Cutting edge chemistry

Cutting edge chemistry

Compiled by Ted Lister

Edited by Susan Aldridge, John Johnston and Colin Osborne

Designed by Black Pearl Consultancy Ltd, London

Picture research by Black Pearl Consultancy Ltd, London

Printed by Butler & Tanner Ltd, Somerset

Published by the Royal Society of Chemistry

For further information on other educational activities undertaken by the Royal Society of Chemistry write to:

Education Department

Royal Society of Chemistry

Burlington House, Piccadilly

London W1V 0BN

United Kingdom

www.rsc.org

www.chemsoc.org

ISBN 0-85404-914-2

British Library Cataloguing in Publication Data.

A catalogue for this book is available from the British Library.

Cover pictures courtesy of PhotoDisc and Swatch AG.

Royal Society of Chemistry

The Royal Society of Chemistry (RSC) is the learned Society for chemistry and the professional body for chemists in the UK, with over 46 000 members worldwide. It was established as the world's first Chemical Society, in London in 1841. The Society is charged by Royal Charter, with the advancement of chemical science and its application, and to serve the public interest by acting as an adviser on matters relating to the science and practice of chemistry. The Society has a comprehensive educational programme for both teachers and students in schools and colleges. This includes providing curriculum resources, in-service training and careers advice. For further information about all the Society's education activities, contact **education@rsc.org**.

Contents

Introduction

What is chemistry?

What is chemistry? In the words of one well-known chemist, chemistry is about making forms of matter that have never existed before. These new substances range from plastics and detergents to contraceptives and anticancer medicines and have had an extraordinary impact on our lives. We now take so many modern products for granted that we forget that they would not exist without the chemical knowledge used to make them.

Although designing and making new molecules is at the heart of chemistry, another important aspect is analysing substances and working out how and why chemical reactions happen. Chemistry has thus contributed to our quality of life by providing the means for quality control in manufacturing, for monitoring the environment and assessing our health needs, and even for crime detection. On a more philosophical level, chemistry is also about seeking a deeper understanding of our place in the grand scheme of things – by revealing the molecular complexity of the world around us and its subtle relationships with ourselves and other living organisms.

Did the study of chemistry start at the campfire?
The National History Museum, London.

The Alchemist.
Reproduced courtesy of the Library and Information Centre, Royal Society of Chemistry.

How chemistry developed

The earliest beginnings

Now that we are starting a new millennium, it is worth looking back at the rise of chemistry and how it has affected our social and cultural development. Mankind's first chemical reactions were probably the use of fire to cook and later to extract metals. Much of the first chemistry was probably accidental. It has been suggested that primitive forms of both soap and glass were first made about 2000 years ago – soap from a mixture of goat tallow and wood ash, and glass from sand, seaweed and salt in the ashes of camp fires on the beach.

Philosophers and alchemists

Two thousand years ago, of course, people had no idea how one material changed into another. The Ancient Greeks were very interested in understanding the material nature of the world but they took the wrong course. Democritus suggested that matter was made of indivisible atoms, but unfortunately his ideas were swamped by those of other philosophers such as Aristotle who thought that matter consisted of the so-called four elements Earth, Air, Fire and Water. In the Middle Ages, the alchemists (the predecessors of modern chemists), supported Aristotle and spent much of their time trying to convert base metals such as lead into gold. Not surprisingly, they were unsuccessful. Nevertheless they did develop methods that were to form the basis of real chemistry such as filtration, crystallisation, and distillation, and they studied a wide range of chemical reactions.

The end of the 18th century

Chemistry as we know it started towards the end of the 18th century. Largely because of the work of the Frenchman Antoine Lavoisier, chemists accepted fundamental ideas such as that combustion was a reaction with oxygen. Lavoisier also put chemical research on a firmer basis by stressing the importance of quantitative measurement. This led to the modern concept of an element as a substance that cannot be further divided into parts. At this time, the first textbook to interpret chemistry using these new ideas appeared.

The new chemical thinking coincided with the beginning of the Industrial Revolution. In the UK the textile mills of Northern England created the need for new chemistry to help convert fibre into finished cloth. Soap-making and glass-making also expanded. Thus a demand grew up for chemicals on a scale not satisfied by traditional methods. Sulfuric acid, required for textile finishing and for making soda and alkali, was produced by the 'lead chamber' process. Bleach was required for textile finishing, and Charles Tennant together with Charles Mackintosh, produced bleaching powder by absorbing chlorine gas in lime. The alkali industry also started in earnest at this time with a process introduced in 1790 by Nicolas Leblanc. This process produced sodium carbonate (for soap and glass-making), previously obtained by burning the barilla plant.

Antoine Lavoisier. Reproduced courtesy of the Library and Information Centre, Royal Society of Chemistry.

TABLE II.

THE ATOMIC WEIGHTS OF THE ELEMENTS

Distribution of the Elements in Periods

Groups	Higher Salt-forming Oxides	Typical or first small period	Large Periods 1st	2nd	3rd	4th	5th
I.	R_2O	Li =7	K 39	Rb 85	Cs 133	—	—
II.	RO	Be =9	Ca 40	S 87	Ba 137	—	—
III.	R_2O_3	B =11	Sc 44	Y 89	La 138	Yb 173	—
IV.	RO_2	C =12	Ti 48	Zr 90	Ce 140	—	Th 232
V.	R_2O_5	N =14	V 51	Nb 94	—	Ta 182	—
VI.	RO_3	O =16	Cr 52	Mo 96	—	W 184	Ur240
VII.	R_2O_7	F =19	Mn 55	—	—	—	—
VIII.			Fe 56	Ru 103	—	Os 191	—
			Co 58.5	Rh 104	—	Ir 193	—
			Ni 59	Pd 106	—	Pt 96	—
I.	R_2O	H=1. Na =23	Cu 63	Ag 108	—	Au 198	—
II.	RO	Mg=24	Zn 65	Cd 112	—	Hg 200	—
III.	R_2O_3	Al =27	Ga 70	In 113	—	Tl 204	—
IV.	RO_2	Si =28	Ge 72	Sn 118	—	Pb 206	—
V.	R_2O_5	P =31	As 75	Sb 120	—	Bi 208	—
VI.	RO_3	S =32	Se 79	Te 125	—	—	—
VII.	R_2O_7	Cl =35.5	Br 80	I 127	—	—	—
		2nd small Period	1st	2nd	3rd	4th	5th
			Large Periods				

An early version of the Periodic Table.

The end of the 19th century

Chemistry really took off in the 19th century. It is interesting to summarise a few things that were known and understood at the end of the century, and, equally revealing, what was not known.

The Periodic Table and atomic theory

The Periodic Table, the foundation of our understanding of chemical behaviour, had been introduced by Dmitri Mendeleef and had shown its power by predicting the existence of then-unknown elements which were later isolated. The figure above shows the Periodic Table as it was in 1900 (and below for comparison the Periodic Table at the end of the 20th century). The atomic theory was, after some skirmishes, firmly established and atomic masses were accurately known for most elements.

1 H hydrogen																		2 He helium
3 Li lithium	4 Be beryllium												5 B boron	6 C carbon	7 N nitrogen	8 O oxygen	9 F fluorine	10 Ne neon
11 Na sodium	12 Mg magnesium												13 Al aluminium	14 Si silicon	15 P phosphorus	16 S sulfur	17 Cl chlorine	18 Ar argon
19 K potassium	20 Ca calcium		21 Sc scandium	22 Ti titanium	23 V vanadium	24 Cr chromium	25 Mn manganese	26 Fe iron	27 Co cobalt	28 Ni nickel	29 Cu copper	30 Zn zinc	31 Ga gallium	32 Ge germanium	33 As arsenic	34 Se selenium	35 Br bromine	36 Kr krypton
37 Rb rubidium	38 Sr strontium		39 Y yttrium	40 Zr zirconium	41 Nb niobium	42 Mo molybdenum	43 Tc technetium	44 Ru ruthenium	45 Rh rhodium	46 Pd palladium	47 Ag silver	48 Cd cadmium	49 In indium	50 Sn tin	51 Sb antimony	52 Te tellurium	53 I iodine	54 Xe xenon
55 Cs caesium	56 Ba barium	57-70 *	71 Lu lutetium	72 Hf hafnium	73 Ta tantalum	74 W tungsten	75 Re rhenium	76 Os osmium	77 Ir iridium	78 Pt platinum	79 Au gold	80 Hg mercury	81 Tl thallium	82 Pb lead	83 Bi bismuth	84 Po polonium	85 At astatine	86 Rn radon
87 Fr francium	88 Ra radium	89-102 **	103 Lr lawrencium	104 Rf rutherfordium	105 Db dubnium	106 Sg seaborgium	107 Bh bohrium	108 Hs hassium	109 Mt meitnerium	110 Uun ununnilium	111 Uuu unununium	112 Uub ununbium						
* lanthanides		57 La lanthanum	58 Ce cerium	59 Pr praseodymium	60 Nd neodymium	61 Pm promethium	62 Sm samarium	63 Eu europium	64 Gd gadolinium	65 Tb terbium	66 Dy dysprosium	67 Ho holmium	68 Er erbium	69 Tm thulium	70 Yb ytterbium			
** actinides		89 Ac actinium	90 Th thorium	91 Pa protactinium	92 U uranium	93 Np neptunium	94 Pu plutonium	95 Am americium	96 Cm curium	97 Bk berkelium	98 Cf californium	99 Es einsteinium	100 Fm fermium	101 Md mendelevium	102 No nobelium			

Courtesy of Bayer plc.

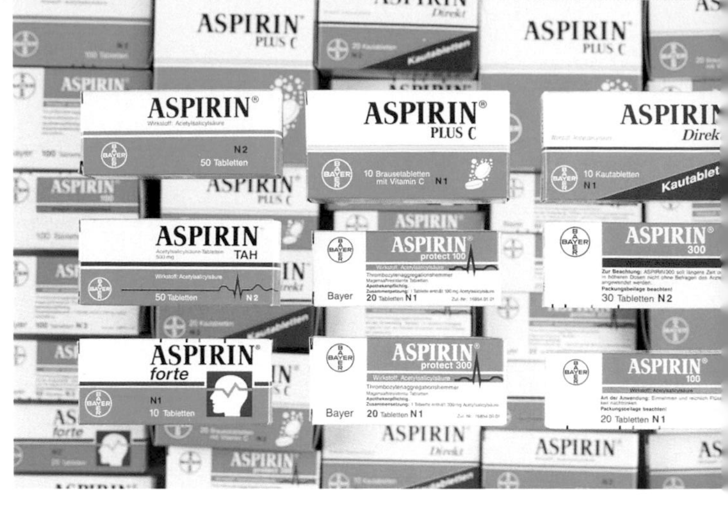

Courtesy of Bayer plc.

Organic chemistry

The mysterious 'life force' supposedly associated with organic material had been dispatched as a myth as chemists realised that organic compounds could be made from inorganic starting materials. Researchers were beginning to understand the structure of organic compounds. They realised that carbon was the major constituent of organic molecules and was very often surrounded by four atoms arranged tetrahedrally. With these insights, synthetic organic chemistry produced some wonderful new molecules, including aspirin, first marketed in 1899.

Flame test on barium.
Jerry Mason/Science Photo Library.

The beginnings of modern chemistry

Many areas that we now consider as modern chemistry had rapidly expanded. The Swiss chemist Alfred Werner had begun his pioneering work on showing how metals could bind with groups of atoms to form complex coordination compounds, now so important in catalysis.
A quantitative understanding of chemical reactions had developed in terms of rates of reactions (kinetics) and the energetic processes that determine whether a reaction happens (thermodynamics). The relationships between electricity and chemistry (electrochemistry) had been thoroughly investigated, and analytical chemistry, mostly based on 'wet' methods and flame tests, had become sophisticated.

Industrial chemistry

The chemical industry had expanded beyond recognition. One of the greatest developments was in synthetic dyes, the origin of which tells a revealing story. William Henry Perkin, a young man of 18, working under the directorship of August Hofmann, was trying to make quinine by oxidising derivatives of anthracene extracted from coal tar pitch – in modern terminology:

$$2C_{10}H_{13}N + 3[O] \rightarrow C_{20}H_{24}N_2O_2 + H_2O$$

Modern chemists know the structural formulae of the starting material and of the product; they bear no relation to each other, so Perkin's efforts were doomed to failure, and he obtained a brown sludge. However, Perkin tried a simpler starting material, aniline (now called phenylamine), also isolated from coal tar. He obtained a mauve compound which he sent off to a firm of dyers. This was the foundation of the dye industry based on synthetic chemistry rather than natural materials.

This story is revealing in two ways. First, it shows the importance of good fortune. Secondly, it shows that the accident must be appreciated by someone – as Louis Pasteur put it: 'Where observation is concerned, chance favours the prepared mind'.

Other major industrial developments included: the replacement of the Leblanc process for producing sodium carbonate by the Solvay process; the replacement of the lead chamber process for sulfuric acid by the contact process; the discovery of the vulcanisation of rubber by Charles Goodyear (in 1841); and the Castner-Kellner electrochemical method for manufacturing both chlorine and sodium hydroxide.

Chemical techniques

What techniques did chemists then have available? These were very limited. They included conventional laboratory methods such as distillation and crystallisation; and analytical techniques based on weighing precipitates, titrations, combustion or flame methods, and also crude ultraviolet/visible spectroscopy and very crude infrared spectroscopy.

A typical 19th century laboratory.
Reproduced courtesy of the Library and Information Centre, Royal Society of Chemistry.

Mauveine dye.
Science Museum/Science and Society Picture Library.

The limits of chemical knowledge

What was *not* known at the end of the 19th century? There was no understanding of atomic structure in terms of a nucleus and electrons, and thus no understanding of the basis of the Periodic Table. There was no idea of what constitutes a chemical bond. There was no understanding of the mechanism of chemical reactions – the series of steps that lead from reactants to products. On the practical front, there was no polymer industry (except for cellulose and bakelite); Paul Ehrlich had not made his 'magic bullet' drug salvarsan (for treating syphilis), and the sulfanilamide anti-bacterial compounds had not been made; there were no synthetic fertilisers. There were none of the modern physical analytical techniques we have today such as X-ray crystallography, and nuclear magnetic resonance (NMR). Indeed it is a constant source of amazement that the chemists of the 19th century could make so many advances with so few resources.

The idea that atoms have an internal structure was not developed until the early part of the 20th century.

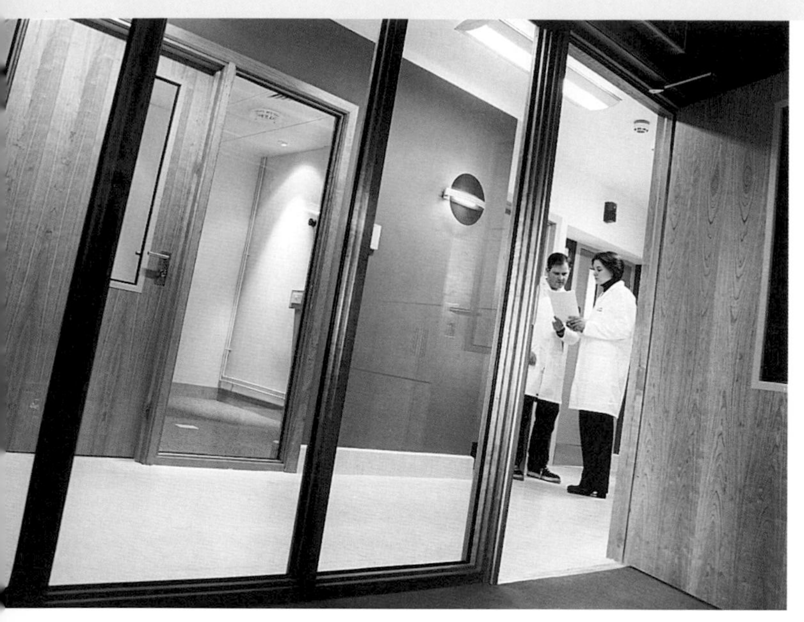

Courtesy of QuChem.

And now, the end of the 20th century

In the 20th century there have been extraordinary strides in chemistry. This book aims to explore some of these advances. It cannot be in any way comprehensive – chemistry is now too large a subject. Instead we have selected some areas of chemistry in which there have been significant developments, particularly in the past 50 years, and which are likely to lead to more scientific breakthroughs in the 21st century. We have also chosen areas that illustrate the overall framework of chemical knowledge and the interdependent relationships between different areas of chemistry. While there is this general theme to the book, each chapter stands on its own. It covers a particular area and is written by chemists doing research at the forefront of that field. The content of the chapters represents, often in a historical context, what they think is exciting and important, and what they think the next developments may be in the next millennium.

Ronald Breslow, recently President of the American Chemical Society, calls chemistry 'The central, useful and creative science'. It is central because it underpins many other scientific disciplines such as biology, geology and materials science. It is useful in providing many of the materials of everyday life, the knowledge to offer better healthcare and better food, and to solve environmental problems. It is creative in that it designs original structures often with new and unique properties. We hope that this book reveals all these aspects.

A quick tour of *Cutting edge chemistry*

The following paragraphs give a flavour of each chapter. The book starts with a chapter on how chemists set about making molecules. *Make me a molecule* gives a flavour of the ingenious strategies that modern synthetic chemists have developed. It emphasises the great contributions made in the early part of this century by an extraordinary group of individuals who were able to synthesise large naturally-occurring molecules using only the limited techniques of the time. Perhaps the most fascinating aspects of modern synthesis are the chemist's ability to tailor-make highly complex molecules with exactly the required three-dimensional geometry, and the automation of laboratory processes which in some areas is rapidly changing the way synthesis is being done.

Synthetic chemistry depends totally on knowing the structure of the end-product. In addition to the 'classical' methods of the 19th century, modern synthetic chemists possess a wide range of techniques, most of them derived from some new physical principle. The chapter on *Analysis and structure of molecules* gives an historical introduction, and outlines what can be learned from the modern techniques – particularly nuclear magnetic resonance (NMR) which has become an indispensable analytical technique for synthetic chemists.

An important aspect of making molecules is the use of catalysts – compounds that can accelerate the rates at which chemical reactions occur. Most of the chemical industry depends on catalysts, and life could not exist without the biological catalysts called enzymes. We have, therefore, included a chapter on catalysis – *Chemical marriage brokers*. Chemists are continually looking for new catalysts which are both selective (make only the desired product), and efficient. They also need to be robust, environmentally-friendly and to be able to work under mild conditions. Enormous effort is expended on studying just how catalysts work at the molecular level to improve their efficiency.

To find better catalysts, and to understand chemical reactivity in general, it is extremely important to know just how a reaction proceeds and what the intermediate stages are. Understanding the details of chemical reactions is now a very important area of chemistry. Many reactions can be extremely fast yet chemists have managed to study them using ingenious techniques. The chapter *Following chemical reactions* describes techniques which have been developed over the past 50 years. Most of the recent work depends on lasers, which has allowed chemists to probe the exact details of how chemical bonds break and reform during reactions – including biological reactions such as photosynthesis in plants.

The chapter on *New science from new materials* selects some examples of new materials – most made only recently – which possess surprising electronic and magnetic properties. These materials have tremendous potential in leading to new types of electronic devices. They include metal oxides whose electrical resistance changes hugely in an applied magnetic field and the famous high-temperature superconductors which lose all electrical resistance at the temperature of liquid nitrogen. Another group of fascinating materials are minute clusters of atoms – nanoclusters – whose properties are neither like isolated atoms nor bulk material. The most famous clusters, however, are the carbon-based fullerenes, such as C_{60} – a molecule shaped like a soccer ball.

Other types of materials that are also common are liquid crystals. As described in *The world of liquid crystals,* these substances consist of molecules that show some kind of orientational order as in a crystal but can move around freely as in a liquid. They were not discovered in the laboratory until the 19th century, although nature long ago invented liquid crystals for making cell membranes and other materials such as beetles' cuticles and spiders' webs. Because certain liquid crystals respond to electric fields, they are used in visual displays for watches and miniature television sets. This modern application of liquid crystals had to wait for synthetic chemistry to produce molecules with the correct tailor-made properties. Among the more important recent developments are liquid crystal polymers such as Kevlar® which are extraordinarily strong.

Perhaps the most visible results of chemical advances in the 20th century have been the synthetic materials, in particular plastics, now used by everyone. The chapter *The age of plastics* plots the progress of polymer research from the time when the early pioneers of polymer chemistry were derided for believing in giant molecules, to the latest developments in light-emitting polymers for electronic devices. As is often the case chance observations have played a part. Polythene, which we now see everywhere, was discovered because some chemists at ICI, studying the reactions of ethene under high pressure and temperature, noticed a completely unanticipated waxy solid in the reaction vessel. Today, there are dozens of different types of plastic, each developed for particular uses.

Courtesy of QuChem.

Photosynthesis is one of several crucially important biological reactions which involve the transfer of energy via the movement of electrons. This is the area of electrochemistry, which makes use of the intimate relationship between electricity and chemistry. The chapter on *Electrochemistry* describes how electricity is produced from chemical reactions in batteries. We all use batteries, whether in cars or personal stereos, and chemists are trying hard to improve them by finding new materials that store energy more efficiently. One of the aims is to develop a lightweight battery with a high storage capacity for the electric car. A competitor to the battery is the fuel cell – a related electrochemical device which depends on a continuous flow of reacting materials.

Our understanding of the way that newly discovered materials behave depends upon theories of chemical bonding and the behaviour of electrons in atoms.
The cornerstone of these theories is quantum mechanics. It is a formidable task to apply quantum ideas to complex atoms and molecules because the equations are difficult to solve. Fortunately today's powerful computers have come to the rescue as shown in the chapter *Computational chemistry and the virtual laboratory.* Chemists can also use computers to display molecular structures as colourful graphics, and also simulate the behaviour of groups of atoms or molecules. These studies have had enormous influence on molecular biology, medicine and materials design.

Perhaps the most important part of modern chemistry is its intimate relationship with biology. Living organisms consist of complex chemicals which behave in a self-organising way. The science of molecular biology has grown out of chemistry and still depends on fundamental chemical ideas and techniques. The final chapter, *The chemistry of life*, traces this relationship by revealing how chemical studies have led to an understanding of living processes and therefore to new medical treatments. The chapter also documents how the structure of DNA was discovered and gave rise to the new science of genetics.

It is worth adding a final comment: what this book is *not* about. It does not have any political agenda and it is not a defence of chemistry. To the average citizen the word 'chemical' conjures up all kinds of environmental and 'unnatural' horrors. We, therefore, offer a gentle reminder that everything around us, including ourselves, is made of chemicals. That is not to say that there are not serious and legitimate issues concerning the production of chemicals and the environment. There is no doubt that the impact of many chemicals has been damaging. The fact that this book does not address these issues directly is not because they are not important – they are – but because we believe that the benefits of chemistry far outweigh the deficits. *Cutting edge chemistry* is simply about why chemistry is exciting and important and why it will be even more so in the future.

To the teacher

This book is aimed at post-16 chemistry students and hopes to give them a picture of cutting edge chemistry research in industry and academia at the beginning of the 21st century. We hope that teachers, too, will find it interesting and that younger students will also be able to use and appreciate at least parts of it.

It draws heavily on the Royal Society of Chemistry publication *The age of the molecule* but has been completely re-edited and re-written to tailor it to post-16 students. One key aim is to show that the principles of chemistry covered in post-16 courses still apply at the cutting edge. However, it is emphatically not another textbook. While we hope that it will be understandable by post-16 students it ranges much wider than current examination specifications and syllabuses.

How to use this book

We hope that teachers will find a variety of uses for this book. Some are listed below but there will be many more.

- **Specific chapters could be used by students as background reading before they tackle corresponding topics in their post-16 course to inspire them and give them a feel for the context of this particular topic.**
- **Alternatively a chapter could be read by students after they have completed a topic on a similar area of chemistry so that they can see how the principles they have studied are used in real life and where the topic leads.**
- **Chapters could be used together with the questions they contain as comprehension exercises or parts of chapters could be used as passages on which to base comprehension questions for examination practice.**
- **The reading of a chapter and answering of the questions within it could be set as a meaningful exercise during the absence (planned or unplanned) of a teacher.**
- **Students could read the book simply for interest and enjoyment using the questions as an aid to understanding as they read.**
- **Teachers themselves might read the book for interest and enjoyment as a means of updating their own knowledge and enlivening their own teaching (both post-16 and pre-16) with up-to-date examples and anecdotes.**

To students

This book is written for you, the post-16 chemistry student, and aims to give you a picture of cutting edge chemistry at the beginning of the 21st century. We hope that it will interest and inspire you and that some of you might go on to contribute to progress in chemistry a little later in this century.

It draws heavily on the Royal Society of Chemistry publication *The age of the molecule* but has been completely re-edited and re-written to tailor it to post-16 students. One key aim is to show that the principles of chemistry covered in your course still apply at the cutting edge.

Questions have been added to the text and have a number of functions.

- **To help you understand the text by making you think through an example rather than just taking it for granted.**
- **To help you see that chemists working at the cutting edge use exactly the same principles that you are learning and using in your school or college course.**
- **To test your understanding of the text.**

Representing molecules

Skeletal notation

There are lots of different ways of representing the formulae of molecules ranging from simple empirical formulae such as C_2H_6O to sophisticated 3D computer graphics. No one method is better than any other – they all have their uses and many different types are used in this book. One that is used a good deal by working organic chemists and is used a lot in this book is skeletal notation. This is useful for complex molecules as it is easy to draw and can give a good idea of the shapes of large molecules – especially those containing rings.

In skeletal notation, carbon atoms are not drawn at all. Straight lines represent carbon-carbon bonds, and carbon atoms are assumed to be at the ends of these lines or where two lines meet. Hydrogen atoms are not drawn either when they are bonded to carbon – each carbon is assumed to form enough carbon-hydrogen bonds to make a total of four bonds. A rough idea of the bond angles is given. In a saturated hydrocarbon chain, the C-C-C angles are 109.5°. Elements other than carbon are drawn as normal.

The figures below show some examples of skeletal notation along with the corresponding displayed (or graphical) formula where all atoms and all bonds are shown.

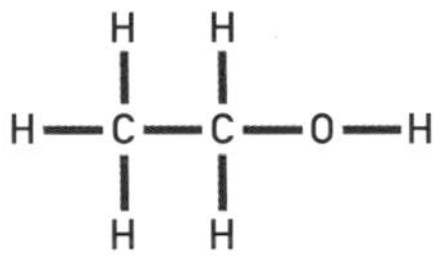

Name	Molecular formula	Graphical (displayed) formula	Skeletal formula
Ethanol	C_2H_6O	Figure 1	Figure 7
Methylpropane	C_4H_{10}	Figure 2	Figure 8
Cyclohexane	C_6H_{12}	Figure 3	Figure 9
Methylcyclohexane	C_7H_{14}	Figure 4	Figure 10
Benzene	C_6H_6	Figure 5	Figure 11
Cholesterol (a steroid)	$C_{27}H_{46}O$	Figure 6	Figure 12

Figure 1.

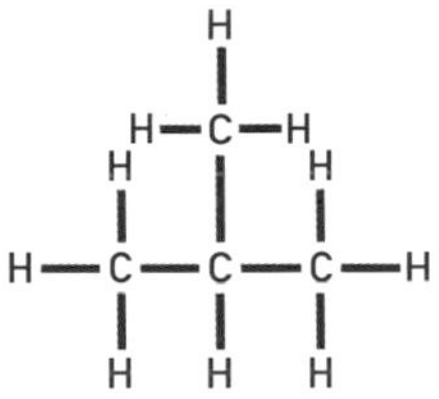

Figure 2.

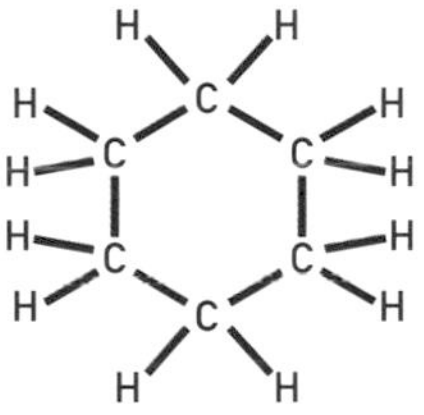

Figure 3.

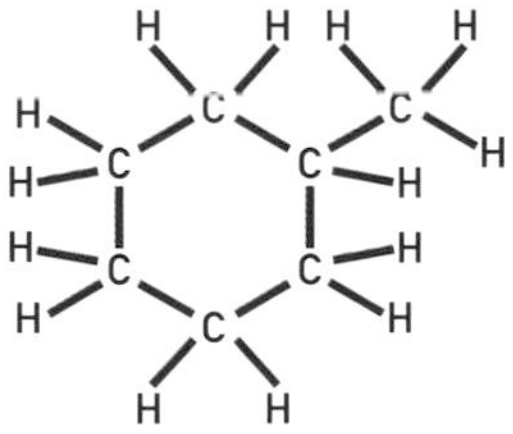

Figure 4.

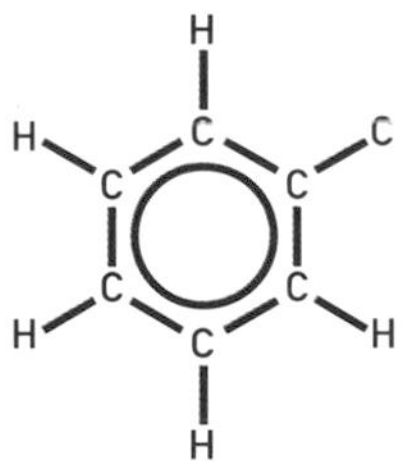

Figure 5.

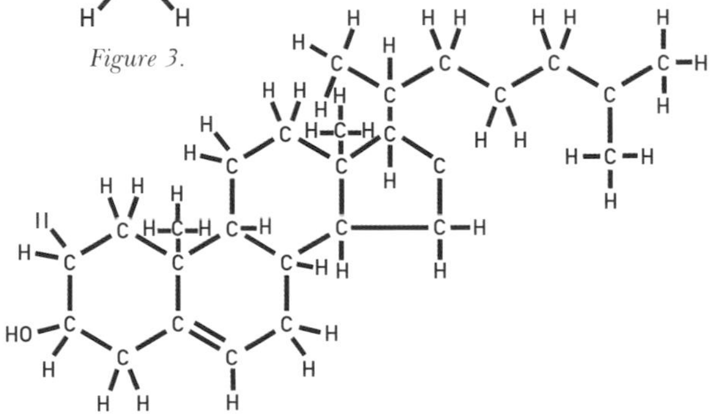

Figure 6.

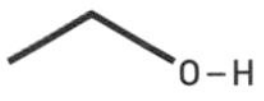

Figure 7.

Figure 8.

Figure 9.

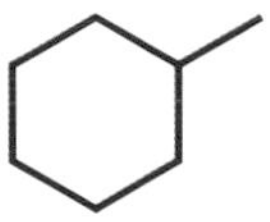

Figure 10.

Figure 11.

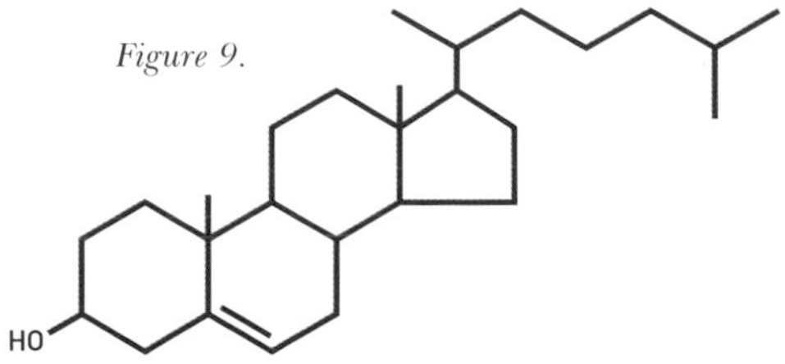

Figure 12.

Wedge bonds

One of the features that you will notice as you use this book is how important the three-dimensional structures of molecules are. These, of course, can be difficult to represent on paper, which is inherently two-dimensional. One shorthand which is used is the convention of wedged or dotted bonds.

A bond drawn ◀ is assumed to be coming out of the paper with the wider end towards you, while one drawn — or ⋯ is assumed to be going back into the paper away from you.

Often in organic chemistry the symbol R is used to represent an organic group. This may be where a group is too complex to draw in full or where a variety of similar groups could equally well be used. For example we might refer to an alcohol as R-OH where R could represent methyl (CH_3-), ethyl (C_2H_5-), propyl (C_3H_7-) *etc.*

Glossary

Most subjects have their own vocabularies, and chemistry is no exception. While reading this book, you will come across many terms which may be new to you such as 'ion selective electrode', 'pharmacophore' and 'laser desorption'. We hope that the text explains these as they crop up – although you might need to make a bit of effort to understand them.

You will also come across a number of technical terms that you probably will have met in your post-16 chemistry like 'ligand', 'cation' and 'excited state'. However, they might not all be at your fingertips so we have included a glossary of these sorts of words in the 'toolbar' at the head of each page.

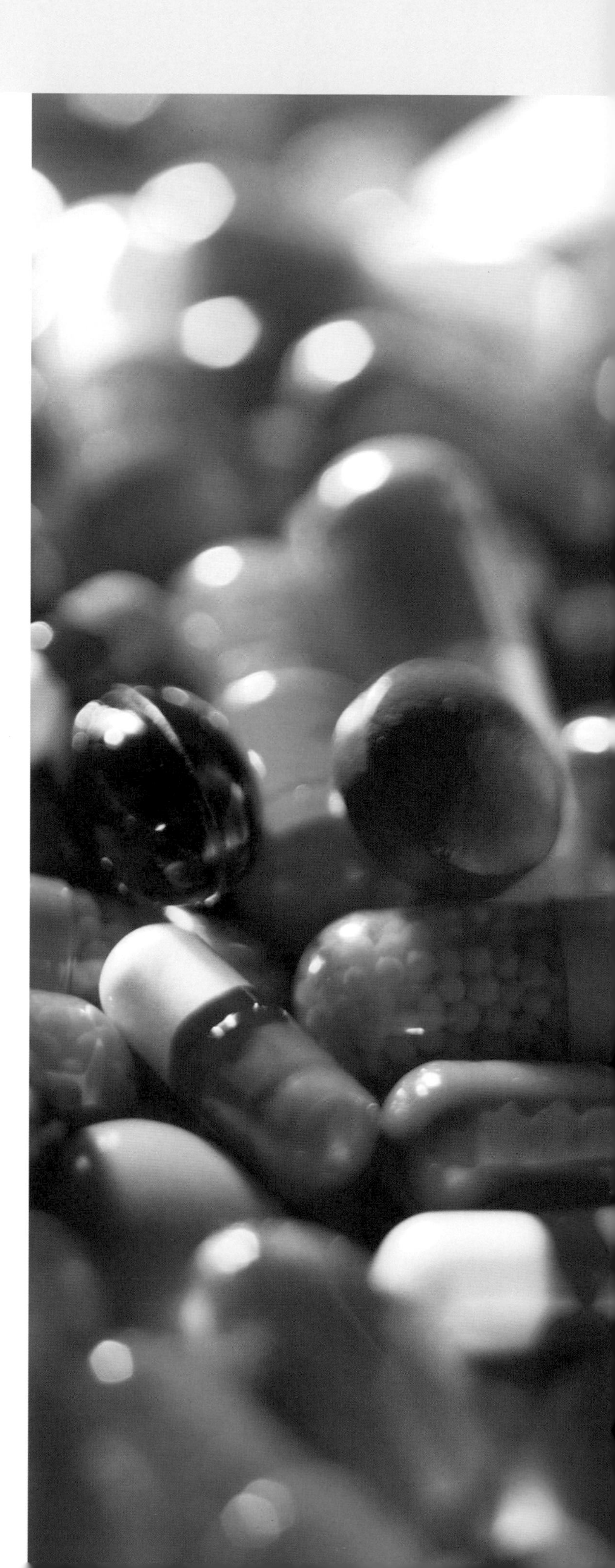

Chemical nomenclature

The system of naming organic compounds used in post-16 courses is normally a systematic one based on the rules set out by The International Union of Pure and Applied Chemistry (IUPAC). This is based on a stem which gives the number of carbon atoms in the longest unbranched chain (or ring) along with prefixes and suffixes to indicate functional groups, and numbers (called locants) to indicate where these functional groups are located along this chain (or ring). So the name 2-bromopropan-1-ol tells us that the longest chain is three carbons long (prop-) and has an alcohol functional group (-ol) on carbon one and a bromine atom (bromo-) on carbon two.

However, universities and industry often use non-systematic names possibly for a number of reasons:

- **a fondness for, and familiarity with, the *non-systematic* names;**
- **for clarity (ethanol and ethanal might be confused when spoken in a noisy factory, while their non-systematic names – ethyl alcohol and acetaldehyde would not); and**
- **because systematic names can be unwieldy – it is easier to ask a pharmacist for aspirin rather than for ethanoyloxybenzenecarboxylic acid, for example.**

In this book we have used both types of name as appropriate. We have tried to give the systematic name where possible but the overall aim has been clarity. Similar considerations have applied to the use of units.

Boxes

Some material is in boxes. In general this means that the chapter still makes sense if you miss out the text in the boxes. This might make the main thrust of the chapter easier to follow on a first reading but we recommend that if you do this, that you should go back to the boxes later.

Acknowledgements

The current book draws heavily on *The Age of the molecule* publication which was managed by Dr Denise Rafferty.
The steering committee consisted of:
Professor J J Turner (Chair)
Dr A D Ashmore
Professor D Gani
Professor J H Holloway
Professor Sir Harry Kroto
Dr B T Pierce
Dr B J Price
Dr D Rafferty
Professor D J Waddington
Professor K Wade

The original authors of the chapters are:

Introduction
Professor James J Turner, *University of Nottingham*

Make me a molecule
Professor Sue E Gibson, *King's College London*
Professor Karl Hale, *University College London*
Professor David A Leigh, *University of Warwick*
Dr Nick Terrett, *Pfizer*
Professor Jonathan Williams, *University of Bath*

Analysis and structure of molecules
Dr Melinda Duer, *University of Cambridge*
Dr Katherine Stott, *University of Cambridge*

Chemical marriage brokers
Dr Phil R Davies, *Cardiff University*
Dr Anthony Haynes, *University of Sheffield*

Following chemical reactions
Dr Helen Fielding, *King's College London*

Electrochemistry
Dr Peter Birkin, *University of Southampton*

The age of plastics
Professor Jim Feast, *University of Durham*

The world of liquid crystals
Dr Corrie Imrie, *University of Aberdeen*

New science from new materials
Dr Paul Attfield, *University of Cambridge*
Dr Roy Johnston, *University of Birmingham*
Professor Sir Harry Kroto, *University of Sussex*
Professor Kosmas Prassides, *University of Sussex*

Computational chemistry and the virtual laboratory
Dr Andrew Leach, *GlaxoWellcome*

The chemistry of life
Professor John Mann, *The Queen's University of Belfast*
Dr Neil Thomas, *University of Nottingham*

All the authors are engaged in research at the cutting edge of chemistry in industry or universities, and their work was edited by ***Nina Hall*** for *The age of the molecule*. Thanks are due to all the authors for reading and commenting yet again on the edited drafts. The drafts were also read by a group of teachers and educationalists mostly from the Royal Society of Chemistry's Committee for Schools and Colleges who made helpful comments.

These were:
Joe Burns, *Coleg Glan Hafran, Cardiff;*
Frank Ellis, *GlaxoWellcome;*
Norman Hooper, *Sexey's School, Somerset;*
Julie Hyde, *The Sheffield College;*
Fiona Lough, *The High School of Glasgow;*
David Moore, *King Edward's School, Oxford;*
Gwen Pilling, *University of York;*
Elaine Wilson, *Homerton College, Cambridge.*

In particular, ***Colin Osborne*** read and commented on two drafts of each chapter and ***John Johnston*** managed the publication process.

The chapter authors would like to acknowledge help from the following:
Professor P N Bartlett
Dr Stef Biagini
Professor Alex Bradshaw
Professor Duncan Bruce
Alex Comely
Dr Paul Dyson
Miguel Gama Goicochea
Professor John Goodby
Professor George Gray
Professor Malcolm Green
Dr Mike Hann
Jerome Jones
Professor G H Kelsall
Professor Peter Maitlis
Professor Ken Packer
Mark Peplow
Professor Stephen Picken
Ellian Rahimian
Professor Peter Raynes
Professor Graham Richards
Professor Wyn Roberts
Dr Glenn Sunley
Professor Sir John Meurig Thomas
Professor Christopher Viney
Nicole Whitcombe

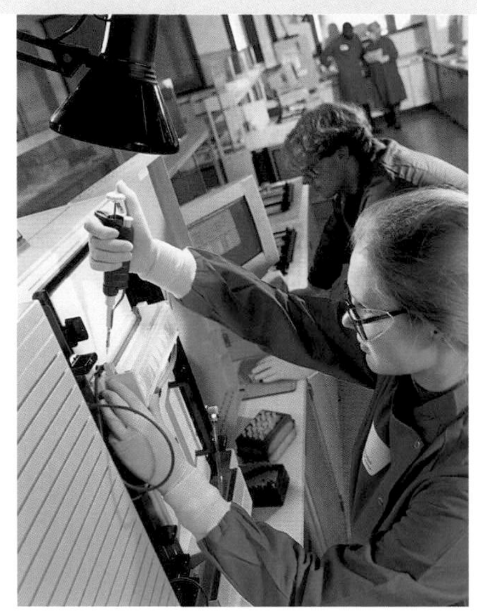

Photograph courtesy of LGC.

Others who have contributed in a variety of ways include:
Professor C Dobson
Professor P P Edwards
Dr P J Dyson
Professor C D Garner
Professor R K Harris
Professor M J Pilling
Professor S M Roberts
Professor P J Sadler
Professor J P Simons
Professor N S Simpkins
Dr J Sloan

Make me a molecule

Chemical synthesis, or making molecules, is both an art and a science which stretches the chemist's knowledge, insight, practical skills and imagination.

Chemists are constantly developing new synthetic methods to make new compounds which not only bring huge social and economic benefits but widen our understanding of nature.

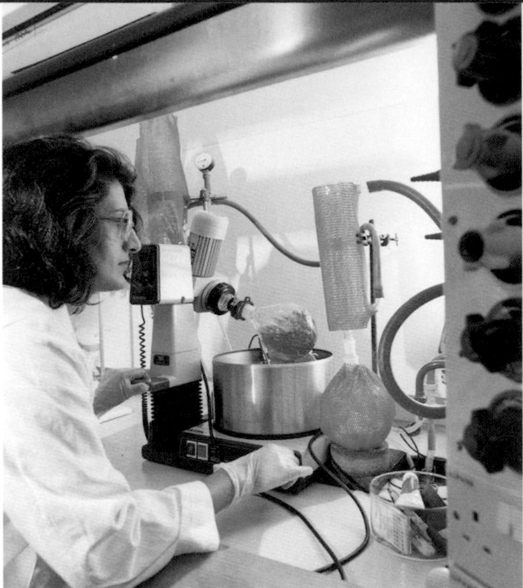
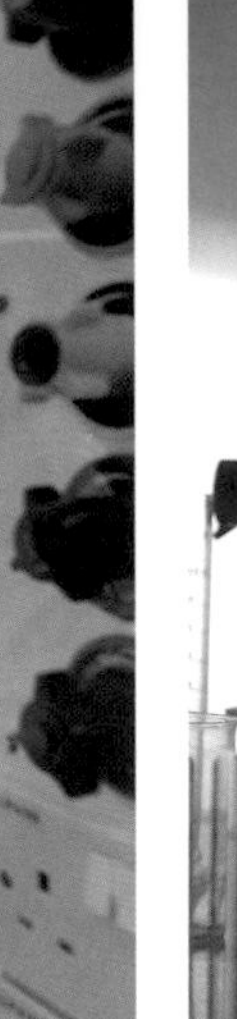

GlaxoWellcome.

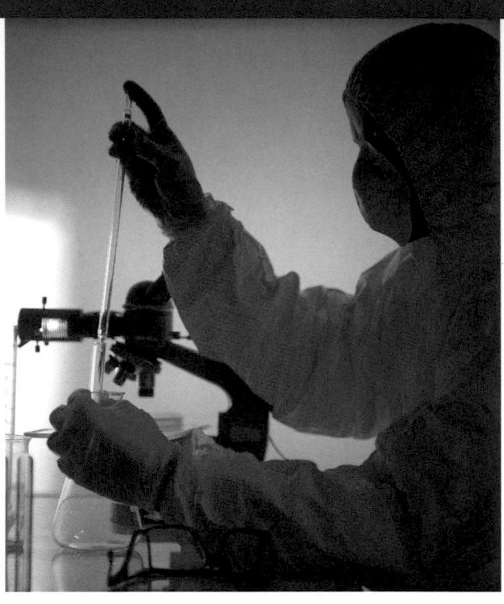

For many people, synthesis is what chemistry is all about, and it is the bedrock on which the chemical and **pharmaceutical** industries are built. Many of the materials we use today are made by chemical synthesis. Chemists create life-saving drugs, pesticides, plastics and specialist materials like liquid crystals.

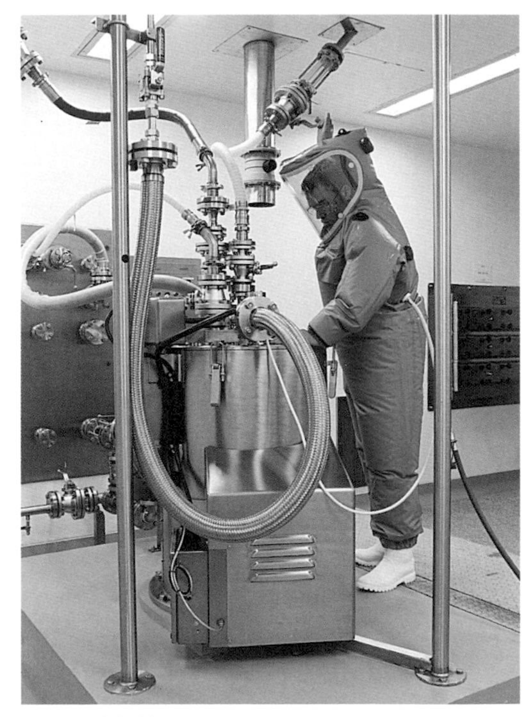

Courtesy of QuChem.

Organic synthesis

Most of the new compounds made today are organic, that is, they have skeletons composed largely of carbon. This element has a unique ability to form strong chemical bonds between carbon atoms and also with the atoms of many other elements such as hydrogen, oxygen, nitrogen, chlorine and sulfur. The electronic structure of a carbon atom is such that it can bond with up to four other atoms, forming single, double and triple bonds. This allows it to form an almost infinite variety of structures – linear and branched chains, rings of all sizes, and more complex bridged structures. Indeed, the chemistry of life is largely a result of the incredible versatility of carbon and is why carbon chemistry is called organic chemistry.

1. Silicon has the same outer electron arrangement as carbon. Explain briefly why Si-Si bonds are weaker than C-C bonds and why bonds between silicon and other elements are weaker than bonds between carbon and these elements.

2. Explain in terms of its electron arrangement why carbon can form up to four covalent bonds.

Glossary

Elimination: A reaction in which a small molecule such as water is lost from the starting material.
Pharmaceutical: To do with the manufacture of medicinal drugs.
Selectivity: The ability of a reaction to make a particular product in preference to any other.
Substitution: A reaction in which one atom or group in a molecule is replaced by another.

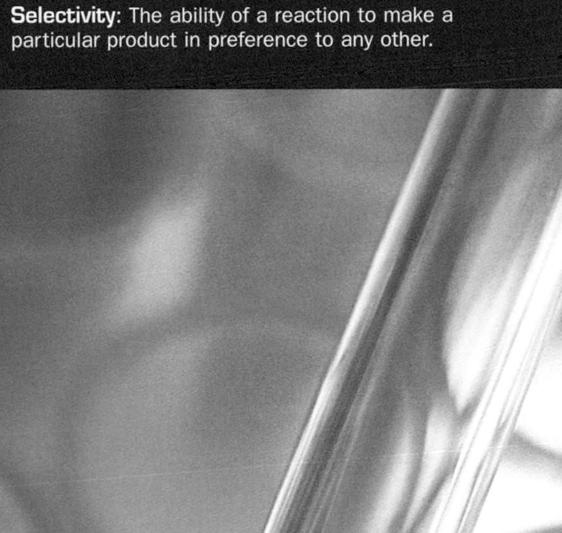

How do chemists go about making new organic molecules?

Synthesis involves breaking and making chemical bonds to create new chemical structures. Much of organic chemistry is concerned with understanding these processes and then using that knowledge to design new kinds of reactions. In this way chemists build up a vast repertoire of reactions.

Often chemists will have a synthetic target in mind such as a possible drug molecule. In this case they have to develop a synthetic strategy – a series of reaction steps – to reach their final compound from readily available starting materials. In the case of large complex molecules, this may involve dozens of steps and a great deal of ingenuity, skill and perseverance.

Organic chemists pride themselves on designing synthetic routes that are both elegant and efficient. This means carrying out the synthesis in the least number of steps, using simple reagents and obtaining the maximum yield of product at each stage. To achieve the latter, **selectivity** is a key idea. Many reactions involve several competing pathways leading to different products. These products are often isomers, in other words, compounds with the same chemical formula but different arrangements of the atoms in space. The chemist tries to tune the conditions of the reaction to obtain as much of the required product as possible.

? 3. The reaction of a haloalkane such as 2-bromopropane with hydroxide ions (OH^-) has two possible pathways leading to different products. The type of reaction which occurs depends on the conditions of the reaction.

(a) In aqueous solution at a little above room temperature, the main reaction is a substitution.

(b) In ethanol solution at 80 °C (353 K) under reflux, the main reaction is an elimination.

Suggest the main product in (a) and in (b).

The basic concepts of synthesis

There are two key concepts in organic chemistry that synthetic chemists work from. One is the notion of a functional group – a group of atoms that tends to behave as a single chemical entity in a reaction. Examples are the carbonyl group (>C=O) or a primary amine group ($-NH_2$). Functional groups undergo particular sets of chemical reactions.

❓ 4. List eight other organic functional groups.

The other vitally important idea is that the carbon atom forms bonds in a definite three-dimensional array. When a carbon atom forms four single bonds, the bonds point to the corners of a tetrahedron (a triangular pyramid), Figure 1. This means that structures of organic compounds must be considered in three dimensions (this is called their **stereochemistry**).

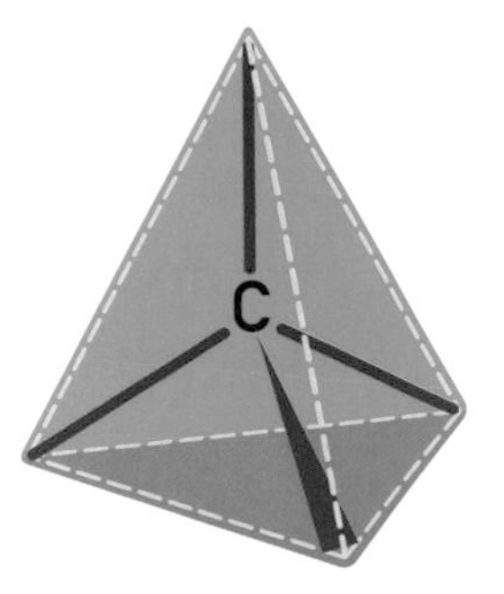

Figure 1. When carbon forms four single bonds, they point to the corners of the tetrahedron (a triangular based pyramid).

❓ 5. Explain why the four single bonds formed by a carbon atom point tetrahedrally. What H-C-H angle results?

Also of importance is the fact that single bonds can rotate, leading to different shapes or **conformations** for the same molecule. An example is the molecule cyclohexane (C_6H_{12}) – a six-membered carbon ring that can adopt the so-called boat or chair conformations, Figure 2. These conformations become significant when considering large natural molecules which are made up of such rings.

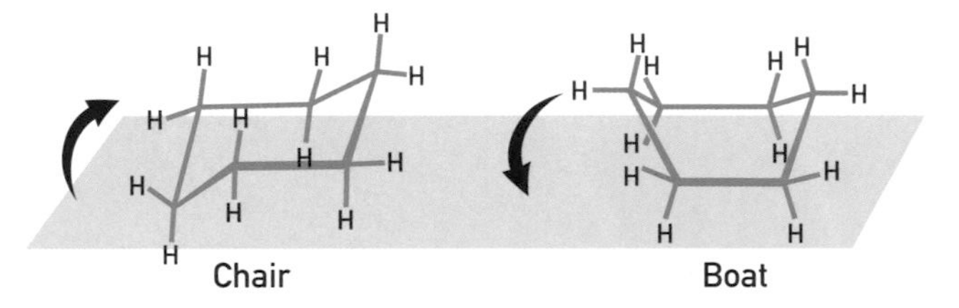

Figure 2. The 'chair' and 'boat' conformations of cyclohexane. They can be interconverted simply by rotating the bonds.

❓ 6. Make a model of cyclohexane with a ball and stick molecular modelling kit. Check that it can easily twist from the boat to the chair conformations without any bonds being broken.

Another consequence of the three-dimensional geometry of carbon compounds is that molecules with the same molecular formula can have their atoms arranged differently in space. In particular, a molecule with a carbon atom which has four different functional groups attached to it (called a chiral centre) can exist as two isomers which are mirror-images but which are non-superimposable, like right and left hands, Figure 3. This property of handedness is called chirality (pronounced *kiy-rality*), and is of great significance in synthesis.

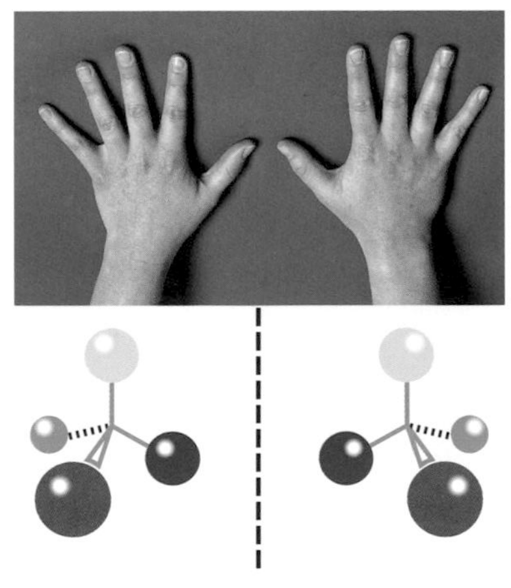

Figure 3. Molecules often have a handedness. Like our hands, the molecules can be mirror images, but not identical.

❓ 7. Use a ball and stick molecular modelling kit to make models of the two molecules in Figure 3. Convince yourself that they are not identical. What happens if on one of the models you exchange the positions of two of the atoms? Does it make a difference which two you exchange?

Glossary

Conformations: The set of different shapes that can be adopted by a molecule by rotation about the single bonds.
Enantiomers: Pairs of molecules (also called optical isomers) which have the same molecular formula but are non-identical mirror images.
Optically active: Describes molecules which can rotate the plane of polarisation of polarised light.
Stereochemistry: Aspects of chemistry that involve three-dimensional considerations in the spatial arrangement of atoms.

Chiral forms of molecules, or **enantiomers,** were discovered in the earliest days of organic chemistry. In 1848, Louis Pasteur showed that salts of tartaric acid crystallised into mirror-image forms which could be separated by picking out the differently-shaped crystals using a pair of tweezers. When Pasteur re-dissolved the crystals he noticed a remarkable thing. One solution made the plane of polarisation of polarised light rotate to the right, while the other rotated it to the left. For this reason, simple chiral isomers like the tartaric acid salts are often called optical isomers and are said to be **optically active**. Twenty years later, Jacobus Henrikus van't Hoff and Josephe-Achille Le Bel explained the origins of the right and left-handed forms of the crystals in terms of mirror-image molecules based on the tetrahedral nature of carbon bonding.

Louis Pasteur was able to separate crystals of tartaric acid into two mirror images. Note that a wedge shape implies a bond coming out of the page and a dashed line one going into the page.
Reproduced courtesy of the Library and Information Centre, Royal Society of Chemistry.

Making complex molecules

It was during the mid-1800s that organic chemistry began to be set on a sound structural framework, allowing chemists for the first time to plan organic syntheses. Some of the landmarks in the history of synthesis are described below.

Fischer synthesises D-glucose

Using the new stereochemical ideas, the German chemist Emil Fischer set about trying to define the absolute configuration (the exact arrangement of atoms in space) of a set of organic molecules that were of great interest in the late 19th century, the optically active sugars. In the process Fischer achieved, in 1890, the first synthesis of a truly complicated organic molecule, the sugar molecule D-glucose (D identifies the particular optical isomer), Figure 4.

This synthesis was important for a number of reasons. First, it yielded D-glucose as a single optical isomer. Secondly, it introduced a number of new reactions into organic chemistry which enabled the synthesis to be completed. Thirdly, it employed multiple sequences of reactions that assembled the sugar rapidly and efficiently all in one step, and fourthly, it proved the structure of D-glucose beyond all doubt. This work is regarded as the catalyst for the development of synthetic organic chemistry in the 20th century.

8. D-glucose has four chiral centres. Mark each of them with a * on a copy of Figure 4.

Emil Fischer
Reproduced courtesy of the Library and Information Centre, Royal Society of Chemistry.

O=C–H
H–C–OH
HO–C–H
H–C–OH
H–C–OH
CH_2OH

Figure 4. One of the forms of D-glucose – it also exists as a ring.

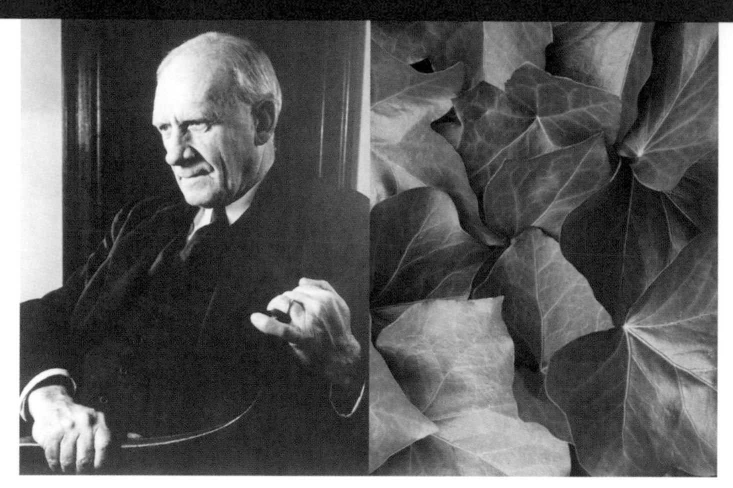

Robert Robinson. *Alkaloids are found in plants.*

Robert Robinson and tropinone

Following Fischer's ground-breaking work, synthetic organic chemists made progress in creating ever more complex structures. In the early part of the 20th century a British chemist, Robert Robinson, at the University of Oxford, synthesised a range of complicated natural products (compounds found in nature), including the alkaloid tropinone. Alkaloids are a group of nitrogen-containing chemicals found in plants, and include morphine.

Tropinone is a seven-sided carbon ring with a carbon-oxygen double bond on one side and a nitrogen-atom 'bridge' across the ring. Based on his own view of how Nature prepares this molecule, Robinson designed a synthesis that involves just one step, Figure 5.

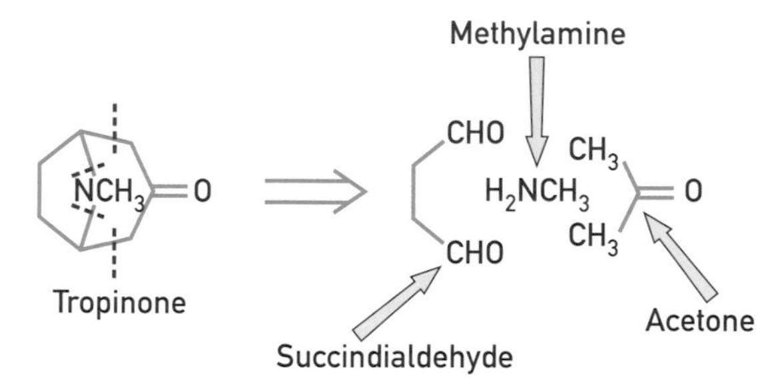

Figure 5. Robert Robinson's retrosynthetic analysis of tropinone.

Robinson's route to tropinone was hailed as revolutionary. It made the previous 19-step synthesis of tropinone, by German chemist Richard Willstätter, seem obsolete and cumbersome. More importantly, it alerted chemists to a better method of planning a synthetic route. This was to look at the target molecule and try to imagine how the molecule could be constructed from simpler chemical units. You can then design a strategy starting from those basic building blocks. This approach is now called retrosynthetic analysis. Robinson made full use of retrosynthetic analytical principles during his planning of the tropinone synthesis, a fact that is made clear from his own words in his research paper describing the synthesis of tropinone: 'By imaginary hydrolysis at the points indicated by the dotted lines, the substance (tropinone) may be resolved into succindialdehyde, methylamine and acetone'.

? 9. What does the term 'hydrolysis' mean?

? 10. Systematic naming of organic compounds was not used in Robinson's time. Methylamine still has the same name, but what are the systematic names for succindialdehyde, and acetone?

Prior to Robinson's brilliant achievement with tropinone, organic chemists had almost universally planned their synthetic routes only in a forward direction. They selected a starting material that bore a structural resemblance to the target molecule, and used reactions that could convert that starting compound into the target. In other words, they allowed the starting material, rather than the target molecule, to dictate their synthetic plan. Often this approach would fail or lead to very lengthy syntheses. By introducing this reverse way of thinking into organic chemistry, Robinson shaped the way total synthesis would be done for the remainder of the 20th century.

Christopher Ingold.

Robinson, with Christopher Ingold of University College London, also worked on organic reaction mechanisms – the series of simple steps by which reactions occur. They developed the ideas of electrophiles and nucleophiles – reagents which attack negatively- and positively-charged areas of organic molecules respectively. This helped chemists to predict the outcomes of new reactions they wanted to use. This approach continues to be used today and was based on careful experimental studies of the rates of organic reactions.

Robert Burns Woodward – master of complexity

As a result of a better understanding of the mechanisms of organic reactions, the field of total synthesis really took off by the 1940s and 1950s. The superstar of this period, and probably the greatest American organic chemist of this century, was Robert Burns Woodward of Harvard University. Woodward was able to make complex molecules with a very limited range of reactions and analytical techniques and he usually finished these synthetic ventures in record time.

He received the Nobel Prize for chemistry in 1965 for 'Achievements in the art of organic synthesis'. Like Fischer,

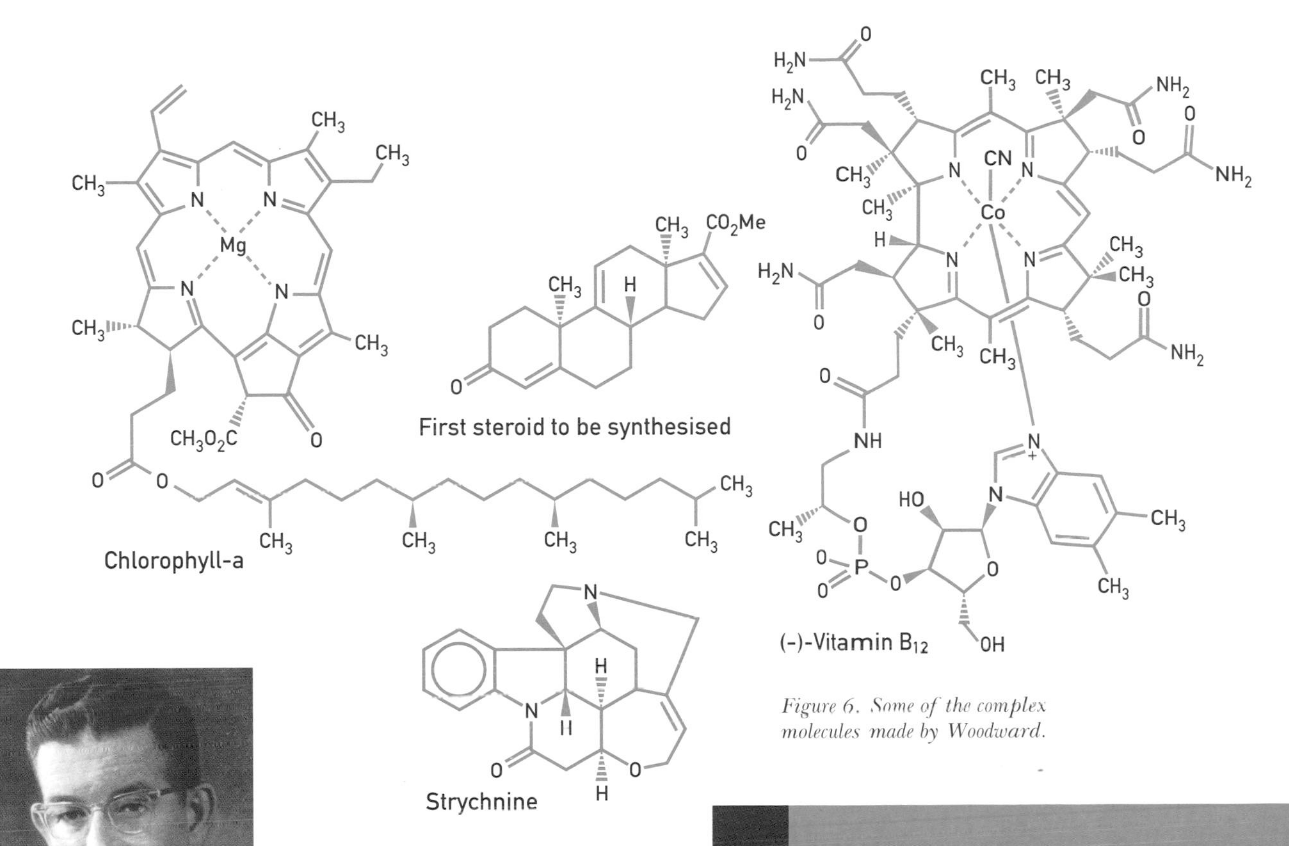

Figure 6. Some of the complex molecules made by Woodward.

Robert Burns Woodward.

Woodward developed much new chemistry in order to complete his syntheses.

The complexity of some of the molecules Woodward synthesised can be seen in Figure 6.

The importance of stereochemistry in synthesis

Stereochemistry describes the relative spatial relationship between the functional groups in a molecule. Woodward was the first organic chemist to make a serious effort to control the stereochemistry in the molecules he was building.

By carefully planning all his synthetic routes with great precision, Woodward would often obtain only one isomer of the product at each stage. The majority of Woodward's syntheses produced the desired target compound as a single enantiomer (provided the target molecule existed in this form).

Woodward's usual approach for obtaining single enantiomers generally entailed synthesising the compound as a 50:50 mixture of opposite enantiomers see Box - *Racemic mixtures*, and then separating that mixture by optical resolution – see Box – *Optical resolution*.

Racemic mixtures

Many organic reactions produce racemic mixtures because they take place via a flat intermediate compound as in the example below. Some nucleophilic substitution reactions take place via a so-called S_N1 mechanism (substitution, nucleophilic, involving one molecule only in the slowest step). The group Y leaves the molecule along with the electron pair in the C-Y bond. This leaves an intermediate positive ion in which the central carbon atom has only three electron pairs in its outer shell. It is therefore a trigonal planar shape. The nucleophile is equally likely to attack the intermediate from either side, leading to a mixture of the two possible enantiomers.

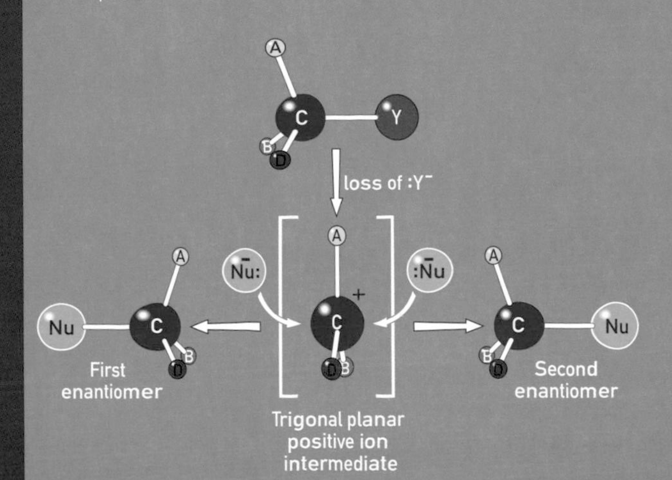

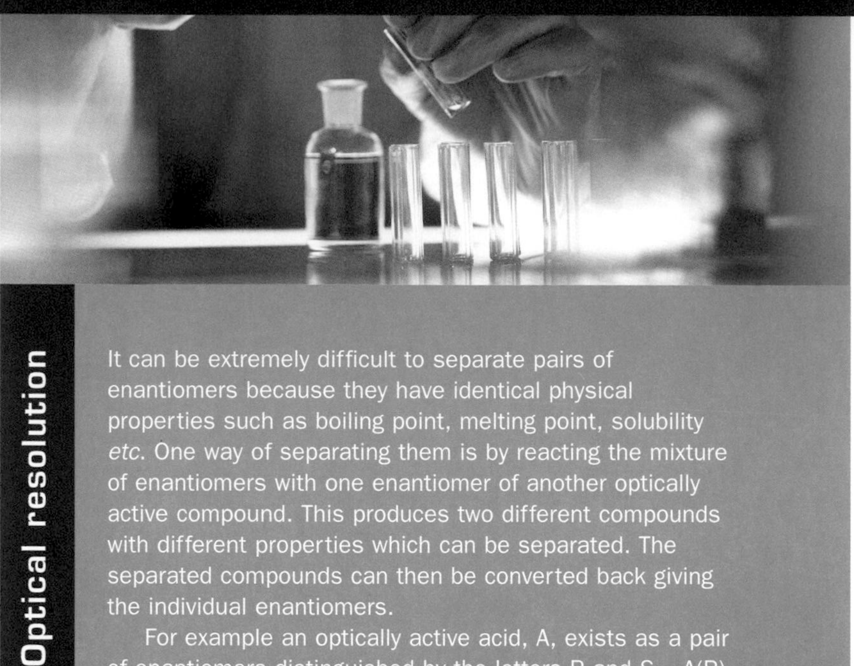

Optical resolution

It can be extremely difficult to separate pairs of enantiomers because they have identical physical properties such as boiling point, melting point, solubility *etc.* One way of separating them is by reacting the mixture of enantiomers with one enantiomer of another optically active compound. This produces two different compounds with different properties which can be separated. The separated compounds can then be converted back giving the individual enantiomers.

For example an optically active acid, A, exists as a pair of enantiomers distinguished by the letters R and S – A(R) and A(S). Similarly an optically active base, B, exists as B(R) and B(S). We can react a racemic mixture of A(R) and A(S) with just one enantiomer of the base, say B(S). This gives two different salts A(R)B(S) and A(S)B(S) with different solubilities. These can be separated by, say, fractional crystallisation. Excess acid can then be added to each of the separated salts to convert them back into the separate free acids.

Indeed, Woodward's research team was able to make a synthetic version of the naturally-occurring compound reserpine, used to treat high blood pressure, controlling the chirality at six carbon positions in this molecule – of which there are 64 (2^6) possible isomers, Figure 7.

*Figure 7. Reserpine. The six chiral centres are marked *.*

By the mid-1970s, most organic chemists preferred to employ a 'chiral pool' of starting materials to carry out their enantiospecific syntheses (*ie* reactions which produced one of a pair of optical isomers rather than the other). The chiral pool is a term used to describe the set of readily available chemical starting materials that exist in Nature as individual pure enantiomers. D-glucose (Figure 4) and (-)-carvone, Figure 8, are two examples of naturally available chiral materials.

Figure 8. (-)-Carvone.

Naming enantiomers

There are (at least) three ways of naming compounds to distinguish between pairs of enantiomers. The +/- notation refers to the direction in which the molecules rotate the plane of polarisation of polarised light. The other two notations, D/L and R/S refer to the actual arrangement of the atoms in space. D/L compares the arrangement to a parent compound and R/S refers to a set of rules based on the atomic numbers of the atoms bonded to the chiral carbon. For example, the enantiomer of 1,2-dihydroxypropanal drawn below is D, R and +. The other enantiomer would be L, S and –.

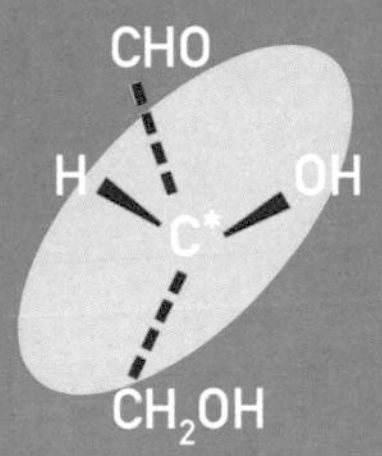

CHO
H C* OH
CH_2OH

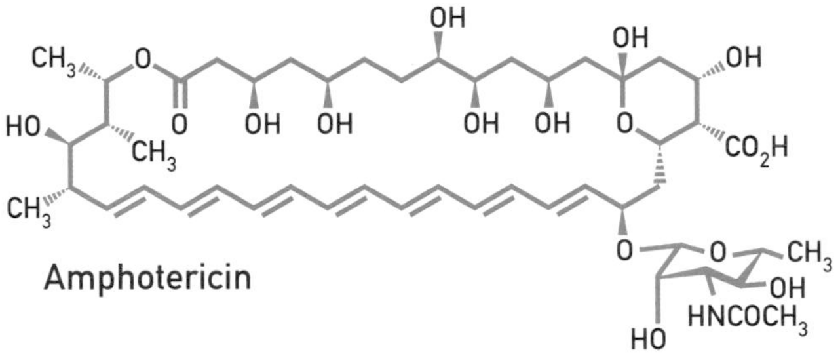

In the 1970s and 1980s, many eminent organic chemists made extensive use of the chiral pool to do their complex molecule synthesis work. Figure 9 shows some of the molecules that were synthesised in this era.

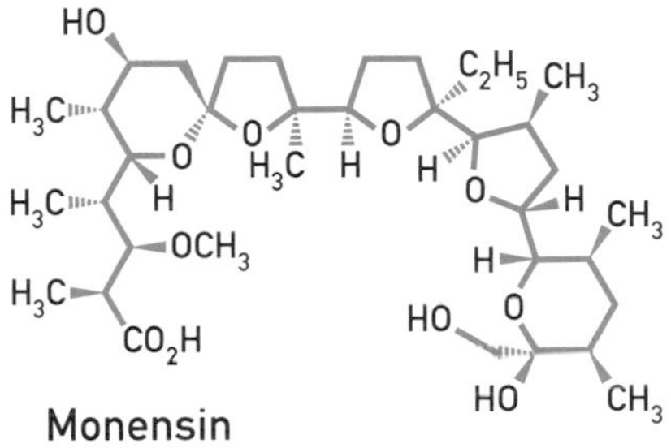

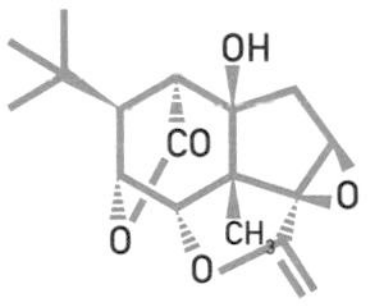

Amphotericin

Monensin

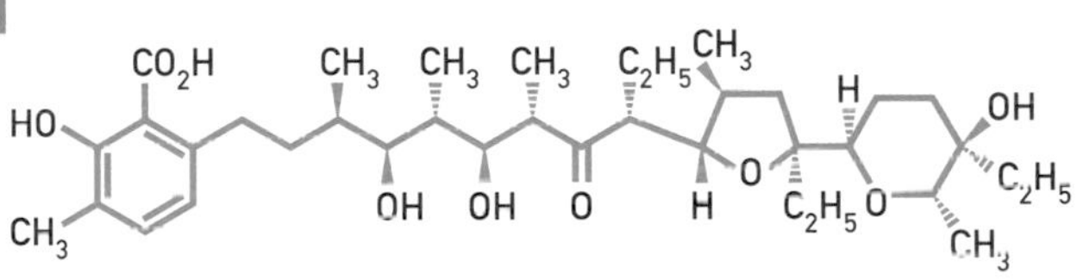

Lasalocid A

Picrotin

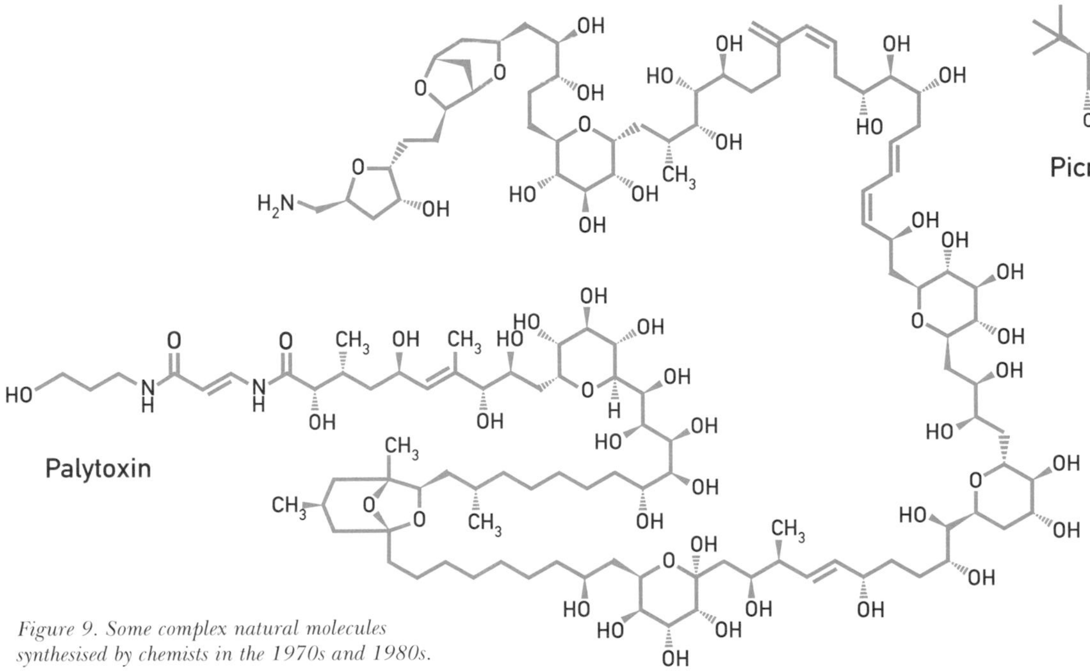

Palytoxin

Figure 9. Some complex natural molecules synthesised by chemists in the 1970s and 1980s.

Mirror molecules and life

Many of the compounds associated with living organisms are chiral – including vital biochemicals such as DNA, enzymes, antibodies and hormones. Each enantiomer may have distinctly different *biological* characteristics. This is despite the fact that they have the same *physical and chemical* properties, such as melting point or solubility in solvents, and give the same spectra. This is true of limonene, a compound which is formed naturally in both chiral forms. One of the enantiomers S-(-)-limonene smells of lemons, while the mirror-image compound R-(+)-limonene smells of oranges. The compound (+)-nootkatone, which is responsible for the smell of grapefruit, smells 750 times more strongly than the (-) enantiomer.

It is thought that the reason we can distinguish the smells of these enantiomers is that our nasal receptors are also made up from chiral molecules that recognise the difference. In fact insects, which have a phenomenal sense of smell, sometimes use chiral chemical messengers (pheromones) as sex attractants. Recently, chemists discovered that one form of the insect pheromone olean attracts male fruit flies, while its mirror image works on the female of the species.

Not surprisingly, because biology is so sensitive to chirality, the activity of drugs also depends on which chiral form is used. However, the significance was not always fully appreciated. In the early 1960s, the drug Thalidomide was prescribed to alleviate morning sickness in pregnant women. Tragically, the drug also caused deformities in the limbs of children born to many of these women. The drug was supplied as a racemic mixture and it seems that one enantiomer of Thalidomide was beneficial while the other caused the birth defects.

There is considerable controversy about the details of this argument, partly because the two mirror image forms of Thalidomide can interconvert easily in the body. Even so, pharmaceutical companies now make sure that both forms of a chiral drug are tested for their biological activity and toxicity before they are marketed.

One current chiral drug recently tested is levobupivacaine, a long-acting anaesthetic which would be ideal for use in dental surgery and for treating deep post-operative pain. At the moment, the only long-lasting local anaesthetic available is the racemic version of bupivacaine. This has limited use because it is toxic to the cardiovascular and central nervous systems. The 'left-handed' version of bupivacaine, levobupivacaine, however, shows considerably reduced heart and neural toxicity and can be used at relatively low dose rates.

Making molecules with mirror images

While today the chiral pool continues to be a major source of chiral starting materials for organic chemists, many are developing single enantiomer reagents which can react with molecules without chiral centres to create new molecules that exist in single enantiomeric form. Such a process is called **asymmetric synthesis**, and this approach is now being used successfully to make molecules for the pharmaceutical industry, for example. This is because different enantiomeric forms of the same molecule may have very different biological properties – see Box – *Mirror molecules and life*. Today, it is usual to isolate the desired enantiomeric form of a compound. Although racemic mixtures can be separated, half the product is then wasted. Companies prefer, therefore, to find efficient methods of asymmetric synthesis.

A non-chiral molecule can be converted into a chiral molecule by simple chemical steps with the aid of a chemical unit called a **chiral auxiliary**. For example, Figure 10 shows a

Propanoic acid
chemical steps
Both handed forms of product
put auxiliary on
chemical steps
take auxiliary off
One handed form only

Figure 10. Adding a chiral auxiliary to flat molecules controls which handed form is made.

reaction starting with propanoic acid, which does not exist in right- and left-handed forms. Attaching an auxiliary to propanoic acid, creates the stereochemical conditions that force the chemical steps to follow a certain geometrical path. Once the handedness of the new molecule has been set, the auxiliary can be taken off (or better, recycled) leaving behind the product molecule in a chiral form.

One drawback of attaching auxiliaries to flat organic molecules, is that an extra step is needed to add the auxiliary, and then another to remove it. An alternative approach uses carefully

Glossary

Asymmetric synthesis: The preparation of a compound (which exists as optical isomers) enriched in one isomer – generally using an optically inactive starting material.
Chiral auxiliary: A single optical isomer which is attached to an optically inactive molecule to control the stereochemistry of subsequent reactions and govern which optical isomer of the product is made.
Enzymes: Proteins which act as biological catalysts.

designed reagents to control the handedness of the reaction. Herbert Brown at Purdue University in the US, was awarded a Nobel prize in 1979 for his work involving organic reactions using boron-containing reagents. His research group has developed the reagent known as Ipc_2BCl which can be used to convert carbonyl groups stereoselectively into alcohols. This can be seen in the reaction in Figure 11. When the C=O is converted into CHOH, only the isomer in which the -OH group sticks out of the paper is formed.

Figure 11. Herbert Brown's boron-containing reagent, Ipc_2BCl is used to convert carbonyl groups to alcohols in a spatially selective way.

11. What type of reaction is the conversion of a carbonyl group into an alcohol? Name at least three reagents (other than Ipc_2BCl) which will bring about this conversion (without controlling the handedness of the reaction).

Ideally, a chiral agent should behave as a catalyst. This means it is needed in only small amounts and is regenerated at the end of the reaction. Nature controls the handedness of molecules by using **enzymes** to catalyse reactions. When enzymes change geometrically 'flat' molecules into three-dimensional chiral molecules they produce only one of the possible isomers.

Recently chemists have designed three-dimensional catalysts (affectionately called 'molecular robots') which can provide enzyme-like levels of selectivity. The advantage of being able to use a catalyst is that a very small amount of handed information can generate a large amount of product.

One of the most important catalytic reactions is the addition of hydrogen to an alkene. One example is the catalysed addition of hydrogen to the alkene shown in Figure 12, where with superb selectivity one chiral form of the anti-inflammatory drug Naproxen is synthesised. The handedness comes from a ruthenium-based catalyst which is attached to a chiral molecule called 'BINAP'.

Figure 12. Ruthenium-based catalysts have been used to add hydrogen selectively to give pharmaceutical compounds.

Chemists are now applying these new asymmetric synthetic strategies to many complicated molecules. For example, the anticancer drug Taxol has a particularly complex molecular structure, Figure 13, but its total synthesis has been achieved by a number of research groups. This may be significant because Taxol can be isolated from yew trees, but in such small quantities that there is not enough compound available to treat many cancer patients effectively.

Figure 13. The anticancer drug Taxol.

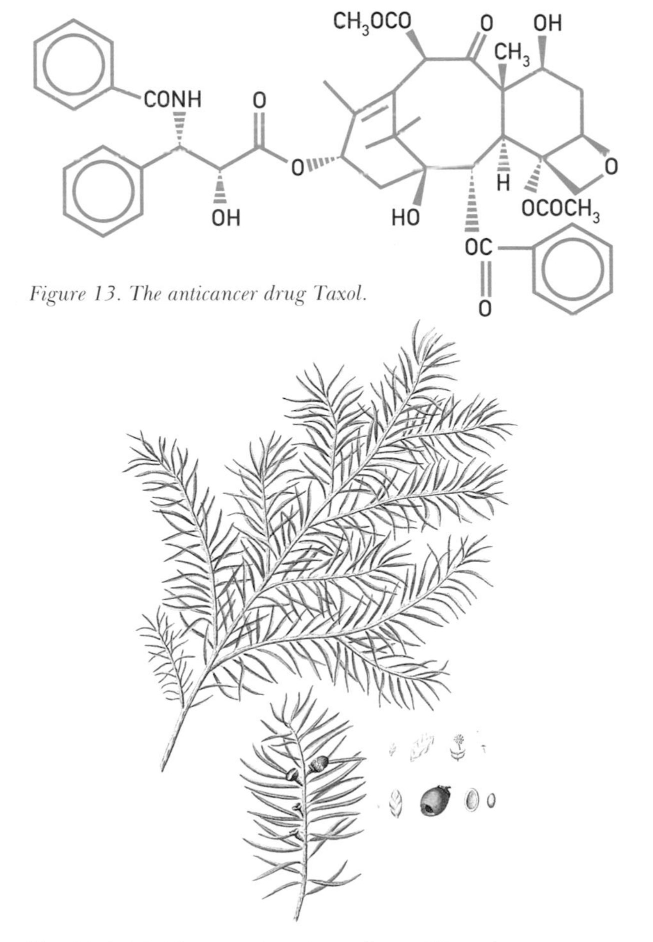

Taxol is isolated from yew trees in small quantities only. Courtesy of Royal Botanic Gardens, Kew.

Organometallic compounds in synthesis

It is only relatively recently that organometallic chemistry (the chemistry of compounds containing carbon-metal bonds) has come to the forefront of organic synthesis. However, since the 1970s, many new types of organometallic reagents and catalysts have been developed and now organometallic compounds are used by organic chemists in a number of ways. We shall look at just two examples.

Some organometallic compounds are used as convenient sources of highly unstable organic molecules. For example, cyclobutadiene, a ring formed from four carbon atoms and four hydrogen atoms, is so strained and thus unstable that it does not exist at normal temperatures and pressures – it reacts with itself to form a larger ring. However, a complex formed with iron tricarbonyl, Figure 14, is a relatively stable and easily-manipulated compound. (Carbonyls are molecules of carbon monoxide bound to a metal through their carbon atoms.) Because the iron and the cyclobutadiene are readily separated, the iron complex provides organic chemists with a convenient source of cyclobutadiene.

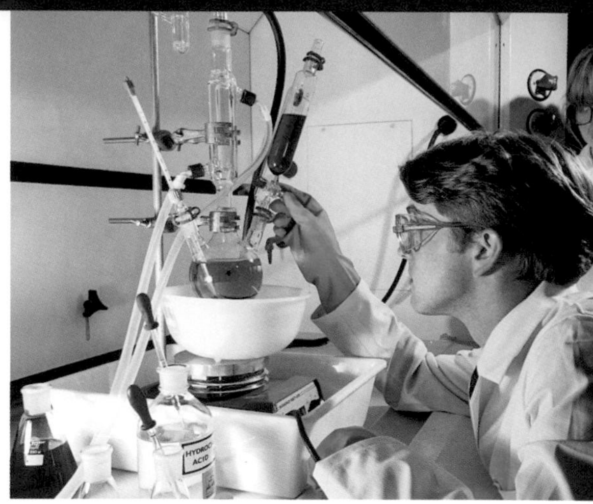

GlaxoWellcome.

12. What is the C–C=C angle in cyclobutadiene? Use this to explain the phrase 'cyclobutadiene ..... is so strained and thus unstable that it does not exist at normal temperatures and pressures'.

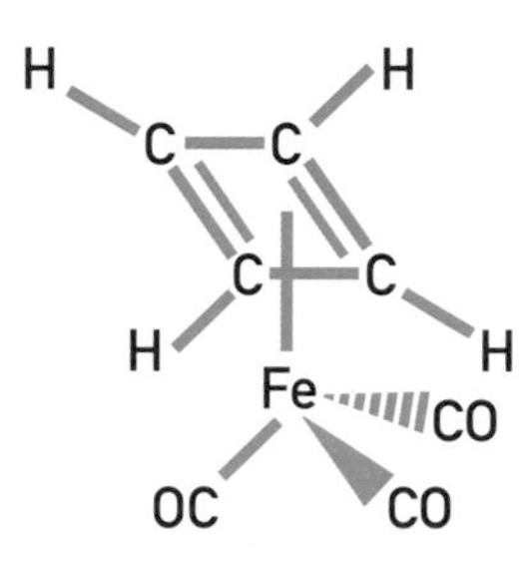

Figure 14. Unstable cyclobutadiene forms a stable complex with iron tricarbonyl, making it easier to manipulate.

Organic chemists also use organometallic compounds to bring about changes in the reactivity of organic molecules. For example, benzene is surrounded by a cloud of negative electrons and so it is attacked by positively-charged **species**. Attaching a metal fragment (based on chromium) to benzene to give the compound in Figure 15, sucks the electrons away from the benzene and changes its preference for positive species to one for negative species. So, by attaching metals to organic molecules, some of the traditional rules governing how organic molecules may be joined together are broken. This provides organic chemists with greater flexibility when they design syntheses.

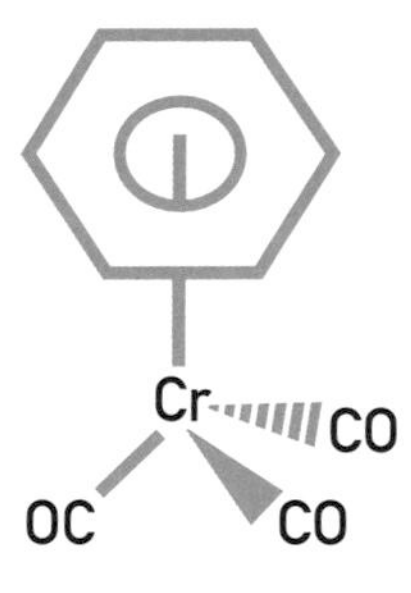

Figure 15. The chromium atom bonded to the benzene ring sucks electrons away from the ring.

13. What terms are normally used for (a) the positively charged species that benzene normally interacts with; and (b) the negatively charged species that it interacts with when bonded to chromium?

Glossary

Species: A general term used to refer to atoms, molecules or ions.

Combinatorial chemistry

Pharmaceutical companies need to make large numbers of molecules which they can screen for activity as drugs. This can be laborious and time consuming using traditional methods.

Currently, however, a group of new synthetic techniques called combinatorial chemistry is being developed to aid the medicinal chemist. They are methods by which large numbers of compounds (called libraries of compounds) can be rapidly synthesised for screening for biological activity. Reactions can either be done in solution or on insoluble resin beads, a technique known as solid phase chemistry.

The solid phase technique has its origins in a method developed in the 1960s by the Nobel prize-winner Bruce Merrifield for making peptides (short chains of amino acids), see Chapter 10 – *The chemisty of life.*

The first amino acid is attached to an insoluble solid support and further amino acids are coupled onto it one after the other with the excess reagents washed away after each step. At the end of the synthesis, the final peptide is removed from the support.

A neat trick which has been developed to give a huge increase in productivity is to make the peptides on tiny polystyrene resin beads only 100 micrometres across, and to use a process of 'mix and split' to make large numbers of peptide products.

Mix and split

The process is illustrated schematically in Figure 16 for a simple case using just three amino acids, represented by an ellipse, a triangle and a square, and three reaction vessels. In step 1, a quantity of resin beads is split into three portions and a different amino acid attached to each of the portions. After this step, all of the resin beads are mixed, and again split into three portions for step 2 – the addition of the next amino acids (a process called coupling). Each portion now contains beads with two amino acids attached, and all nine possible combinations of dipeptide are generated. A further cycle of mix and split followed by amino acid coupling (step 3) produces a total of 27 tripeptides. The process can be continued.

The size of the peptide 'library' generated is X^n ,where X is the number of amino acids used and n the number of coupling steps.Thus, using 20 amino acids in each cycle permits the production of 400 dipeptides, 8000 tripeptides or 160,000 tetrapeptides and so on. At the end of the synthesis, the combinatorial library consists of a large number of resin beads, with many molecules of only one specific peptide sequence attached to each bead, Figure 17. An individual bead can then be picked out and its specific peptide released.

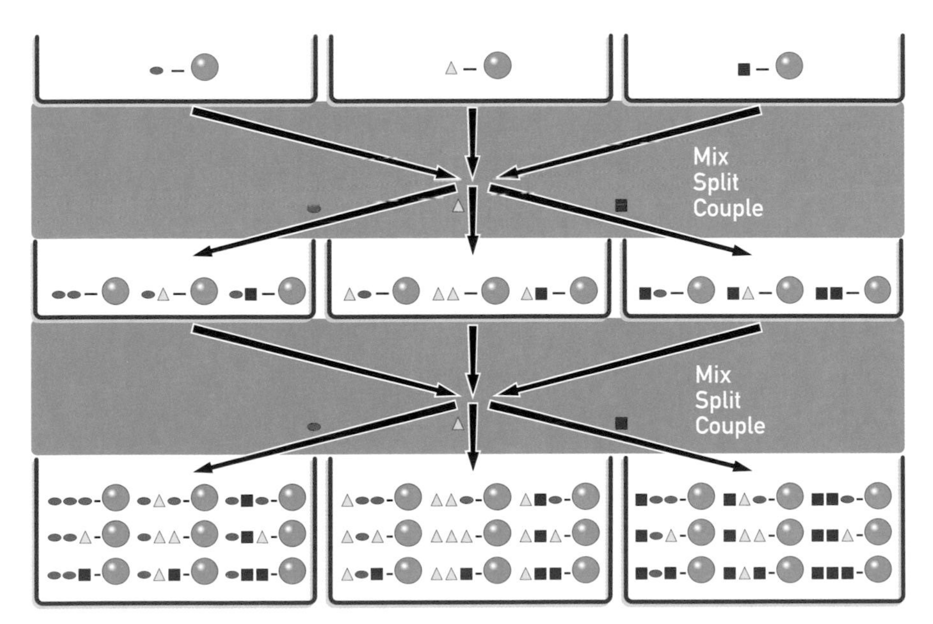

Figure 16. The mix-and-split process. Chemical building blocks (shown here as an ellipse, triangle and square representing monomers such as amino acids used to make peptides) are attached to beads, which are mixed and split into three groups for a second coupling with the same three monomers. The procedure is repeated to give the 27-compound library shown here.

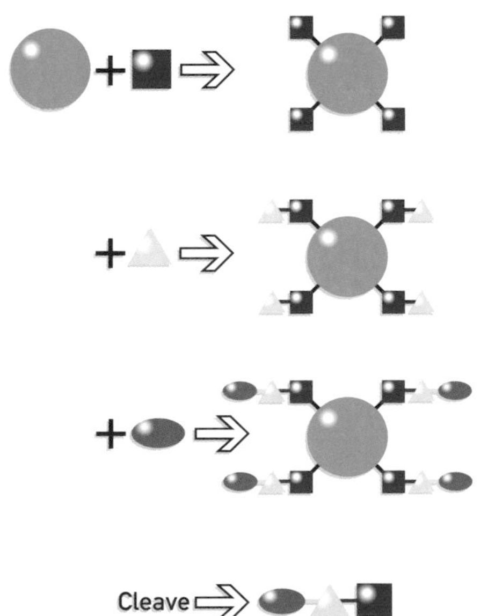

Figure 17. Solid-phase chemistry allows the synthesis of compounds attached to a resin bead. A final cleavage step generates the product in solution.

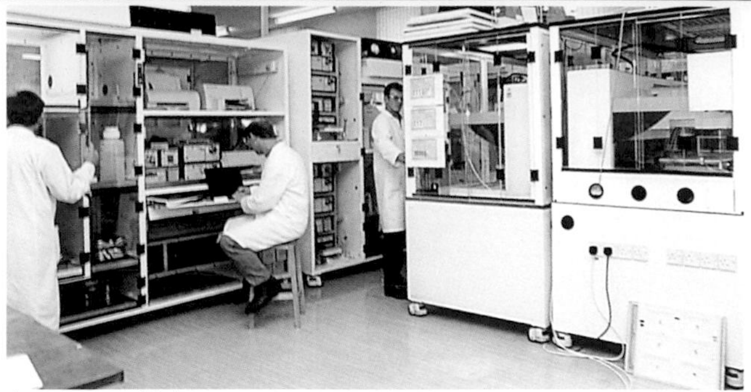

An automated laboratory.
Courtesy of Pfizer.

The combinatorial library process has many advantages.

- **It generates huge numbers of peptides in only a few steps.**
- **Many different reactions involving the same reagents can be combined in the same reaction flask.**
- **A large excess of the second (and subsequent) amino acids can be used to drive the equilibria towards the product (Le Chatelier's principle).**
- **The excess amino acid can simply be washed away while the product remains on the resin.**
- **Purification of the final product is easy. After removing the final peptide from its support, the resin beads can simply be filtered off from the final reaction mixture and washed.**
- **The process can be automated using synthetic 'robots' – essentially computer-controlled syringes which add reagents.**
- **Sufficient material can be made on a single bead for preliminary testing for biological activity in vitro to see if it affects enzymes or binds to receptors in cells. There is also sufficient material to confirm the product's identity by a variety of instrumental techniques (See Chapter 2 – *Analysis and structure of molecules*).**

Fortunately, the technique can easily be adapted to make libraries of non-peptide molecules because peptides do not make good drugs. Today just about every major pharmaceutical company has a group dedicated to the production of novel organic molecules using combinatorial chemistry.

14. Suggest what might happen to a peptide in the human gut which would make it ineffective as a drug.

Glossary

In vitro: Literally 'in glass' *ie* in test-tube reactions rather than on living organisms.

Supramolecular chemistry: Chemistry relating to the structure, properties and characteristics of assemblies of two or more molecules held together by intermolecular forces.

Courtesy of QuChem.

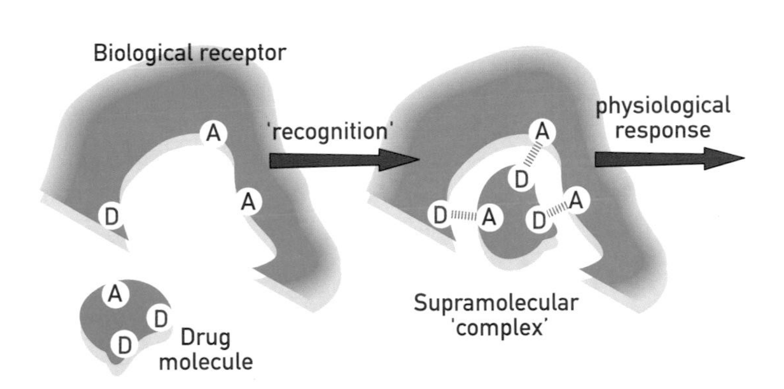

Figure 18. The binding between drug and biological receptors through hydrogen bonding, shown by D = hydrogen donor sites and A = hydrogen acceptor sites.

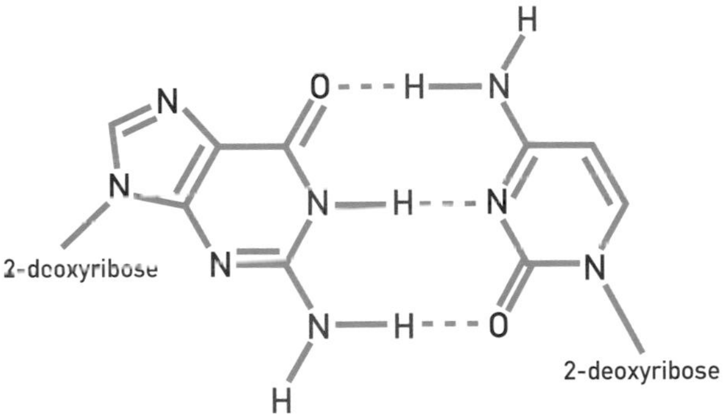

Chemistry beyond the molecule

Over the past 30 years, there has been increasing interest in preparing chemical systems that mimic the behaviour of natural ones. This is because Nature carries out so many difficult chemical processes with such efficiency. One of the crucial characteristics of biological molecules is that they often interact through weak electrostatic attractions such as hydrogen bonding, for example the interaction between a receptor protein and a small drug molecule which brings about a response by the body. This is called molecular recognition, Figure 18.

Large biological molecules such as proteins and the strands of DNA are held in their characteristic three-dimensional shapes by hydrogen bonding – the attractive force between slightly positively charged hydrogen atoms and the negatively charged electrons on oxygen and nitrogen atoms, Figure 19. Without hydrogen bonding, life would not exist.

Understanding how these weak forces control the architecture and behaviour of large molecules and assemblies of molecules is now an exciting new field of chemistry. It is called **supramolecular chemistry** – that is, chemistry beyond the molecule.

Some of the first attempts to copy the supramolecular chemistry of biological molecules were to create artificial receptors whereby ions of alkali metals such as sodium and potassium bind into the centre of a cyclic organic compound in what is called a host-guest complex.

Some of the molecules which can do this are shown in Figure 20.

15. Show on a copy of Figure 20 where you would expect an alkali metal ion to fit into each molecule. What sort of bonding would hold it in place?

2-deoxyribose

2-deoxyribose

Figure 19. Hydrogen bonding between complementary nucleic acid bases guanine (on the left) and cytosine in DNA.

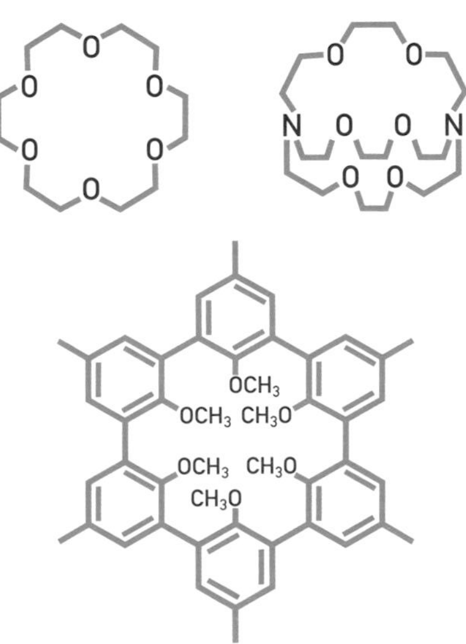

Figure 20. Some molecules which can bind to alkali metal ions.

Donald Cram.
Courtesy of Robin Robin and UCLA.

Boxing clever

In a development of this idea, 1987 Nobel Prizewinner Donald Cram synthesised a 'molecular prison' (Figure 21) in which he trapped the hyper-reactive cyclobutadiene molecule mentioned earlier. He called the molecule a carcerand, from the word 'incarcerate' meaning to imprison. At low temperatures, the bars of the carcerand are closed and nothing can get in or out, but at temperatures greater than 130 °C (403 K), the bars bend and twist and small molecules can be forced in and out. In fact, Cram's team first trapped a molecule called α-pyrone and then used ultraviolet light to decompose it into cyclobutadiene and carbon dioxide which could 'escape' from the cage, Figure 22. Once trapped within its protective shell, the cyclobutadiene was unable to reach other molecules of its type to react with and thus its characteristics and chemistry could be studied at leisure.

Another molecule which has been studied while 'trapped' is 'benzyne' a benzene-like molecule with a carbon-carbon triple bond.

16. Suggest how chemists would be able to show that benzyne has a triple bond?

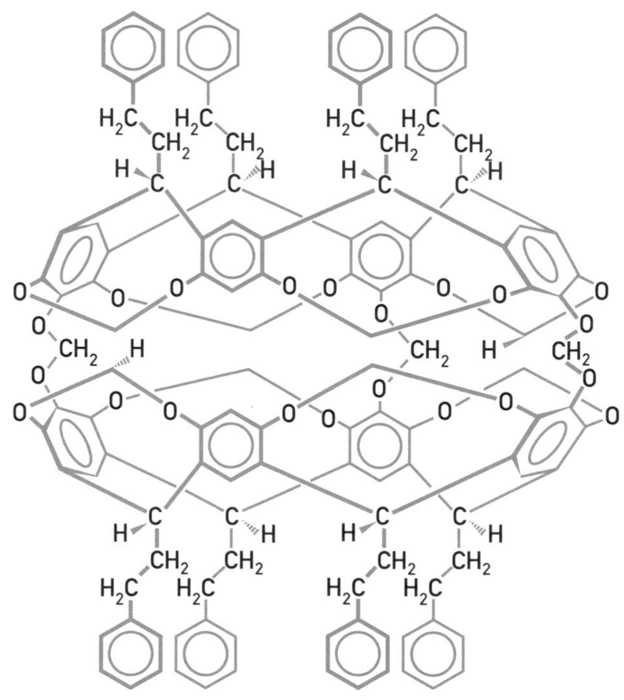

Figure 21. Cram's carcerand in which he imprisoned cyclobutadiene.

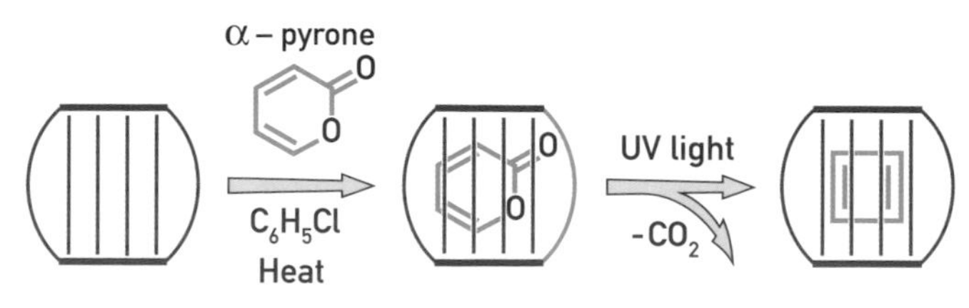

Figure 22. Incarceration of cyclobutadiene.

Extracting caffeine from tea or coffee could be done by host-guest chemistry.

Chemical sensors

The ideas of host-guest molecules have also been used to develop chemical sensors. The molecule (b) in Figure 23 hydrogen bonds very specifically to the molecule (a) which is one of a class of drugs called barbiturates. This could be used to develop a sensor or 'artificial nose', for detecting barbiturates in mixtures of other chemicals.

Another possible application is to construct molecules that can selectively bind strongly to other species and use them for applications such as:

- **removing urea from blood (artificial kidney dialysis);**
- **extracting the vast quantities of uranium and gold naturally present at low concentrations in sea water;**
- **eliminating cadmium, lead and other pollutants from industrial effluents; and**
- **extracting caffeine from tea and coffee.**

17. The stucture of caffeine is shown below. Which atoms in the caffeine molecule might form hydrogen bonds to a sensor molecule?

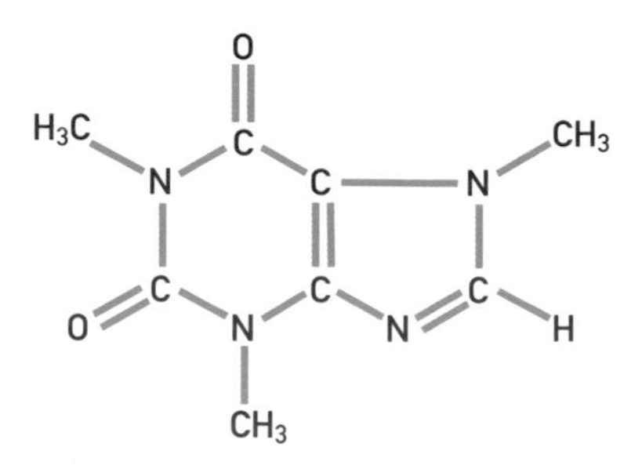

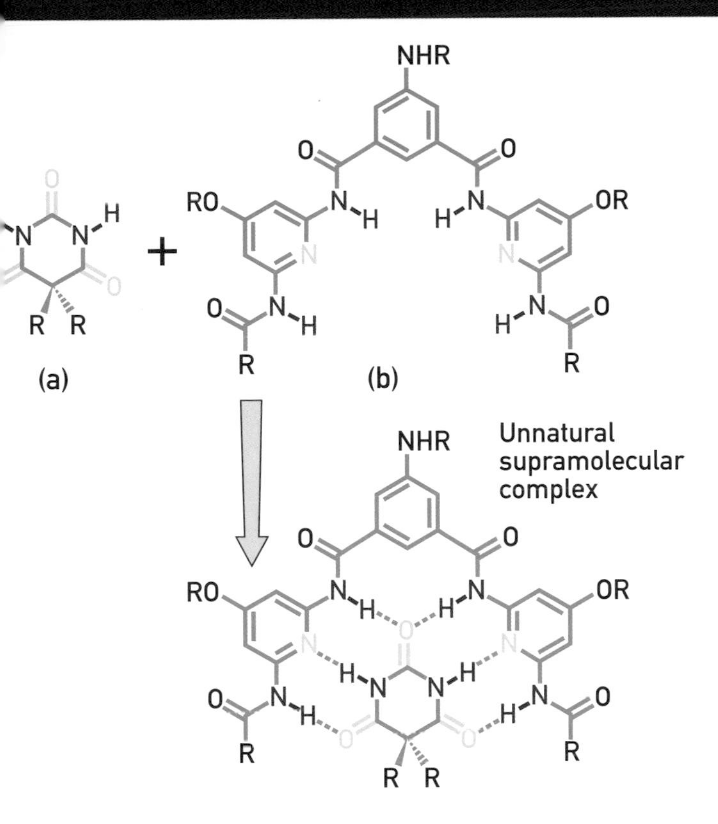

Figure 23. Recognition between barbiturates and a synthetic receptor – a prototype chemical sensor.

The future of synthesis

Synthetic organic chemistry still presents enormous challenges. One is the preparation of large, naturally-occurring molecules. Synthesising them helps chemists understand how these molecules work and also stimulates them to develop new reactions. Increasingly, chemists try to mimic the way nature makes the molecule, using types of reactions that automatically prime the product for the next reaction step. This means that the reagents needed for several steps can all be put together in one flask and left to get on with it – 'one pot synthesis'.

A second area of interest is designing environmentally-friendly syntheses – ones which can be carried out at low temperatures to save energy and in water rather than in traditional solvents to reduce pollution.

Computers and robots will increasingly be used to both design and carry out syntheses and, in years to come, it may be possible to design complex pieces of molecular architecture that will automatically self-assemble from simple building blocks into a useful device. Although that view is science fiction now, the 21st century is sure to reveal many such exciting advances in organic synthesis.

Courtesy of Hornby Hobbies Ltd.

Molecular train sets

One of the characteristics of biological systems is that they tend to 'self assemble' by intermolecular forces – the replication of DNA is a good example. A number of chemists are beginning to take this idea further and design molecules that can act as wheels, cogs, spindles and switches in molecular-sized 'machines', sometimes called nano-machines after their sizes.

One team has made doughnut-shaped molecules which can be moved up and down a linear molecule (like a bead on a thread) in response to signals with protons or electrons, and they have even made a 'molecular train set', where a bead moves from 'station' to 'station' around a circular track. Although practical applications are some way away, this achievement gives a glimpse into one possible future for organic synthesis.

Answers

1. Silicon atoms are bigger than carbon atoms, so the shared electrons in bonds are further from the nuclei which they are holding together and therefore the bonds are weaker. (Although the electrons 'feel' the same shielded nuclear charge.)

2. Carbon has four electrons in its outer shell and 'needs' four more to form a full outer shell.

3. (a) Propan-2-ol (*ie* the OH group has replaced the Br atom).
   (b) Propene (*ie* a molecule of water has been lost).

4. A variety of answers is possible depending on the experience of the student, *eg* alkene, alkyne, haloalkane, alcohol, carboxylic acid, ester, acid halide, amide, nitrile.

5. This allows the four shared electron pairs in the bonds to be as far apart as possible. 109.5°.

6. Yes it can.

7. Exchange of any pair of atoms leads to the two molecules becoming identical.

8.

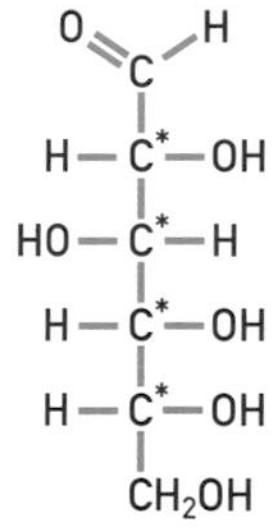

```
 O    H
  \\ /
    C
    |
 H—C*—OH
    |
HO—C*—H
    |
 H—C*—OH
    |
 H—C*—OH
    |
   CH2OH
```

9. Breaking up by reaction with water.

10. Butanedial and propanone.

11. Reduction. Various answers including hydrogen with a suitable catalyst, sodium tetrahydridoborate(III), lithium tetrahydridoaluminate(III).

12. 90°. Ideally, the bond angle would be 120° as in ethene.

13. Electrophiles and nucleophiles respectively.

14. The peptide bonds would be hydrolysed by acid and/or enzymes. (In fact peptides are too polar to pass through cell membranes or the blood-brain barrier which also restricts their use as drugs.)

15. Bonding between the $O^{\delta-}$, $N^{\delta-}$ and the M^+ ion.

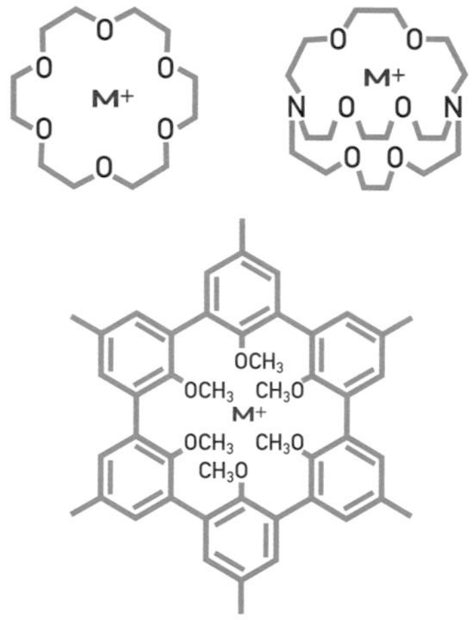

O O O O O O M^+

O O N O O N O O M^+

OCH_3 OCH_3 CH_3O M^+ OCH_3 CH_3O CH_3O

16. One carbon-carbon bond length would be shorter than the rest and thus have a different infrared absorption.

17. The nitrogen and oxygen atoms could form hydrogen bonds to hydrogen atoms on the sensor molecule that had sufficient δ^+ character.

Introductory page pictures: Hornby Hobbies Ltd.

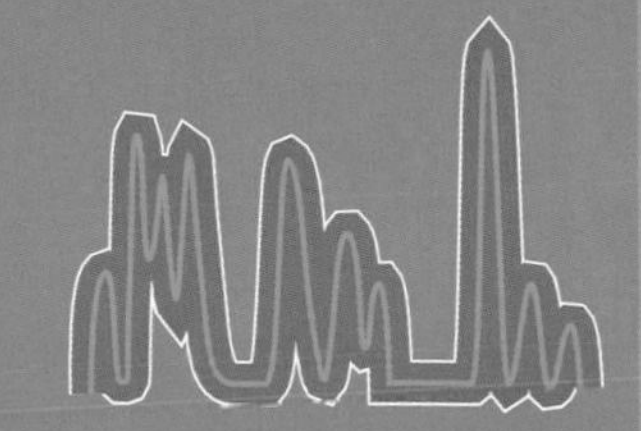

Analysis and structure of molecules

Over the past century, chemists have accumulated a vast armoury of ingenious analytical methods to determine the structures of the chemical substances they produce.

These include:

- 'classical' chemical analysis using a variety of test reactions and techniques;
- **ultraviolet** (UV) and visible spectroscopy;
- **infrared** (IR) spectroscopy;
- mass spectrometry;
- X-ray, neutron and electron **diffraction**; and
- nuclear magnetic resonance (NMR) spectroscopy.

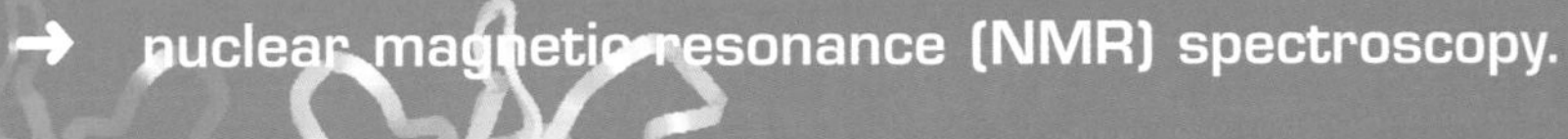

Glossary

Carcinogen: A cancer-causing chemical.
Diffraction: The spreading out of waves as they pass through a gap comparable in size to their wavelength.
Electromagnetic radiation: Energy, consisting of oscillating electric and magnetic fields, which

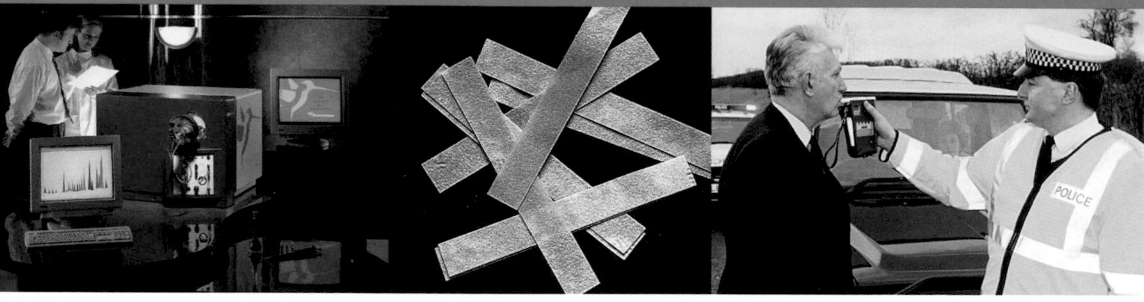

Different methods of chemical analysis.

Courtesy of Micromass UK Ltd. *Charles D Winter/Science Photo Library.* *Courtesy of West Midlands Police.*

Pick up any newspaper and it is likely to contain a news story in which analytical chemistry plays a part: a celebrity is accused of drink-driving or an athlete of taking drugs; river water is found to contain toxic metals or terrorists who have planted a bomb have been incriminated. The outcome of each case involves identifying and measuring traces of a chemical – alcohol, a drug, a metal or an explosive. Mostly, though, the role of analysis is less obvious. When a new 'wonder drug' is announced, its discovery and development will have depended on a whole battery of analytical tools. Sometimes the analytical requirement is more exotic; how much iron is there in dust brought back from the Moon, for example? Over recent decades, chemical analysis has become increasingly sophisticated, taking advantage of computer technology and robotics. It is now possible to detect minute quantities of materials – which can pose problems of interpretation. In 1990 there was a scare over benzene in Perrier water. This arose only because a new analytical technique was introduced that was so sensitive that it could detect minute amounts of the benzene (a **carcinogen**) – down to one part per billion. This is the equivalent of a single sheet of paper in a stack 50 km (30 miles) high! The amounts detected were between 7 and 20 parts per billion. There is now no detectable quantity of benzene in Perrier water.

Perrier water production halted

By Alan Friedman in New York

PERRIER, the French mineral water company, has halted production, and North American sales have been stopped, because of suspected contamination.

Perrier Group of America, Perrier's American importer, around $40m in lost sales, said Mr Ronald Davis, president of Perrier Group of America.

US sales of Perrier's regular and flavoured waters are $160m a year. Mr Davis said on Saturday that the search for the cause of the "chemical month, appeared to be "human error". He described it as "a very freak accident."

Tests of Perrier bottles in North Carolina and Georgia found the benzene level to be 12.3 to 19.9 parts per billion, well above the FDA's permissi-

Courtesy of Financial Times.

Classical chemical analysis

What sorts of analysis do chemists do? Much modern analytical work is done with instruments which probe the structures of substances using **electromagnetic radiation**. However, there is still a place for traditional methods as illustrated by the following story which concerns the first high temperature **superconductors**, discovered in 1986, see Box – *Analysis of superconducting materials.*

Courtesy of QuChem.

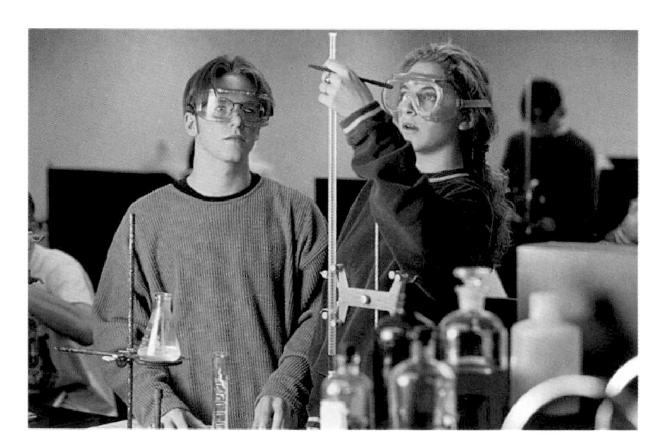

Titration in school.
Tony Stone Images.

can pass through a vacuum.
Infrared: The part of the electromagnetic spectrum with wavelengths a little longer than those of visible light.
Superconductor: A material that can conduct electricity without any resistance.
Ultraviolet: The part of the electromagnetic spectrum with wavelengths a little shorter than those of visible light.

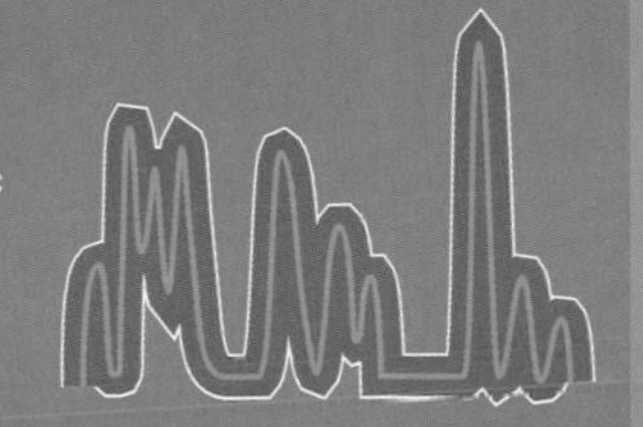

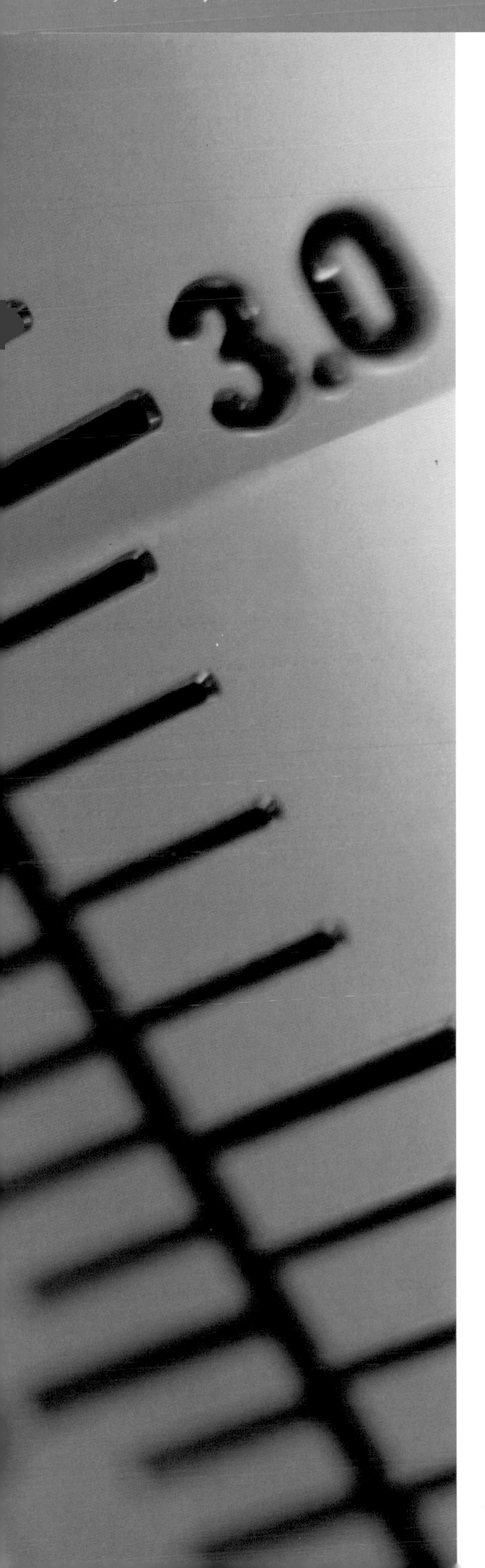

Analysis of superconducting materials

The first high temperature superconductors were a group of oxides of yttrium, barium and copper of formula $YBa_2Cu_3O_x$. The value of x depends on the oxidation state of the copper – in particular whether or not there is any Cu^{3+}.

? 1. What would be the value of x if all the copper were in the Cu^{3+} oxidation state? Yttrium normally forms Y^{3+} ions.

The best way of determining the amount of this unusual oxidation state of copper turns out to be a classical titration method, called iodometry, introduced by Robert Bunsen (of burner fame) in 1853. Copper (either in oxidation state II or III) oxidises iodide ions (I^-) to iodine (I_2), while itself being reduced to copper(I). The amount of iodine produced can then be determined by titration with sodium thiosulfate. Thus, if we know the concentration of sodium thiosulfate solution, we can calculate the oxidation state of copper in the original material, in other words, the value of x.

? 2. How many iodide ions (I^-) could be oxidised to iodine molecules (I_2) by (a) a Cu^{2+} ion (b) a Cu^{3+} ion?

? 3. Write the equation for the reaction of sodium thiosulfate with iodine. How is the end point determined in the titration?

The other method giving comparable accuracy is a traditional thermogravimetric analysis. The $YBa_2Cu_3O_x$ sample is heated under a flow of hydrogen gas which reduces it to a mixture of yttrium oxide (Y_2O_3), barium oxide (BaO) and copper metal; x is found from the percentage mass loss.

? 4. What will be the other product of this reaction? Write an equation for the reaction if the starting material is $YBa_2Cu_3O_8$.

Combustion analysis

Another 'classical' technique which is still used is combustion analysis in which the amounts of carbon, nitrogen, hydrogen and sulfur in a compound are found by burning the compound in a stream of oxygen and measuring the amounts of products formed. This technique is used to confirm the formulae of organic compounds and can be automated.

? 5. Suggest the products of combustion analysis of a compound containing the elements carbon, nitrogen, hydrogen and sulfur.

Glossary

Adsorb: Form weak chemical bonds with a surface.
Atomic spectroscopy: The study of the electromagnetic radiation emitted and absorbed by substances.
Electrode: An electrically conducting material through which electrons enter or leave an

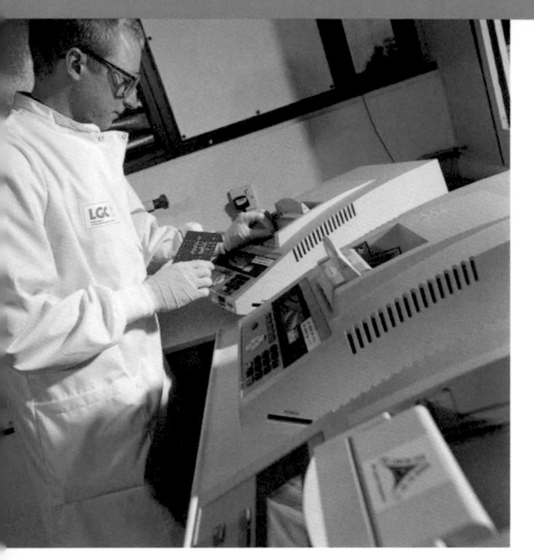

An analyst working with modern analytical instruments.
Photograph courtesy of LGC.

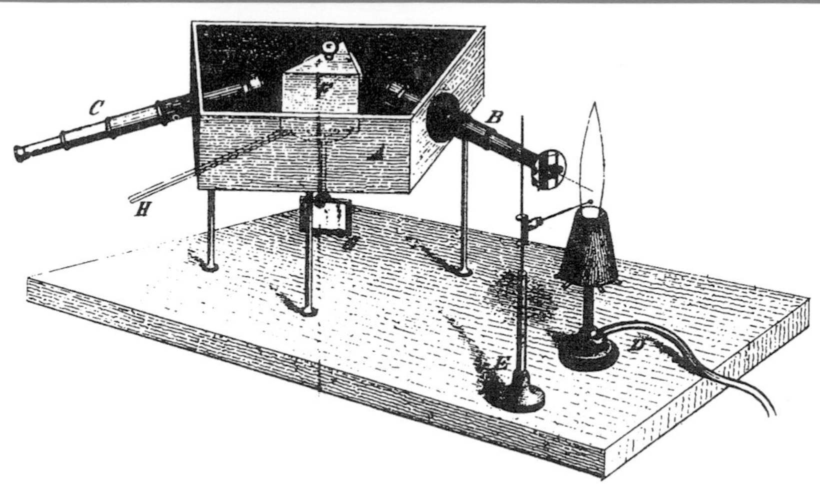

A 19th century spectroscope.
The History of Chemistry, Hudson, Macmillan.

The rise of physical methods

The most spectacular advances in analytical science, however, have followed the discovery of some new physical phenomenon. An example with a long history is **atomic spectroscopy**. In the 19th century, scientists realised that when some compounds are heated in flames, they emit light with a colour characteristic of certain constituent elements. For instance, compounds containing sodium show a bright yellow flame (this is why your gas cooker flame burns yellow when salty water boils over). If the emitted light is then passed through a spectroscope, which acts like a prism, it is split into bands of colours – the spectrum. The bands, or lines, can be used to identify the elements. Atomic spectroscopy enormously enhanced analytical chemistry. At first, the method was used for qualitative analysis (*what is present?*) as in simple flame tests. Now it can be used for quantitative analysis (*how much is present?*) and there are several variants of the technique which, when combined with modern technology, give extremely sensitive and precise results. For instance, when the sample is heated in an inductively coupled **plasma** (ICP), rather than a simple flame, it is possible to detect 0.08 millionth of a gram of magnesium in 1 dm^3 of solution. An ICP is a stream of ionised argon atoms heated up to 10 000 K by a radio frequency generator.

Very small amounts of substances can also be measured using electrochemistry. In this case the voltage at an **electrode** is affected by the concentration of ions around it. The voltage of an ion selective electrode depends on the concentration of only one species of ion. So it gives a direct and easy method of measuring its concentration, especially in the presence of other ions. The crucial part of an ion-selective electrode is a **membrane** which permits diffusion of only one species. For example, the concentration of fluoride ions can be measured at levels down to billionths of a gram per dm^3 by using a solid membrane of lanthanum fluoride (LaF_3) doped with europium fluoride (EuF_2) which lets through only fluoride ions. A more common system is the glass electrode which is permeable to H^+ ions only and is used to measure the pH of solutions.

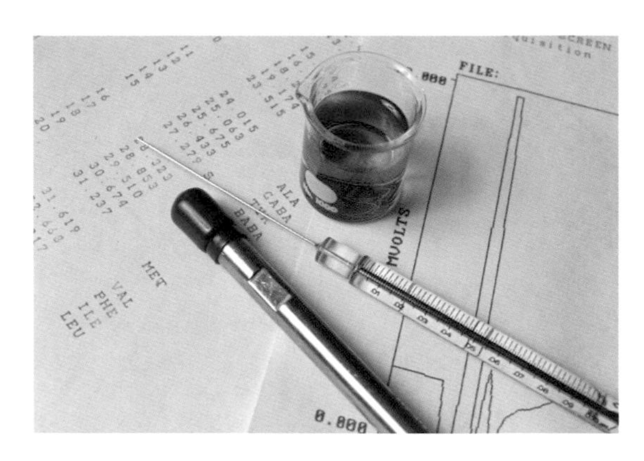

Analysing mixtures

Often, before we can analyse one component of a mixture, we must separate the mixture so that signals from the compound we are interested in are not obscured by those from the other components. One of the most powerful techniques for separating mixtures is the family of methods called chromatography, see Box – *Chromatography*.

Courtesy of QuChem.

electrochemical cell and where electrochemical reactions take place.
Membrane : A sheet of material that separates two phases such as a liquid and a gas.
Optical isomers: Pairs of molecules which are non-identical mirror images of one another – they have virtually identical physical properties which makes them very difficult to separate.
Phase: Different phases of a system are those separated by a distinct boundary, such as liquid and solid.
Plasma: A gas consisting of positive ions and electrons.

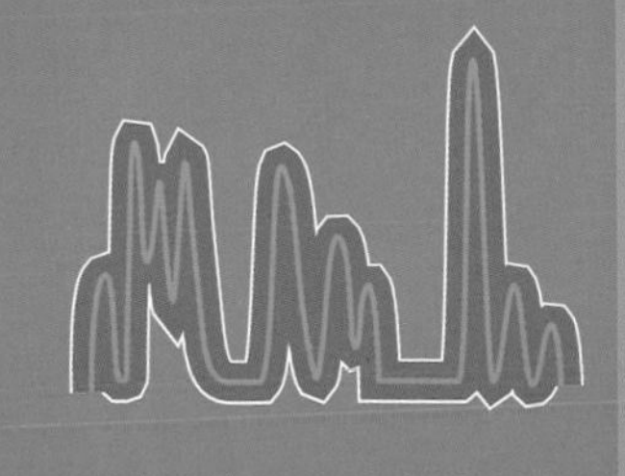

Chromatography

Spill fruit juice on a white tee-shirt and you can often see chromatography at work. As the stain spreads, it may separate into circles of different colours showing that it is in fact a mixture. The botanist Mikhail Semenovich Tswett coined the term chromatography ('coloured writing') in 1906. He passed a solution of plant pigments down a glass column packed with calcium carbonate particles, and then poured solvent down the column. Different coloured bands appeared as the components of the mixture were separated. The different coloured bands travelled separately down the column and were collected in the receiving vessel one by one. Thus the components of the original mixture were separated.

Chromatography depends on the presence of a stationary **phase** (often a solid) and a mobile phase (a liquid or a gas). In Tswett's experiments, the stationary phase was calcium carbonate and the mobile phase was the solvent moving down the column. Suppose the mixture has only two components, X and Y. If X **adsorbs** more strongly than Y on calcium carbonate before being washed off, then it will pass more slowly through the column than Y and the two can be separated.

A powerful development of this technique is known as high performance liquid chromatography (HPLC). If the particles of the stationary phase in the column are very small then separation can be very efficient, but the rate at which the solution passes through is very slow. However, the solution can be forced through the column by applying high pressure, and this results in extremely efficient separation. With a suitable stationary phase, even **optical isomers** can be separated.

Chromatography is not confined to a solid stationary phase and a liquid mobile phase. One of the major advances in chromatography was devised by two British chemists Archer Martin and Richard Synge who received the 1952 Nobel Prize for their work. They developed partition chromatography, in which the mobile phase is a liquid, and the stationary phase is also a liquid but which is held as a thin film on a solid support. An important application of this is the separation of amino acids using wet filter paper as the stationary phase and a variety of solvents as the mobile phase.

A very important type of chromatography uses gases as the mobile phase and either a liquid or a solid as the stationary phase in a long, narrow tube. Gas chromatography (GC) is often used in industry to analyse mixtures of volatile organic compounds such as perfumes and flavours in cosmetics and foods, and to monitor air pollution.

If the components of the mixture are not coloured, some method has to be used to detect them after they are separated. Chromatographic methods are often directly combined with an analytical technique such as mass spectrometry. GC-MS (gas chromatography combined with mass spectrometry) and LC-MS (liquid chromatography-mass spectrometry) equipment is found routinely in many laboratories in the fine chemicals and consumer products sectors.

Richard Synge.
©The Nobel Foundation.

Archer Martin.
©The Nobel Foundation.

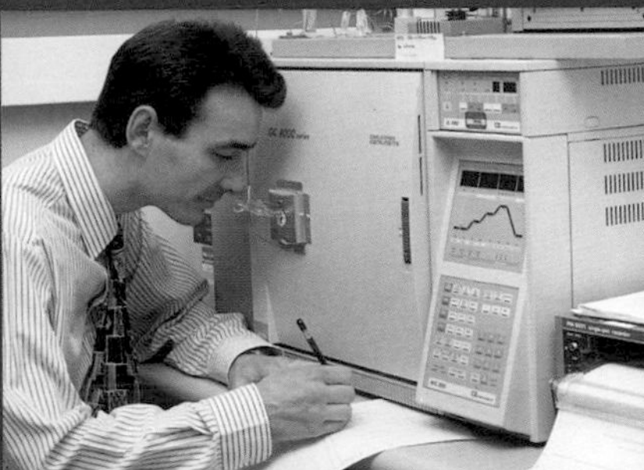

GC 'smelling machine'.
Quest International.

Wavelength/m	10^{-10}	10^{-9}	10^{-8}	10^{-7}	10^{-6}	10^{-5}	10^{-4}	10^{-3}	10^{-2}	10^{-1}	10^{0}	10^{1}	10^{2}	10^{3}
Frequency/s^{-1}			$3x10^{16}$		$3x10^{14}$		$3x10^{12}$		$3x10^{10}$		$3x10^{8}$		$3x10^{6}$	
Radiation type	X-rays		Ultraviolet		Visible		Infrared		Microwave				Radiowaves	
Name of technique	X-ray diffraction		UV/visible spectrophotometry				IR spectro-photometry		Microwave spectroscopy				Nuclear magnetic resonance	
Summary	Arrangement and spacing of atoms in solids		Electrons moving between different energy levels – identifies different elements and also special groups called chromophores				Vibrations of bonds – identifies the types of bonds present in a molecule		Rotation of molecule – gives bond lengths				Flipping of nuclei in a magnetic field – identifies the number and type o hydrogen (and oth atoms in molecule	

Table 1. Analytical techniques which use electromagnetic radiation.

The structures of compounds

The examples of analysis given so far offer no information about chemical structure. They tell us neither the functional groups present, such as the -OH group in ethanol, CH_3CH_2OH, nor the three-dimensional arrangement of the atoms. A glance at almost any chapter in this book will reveal the vital importance to chemistry of knowing molecular structure, from the simplest molecules containing a few atoms to the most intricate biomolecular systems containing thousands.

Chemists use a number of methods to determine the structure of the chemical substances they produce. Most of them involve probing the molecule with either electromagnetic radiation, see Table 1, or particles such as electrons or neutrons. Each analytical technique, offers a certain kind of information, or is aimed at specific kinds of molecules or structures (such as crystals or surfaces). Here we mention a few of the main methods.

Glossary

Constructive interference: The adding together of two waves when their peaks and troughs coincide.

Destructive interference: The cancelling of two waves which occurs when the peaks of one wave coincide with the troughs of the other.

In phase: Two waves are in phase when their peaks and troughs line up.

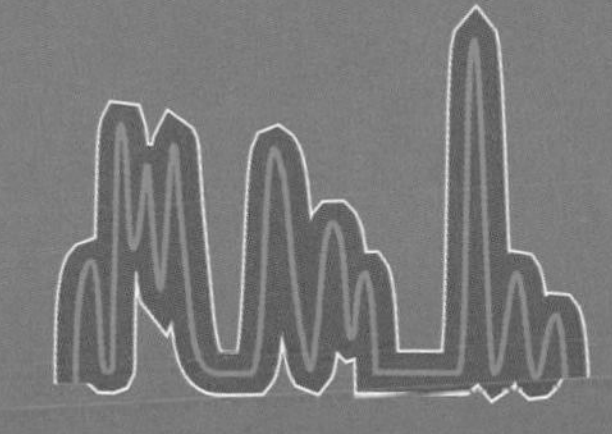

X-ray diffractometer.
Courtesy of Bruker.

X-ray, neutron and electron diffraction

Perhaps the most straightforward way of examining the structure of a compound is to shine light on it and then examine the light which is scattered. (This is how our eyes see the shapes and surfaces of things.) But to discern structure at the level of chemical bonds, the radiation wavelength must be of comparable dimensions to interatomic distances, and that falls within the X-ray region of the electromagnetic spectrum.

The simplest interaction of X-rays is with substances in a crystalline form – through the phenomenon of diffraction. A crystal consists of repeating arrangements of molecules, atoms or ions in a giant three-dimensional array which can be viewed as regularly oriented planes of similar entities. When X-ray waves hit the sample, they are reflected off consecutive planes in such a way that the waves interfere and reinforce or cancel each other periodically to produce a diffraction pattern – a series of peaks and troughs in X-ray intensity. The pattern contains clues to the arrangement of atoms in the crystal and can be interpreted by Bragg's Law – a famous relationship discovered in 1912 by the father and son team of William and Lawrence Bragg.

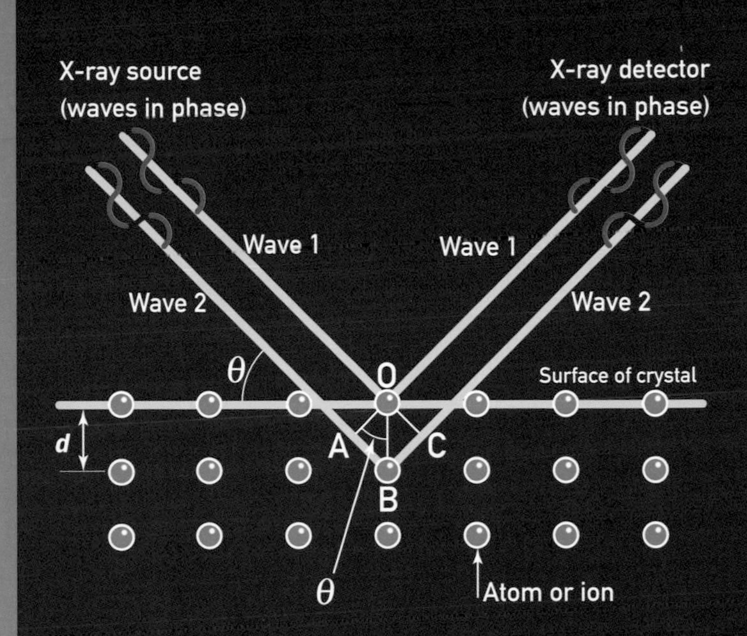

The diagram shows what happens when a beam of X-rays is shone on the surface of a crystal which is composed of a regular array of atoms or ions. The X-rays behave as though they are reflected by the atoms or ions. Those reflected from the second layer of atoms in the crystal travel further than those reflected from the surface atoms, so they 'lag behind'. Unless they emerge **in phase** with the X-rays reflected from the first layer, they will cancel out (**destructive interference**) and no X-rays are detected. Only if the X-rays reflected from the second layer lag behind those from the first by a *whole number of wavelengths*, *n*, does **constructive interference** occur and significant amounts of X-rays can be detected. If we start with the X-ray beam parallel to the surface of the crystal and gradually rotate the crystal to increase the angle θ, the X-rays reflected from the second layer of atoms lag further and further behind those reflected from the surface layer until they are one wavelength behind ($n = 1$) and constructive interference occurs. At this point, they have travelled a distance ABC further.

A little trigonometry on triangles OAB and OAC shows us that

$$\mathbf{ABC = 2}d\,\mathbf{sin}\theta$$

And for the X-rays to be in phase:

$$n\lambda = 2d\sin\theta$$

This is Bragg's Law and it enables us to find the spacing, *d*, between the layers of atoms or ions in the crystal provided we know the wavelength of the X-rays.

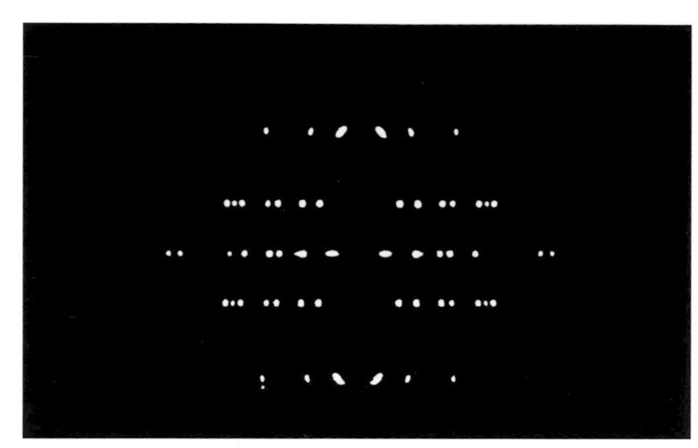

The diffraction pattern of urea is converted mathematically into...

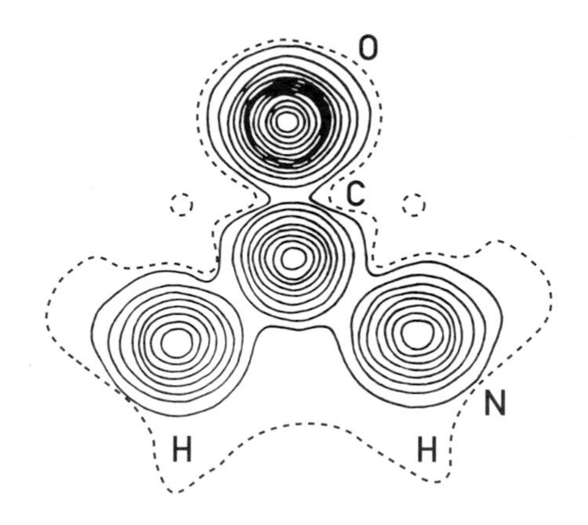

...an electron density distribution which indicates...

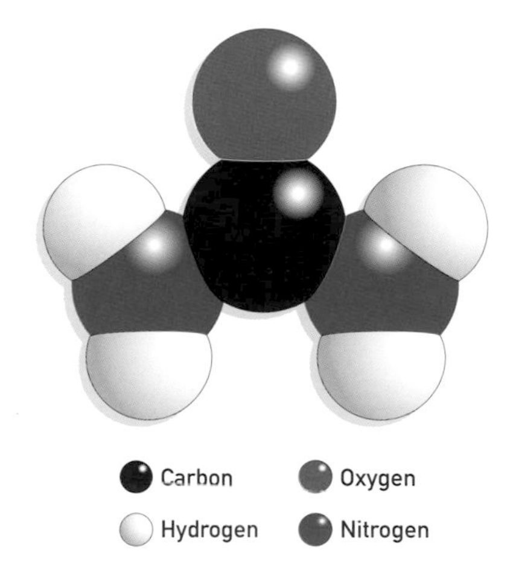

...the positions of the atoms.

The crystal is rotated through all angles to obtain diffraction patterns for all planes, and the patterns are then analysed. Because X-rays are a form of electromagnetic radiation they interact with the clouds of electrons in each atom, so what we actually have is a record of the distribution of **electron density** in the compound. The conversion of the scattering data to structure is actually extremely complicated and involves a great deal of computation. However, increases in computing power over recent decades and new techniques have greatly speeded up X-ray analysis so that quite complex structures can now be solved relatively quickly.

X-ray diffraction, also called X-ray crystallography, provides the most complete structural picture of a molecule. Most university research laboratories have at least one X-ray instrument. X-ray analysis requires a fairly large single crystal (although for some compounds it is possible to determine stucture from X-ray diffraction on powders).

X-ray diffraction has become a particularly powerful tool in molecular biology (See Chapter 10 – *The chemistry of life*). It is used widely and routinely to solve very large three-dimensional structures containing thousands of atoms, such as proteins, **RNA**, **DNA** and other complex polymers. Even the molecular structures of very large assemblies, such as viruses, have been analysed using a particularly intense source of X-rays called synchrotron radiation. This is produced in large ring-shaped machines where electrons circulate at close to the speed of light. The electrons lose energy by emitting extremely intense electromagnetic radiation at all wavelengths. Very narrow beams of X-rays of precise wavelength can be siphoned off at 'beam line' stations around the ring and used for all kinds of chemical analysis including high-resolution X-ray crystallography. UK chemists have access to two main synchrotron sources in Europe – one at Daresbury in Cheshire and the other at Grenoble, France.

The European Synchrotron Radiation Facility in Grenoble produces intense beams of X-rays suitable for X-ray crystallography on complex molecules like proteins.
Courtesy of ARTECHNIQUE.

Glossary

DNA: Deoxyribonucleic acid. The double-helix molecule which stores genetic information.

Electron density: The concentration of electron charge in a particular region of space.

Quantum mechanics: A theory which describes how objects of the size of atoms or smaller behave.

RNA: Ribonucleic aid – a molecule involved in the synthesis of proteins from the information stored by DNA.

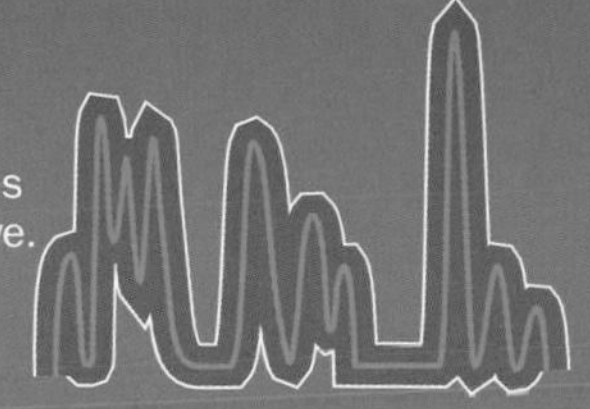

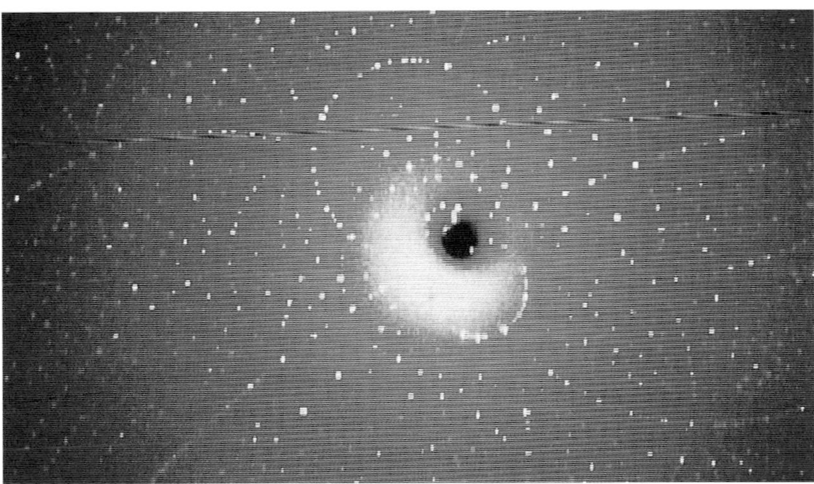

X-ray diffraction pattern of lysozyme.
Phillippe Plailly/Eurelios/Science Photo Library.

Another related type of structural analysis is neutron diffraction, or neutron scattering, which works on the same principle of Bragg's Law. Although neutrons are subatomic particles (found in the atomic nucleus), according to **quantum mechanics**, they also behave like waves, and their wavelengths, like those of X-rays, correspond to interatomic and intermolecular distances.

Neutrons, being electrically neutral, pass through the electron clouds of the atom and interact with the nucleus. This means they provide slightly different information about atomic positions than do X-rays. Although neutrons generally do not achieve the resolution of X-rays, they can locate hydrogen atoms (protons), which usually do not offer enough electron density to be recorded via X-ray diffraction. What is more, it is possible to distinguish between selected hydrogen positions in a molecule by substituting with the hydrogen isotope deuterium, because the two isotopes scatter neutrons very differently.

6. Explain why hydrogen atoms are difficult to locate with X-ray diffraction.

7. Explain why hydrogen and deuterium scatter neutrons very differently.

Because neutron beams are produced from high-energy sources – nuclear reactors or particle accelerators – and neutron diffraction requires very large crystals, the technique is less used by chemists than is X-ray diffraction.

Electrons also behave as waves which are diffracted by matter. If electrons are accelerated through a voltage of 50,000 volts, then their wavelength is about right for probing interatomic distances. Since electrons are charged particles, they cannot penetrate far into solids or liquids but the technique of electron diffraction can be very useful for examining surfaces. However, diffraction of a beam of electrons by a gas can give very detailed structural information. Each atom in the molecule scatters the incoming electron beam; the scattered electrons from neighbouring atoms interfere with each other and the end result is a complicated interference pattern. This can then be converted into a so-called 'radial distribution curve' which gives interatomic distances. An example is shown in Figure 1.

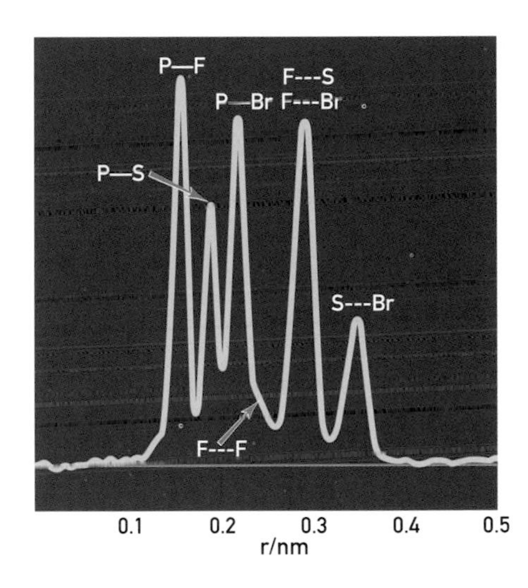

Figure 1(a). A radial distribution curve for $PBrF_2S$, where the bromine, fluorine and sulfur atoms are all bonded to the phosphorus atom, gives the various interatomic distances.
Adapted from Structural Methods in Inorganic Chemistry. E. A. V. Ebsworth, D. W. H. Rankin and S. Cradock, Blackwell, Oxford, 1991.

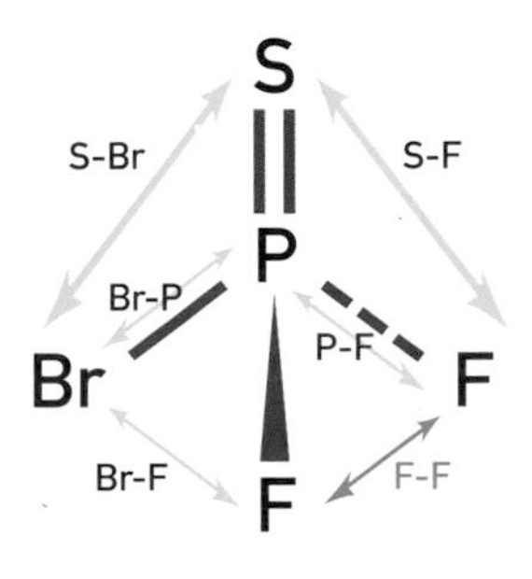

Figure 1(b). The $PBrF_2S$ molecule with the interatomic distances marked.

Figure 2. The UV spectrum of caffeine.

Light spectroscopy

Ultraviolet/visible spectroscopy

We have seen earlier that atomic spectroscopy is a powerful analytical tool. It depends on electrons jumping from one energy level in the atom to another and emitting or absorbing a characteristic frequency of light in the process; some of these transitions occur in the visible region of the spectrum and hence are easily seen. However light spectroscopy is much richer than this. In the first place, molecules can interact with light over the whole spectral range from the far ultraviolet to the far infrared and microwave regions. Secondly, the spectral pattern of the interaction can provide quite detailed information about the structure of the molecule.

In the ultraviolet and visible regions of the spectrum, the interaction is with the electronic energy levels of molecules (as for atomic spectroscopy). This can provide a means of identifying what compound we have (for example, in the case of transition metal **coordination compounds** and some **organic compounds**) and can also be used for quantitative analysis – that is how much of it we have. This is because the amount of light absorbed is related to the concentration of the absorbing species. A typical ultraviolet/visible spectrum is shown in Figure 2.

Ultraviolet/visible spectroscopy can be very sensitive if the compound absorbs light very effectively. For instance, the complex formed between iron ions and *o*-phenanthroline is highly coloured and provides a good method of measuring low concentrations of iron in natural waters.

The ultraviolet spectrum of organic compounds can also show the presence of particular chemical groups (called chromophores), such as those containing double bonds. Much more sophisticated use of ultraviolet/visible spectroscopy can be applied to metal coordination compounds, where the position and intensity of the spectral bands can provide a great deal of information about the electronic structure of the molecule.

Analysts at work.

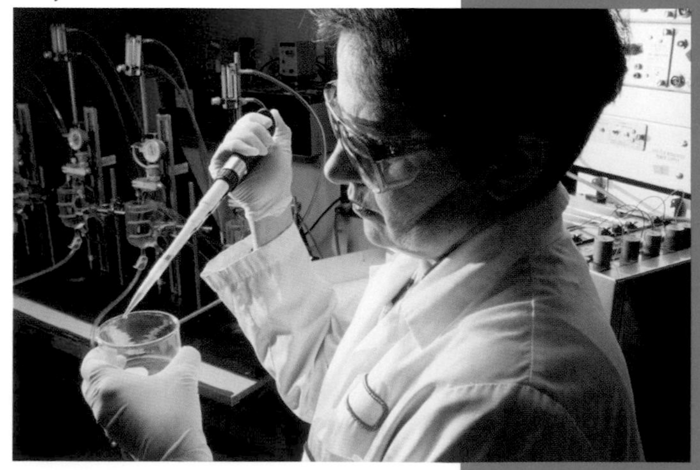

Courtesy of QuChem.

Glossary

Coordination compound: A complex in which molecules called ligands form dative covalent bonds with a transition metal atom or ion.
Organic compound: A compound based on a skeleton of carbon and hydrogen.

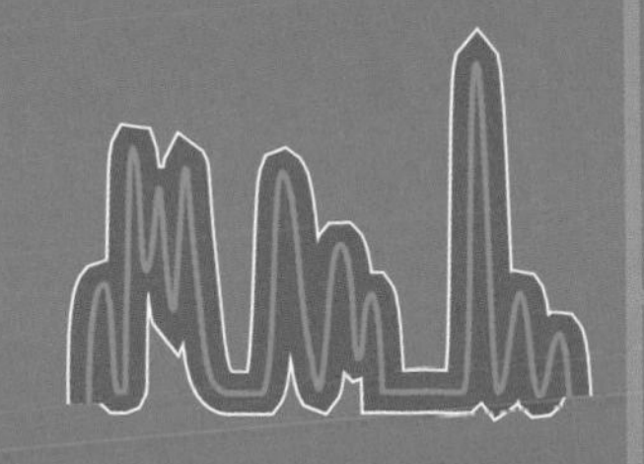

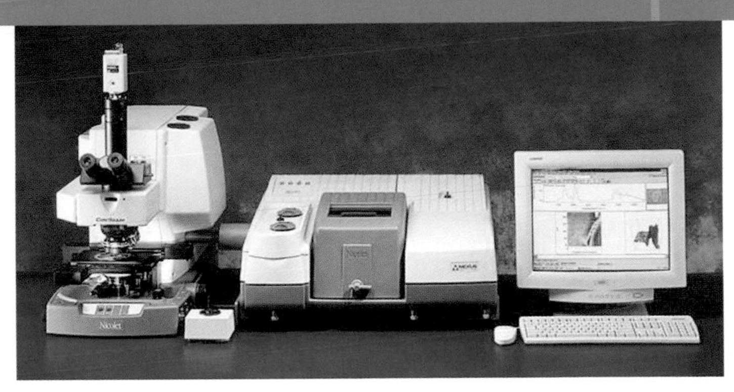

A modern FT-IR spectrometer.
Courtesy of Nicolet Instruments Ltd.

Infrared spectroscopy

It is in the infrared region of the spectrum where most information about chemical bonds and functional groups can be obtained. Molecules are constantly vibrating rapidly, and the energy levels corresponding to these motions occur mostly in the infrared region of the spectrum, in the frequency range 1×10^{14} s^{-1} to 5×10^{12} s^{-1}. This corresponds to wavenumbers (the number of waves in one cm) of 4000 to 200 cm^{-1} in the units commonly used by infrared spectroscopists.

Infrared spectroscopy probes these vibrations. Since the positions of particular bands in the infrared depend on the 'springiness' of individual bonds and on the masses of the appropriate atoms, the infrared spectrum can help identify the various chemical groups in the molecule. This is very useful in organic chemistry, particularly if the infrared bands are very intense. For instance, an organic carbonyl (C=O) group gives a very intense band at about 1700 cm^{-1}, Figure 3.

Traditionally infrared spectra were obtained by scanning through a range of frequencies. The situation was revolutionised with the development of Fourier transform infrared spectrometers. The technique of Fourier transform is important in many areas of analytical science – see Box – *Fourier transform – a useful mathematical trick.*

Although valuable for qualitative and structural analysis, infrared spectroscopy is seldom used for quantitative work, except in specialised areas, partly because it is difficult to get very accurate measurements of intensity.

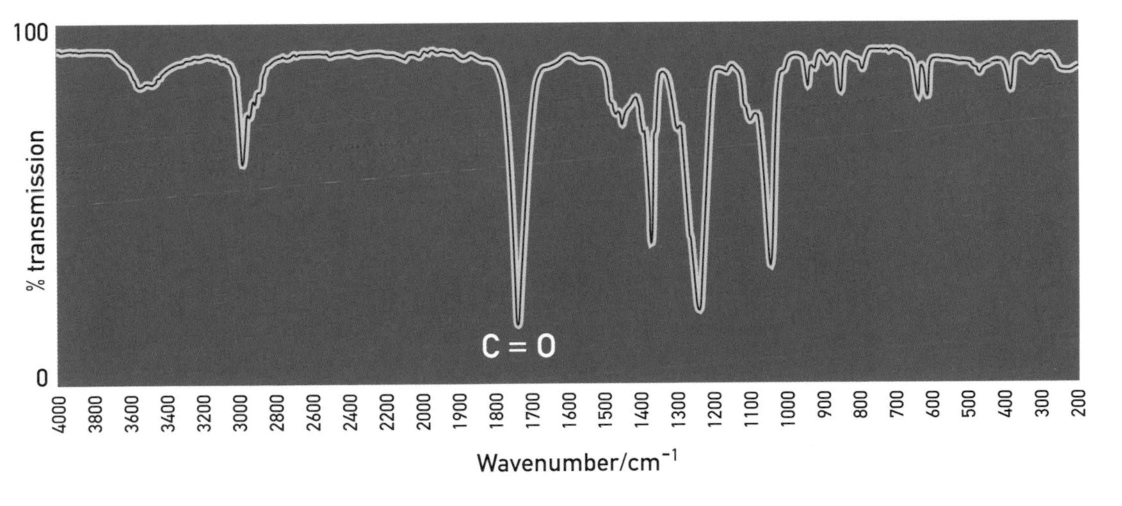

Figure 3. An infrared spectrum of ethyl ethanoate showing a C=O stretch between 1700-1800 cm^{-1}.

A modern FT-IR spectrometer.
Courtesy of Nicolet Instruments Ltd.

An infrared spectrum is normally displayed as a graph of % transmission, (*ie* the amount of infrared which passes through the sample without being absorbed) plotted vertically against frequency (usually expressed as wavenumber) horizontally, Figure 3. Traditionally the spectrum was obtained by using a diffraction grating (or prism) to split the infrared radiation from the source into its component frequencies and rotating the grating so that the frequencies passed through the sample one at a time. A detector measured how much radiation passed through the sample for each frequency. This could be quite a slow process and much of the time is in fact wasted as the detector is collecting data about areas of the spectrum where no radiation is being absorbed.

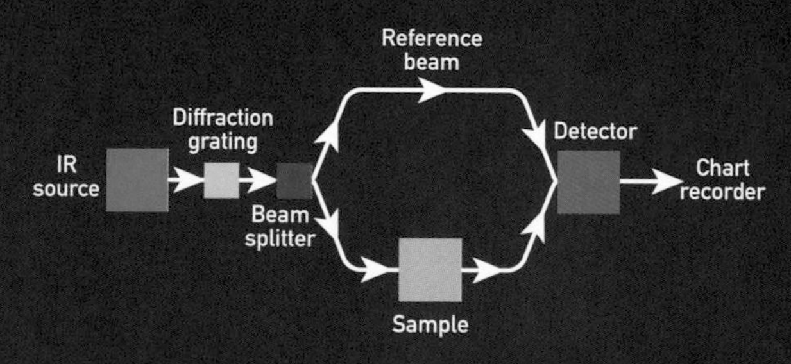

A traditional 'double beam' infrared instrument. The instrument compares the intensity of the reference beam with that which passes through the sample.

Modern instruments do away with the grating and pass a pulse of radiation containing all the required frequencies through the sample at once. After passing through the sample the waves of different frequencies interfere with one another (that is they add together or cancel depending whether or not they are in phase). The detector records the interference pattern which is produced. This contains all the information we need to produce the traditional graph which chemists are used to interpreting. A computer is used to mathematically transform the interference pattern into the graph by a mathematical technique called Fourier transformation after the French mathematician Jean-Baptiste Fourier.

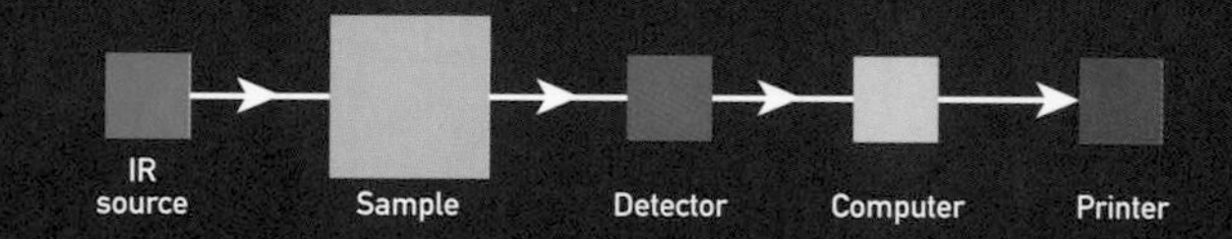

A Fourier transform infrared instrument.

Fourier transform (FT) instruments record the spectrum quickly. They also have the advantage that if the spectrum signal is weak, several pulses of radiation can be used and the results from them added together to give the final spectrum, so FT instruments can be made to be more sensitive than traditional IR spectrometers.

The FT technique can be used for ultraviolet/visible spectra, mass spectra, nuclear magnetic resonance and microwave spectroscopy as well as infrared.

Microwave spectroscopy

As well as vibrating, molecules also rotate – especially in liquids and gases where the molecules can readily tumble around (although molecules can rotate in solids as well). This leads to energy levels associated with rotation. These occur in the **microwave** part of the spectrum. The technique of microwave spectroscopy is most useful in gases where the levels are well-defined and separated by energies that depend on the moment of inertia of the molecule – a property which measures how the mass of a molecule is distributed. The molecule must have an overall **dipole moment** to give a microwave spectrum, which means an asymmetrical distribution of electric charge is needed. Methane, for example, has no microwave spectrum. Although each bond has a dipole, the dipoles cancel out because of the tetrahedral shape of the molecule.

However, where there is a dipole moment, microwave spectra can provide very accurate values for moments of inertia, bond lengths and, therefore, structure.

? 8. (a) Which of the following molecules have a dipole moment?
i) water, H_2O;
ii) oxygen, O_2;
iii) carbon monoxide, CO;
iv) ammonia, NH_3.
Use diagrams to explain your answers.

(b) Explain why sulfur dioxide (SO_2) has a dipole moment while carbon dioxide (CO_2) does not.

Glossary

Dipole moment: A molecule has a dipole moment if it has an asymmetric distribution of charge.
Microwave: Part of the electromagnetic spectrum with wavelengths around 1 cm.

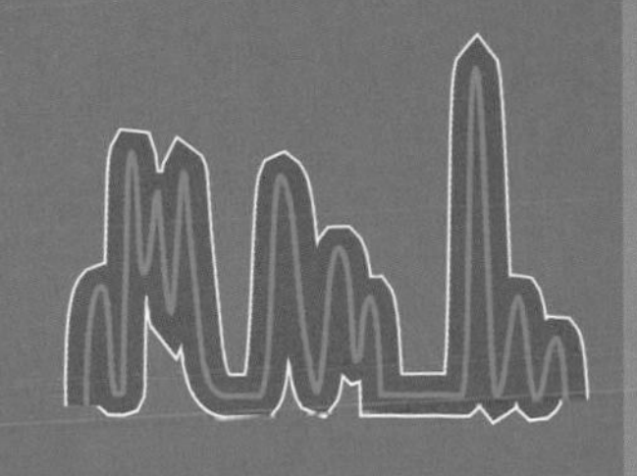

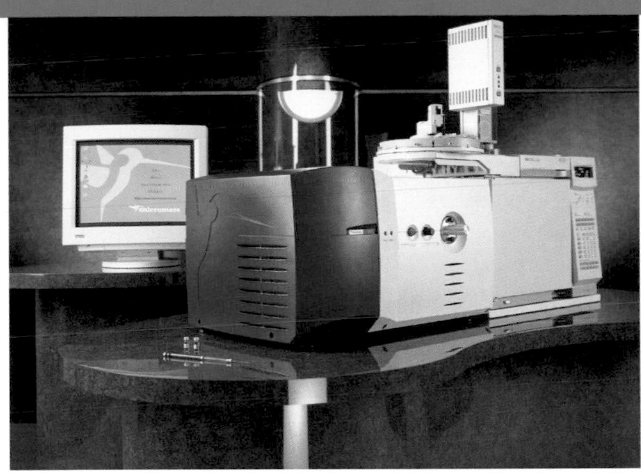

A GC-MS instrument.
Courtesy of Micromass UK Ltd.

Mass spectrometry

Another technique that is regularly used by organic chemists is mass spectrometry. In one form of mass spectrometer, the molecules in the sample are first vaporised and broken into charged fragments (ions) using an electron beam. They are then accelerated and separated by electric and magnetic fields before reaching a suitable detector. The ratio of the charge to the mass of the fragments is then recorded. This allows the mass of the molecule to be accurately determined as well as the nature of the atoms in the fragments. The pattern of fragmentation also depends on the most energetically favourable routes for decomposition and also helps analysts to work out the structure of the molecule.

Mass spectrometry is a technique that is rapidly growing in popularity because of improved methods of vaporising molecules. Until recently, it could not be applied to large molecules such as proteins because they would decompose, losing their identity completely. Now, however, there are two new methods for getting proteins into a sort of 'gaseous' state. One, called electrospray, involves spraying a solution of the protein into the instrument through a syringe equipped with a needle at high electric potential. The electric field breaks up the liquid into a fog of charged droplets which then eject protein molecules carrying many charges. The resulting 'protein ions' can then be analysed in the mass spectrometer. In the other method, laser desorption, the sample protein is mixed with a matrix of smaller molecules such as urea. Then a laser pulse is fired with the right energy and wavelength to vaporise the matrix material without destroying the protein. The matrix and protein, which picks up charge from the matrix, are then propelled towards the detector. The mass of the charged protein is obtained from an accurate measurement of the time required for the protein to travel to the detector – heavier proteins will take longer. Once the mass of the protein has been measured, it can be broken into fragments in a controlled way and its mass spectrum measured.

Nuclear magnetic resonance

Perhaps the chemist's favourite structural analytical tool is nuclear magnetic resonance (NMR) spectroscopy, and it is on this technique that we are going to concentrate for the rest of the chapter. It is a method that depends on an unusual combination of nuclear physics and chemistry.

Nuclear magnetic resonance depends on the magnetic properties of certain nuclei and the subtle influence of their chemical environment. Since the 1960s it has been used by organic chemists to help solve the structures of relatively simple molecules. As the technique has developed, NMR has been applied to ever more complex structures including large biological molecules, complex solids and living tissue. Today it can be used to deduce the network of bonds in a molecule, identify atoms close in space, and probe the complicated way in which the molecule moves and flexes over time. Nuclear magnetic resonance is not limited to a single state of matter, but is used on solids, liquid crystals, liquids and solutions. For all these reasons, NMR occupies a unique position in analytical chemistry.

How does NMR work?

As the name suggests, NMR is concerned with the nuclei of atoms. Atomic nuclei have a property called 'spin' which gives them magnetic properties. Some nuclei (ones with odd numbers of protons and neutrons such as ^{1}H, ^{13}C, ^{19}F and ^{31}P) have a value of spin which makes them behave like tiny bar magnets. Some of the 'nuclear bar magnets' will line up in the same direction as an external magnetic field (*ie* parallel to the field) and some line up in the opposite direction (antiparallel). There are slightly more in the parallel direction than the antiparallel direction as the parallel configuration is of lower energy. So the external field magnetises the sample.

If electromagnetic radiation of radio frequency is now applied to the sample, some of these nuclear magnets absorb energy and 'flip' from the parallel to the antiparallel position, Figure 4. The energy, and therefore frequency, of the radiation required depends on the strength of the applied magnetic field and also the strength of the nuclear magnet. This frequency is called the resonance frequency.

Nuclear magnetic resonance experiments were first carried out shortly after World War II by physicists using electronic components salvaged from radar sets. They hoped to measure the magnetic moments of nuclei – the strength of their magnetism. However, they came across a problem – they got different values for the same nucleus in different chemical compounds. They realised that this was because the nuclei were being shielded from the applied magnetic field by the electrons, which also have spin and therefore a magnetic moment. So the experiment was unable to measure nuclear magnetic moments but it was able to provide information about the electrons – the very stuff of chemistry. So the technique was taken over by chemists and went on to become probably the most useful method for determining chemical structure.

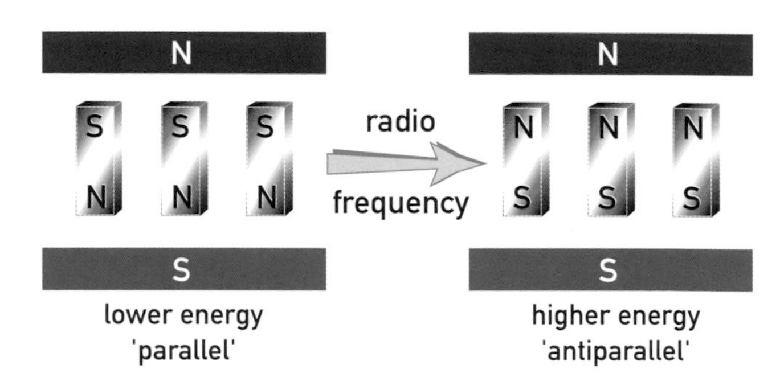

Figure 4. Nuclear magnets can 'flip' from parallel to antiparallel alignments in a magnetic field when radiofrequency radiation is supplied.

The chemical shift

The distribution of electrons around it affects the resonance frequency of a particular magnetic nucleus. If the resonance frequency of *all* hydrogen nuclei (which are single protons), for example, were the same, then NMR would not be much use as an analytical tool since there would only ever be one peak in an NMR spectrum. However, the surrounding electrons set up their own magnetic field which usually opposes the external field. This effectively shields the nucleus, so that it 'feels' a slightly different field from that of the applied field. Because the exact electronic environment varies from one proton to another in the same compound, their resonance frequencies are different.

9. Consider the hydrogen nuclei in water (H_2O) and methane (CH_4). In which molecule are the hydrogen nuclei surrounded by most electrons? Explain your reasoning.

An NMR spectrum can be recorded for any magnetic nucleus but the spectra for different nuclei are usually recorded separately, so, for instance, the hydrogen (^{1}H) and carbon-13 (^{13}C) spectra of an organic compound are recorded in two separate experiments.

Each spectrum consists of a series of peaks, or lines, Figure 5, corresponding to a particular nucleus in different electronic environments. The frequency of each line is known as the chemical shift, which is the shift in the spectrum from a designated reference standard. In hydrogen, or proton, NMR, the standard is the compound tetramethylsilane, TMS ($Si(CH_3)_4$), a small amount of which is added to the sample before the spectrum is run. Chemical shifts are given the symbol δ and measured in units of parts per million (ppm) from the TMS peak.

10. Draw the displayed formula of TMS. What can you say about the chemical environment of each hydrogen atom?

The intensity (peak area) of each line is proportional to the number of nuclei in the sample in that particular environment. So for instance, a proton-NMR spectrum of ethanol Figure 5 (a), (CH_3CH_2OH) gives three signals; one from the three hydrogens in the methyl (-CH_3) group (relative intensity 3); one from the two hydrogens in the -CH_2- group (relative intensity 2) and one from the single hydrogen bonded to the oxygen (relative intensity 1). In water, on the other hand, both hydrogen atoms are the same, both bonded to oxygen with identical bonds, so only one line is seen, Figure 5(b). It is possible to use the chemical shift to identify the types of protons in a molecule. Some values are given in Table 2.

An 800 MHz NMR machine.
Courtesy of Bruker.

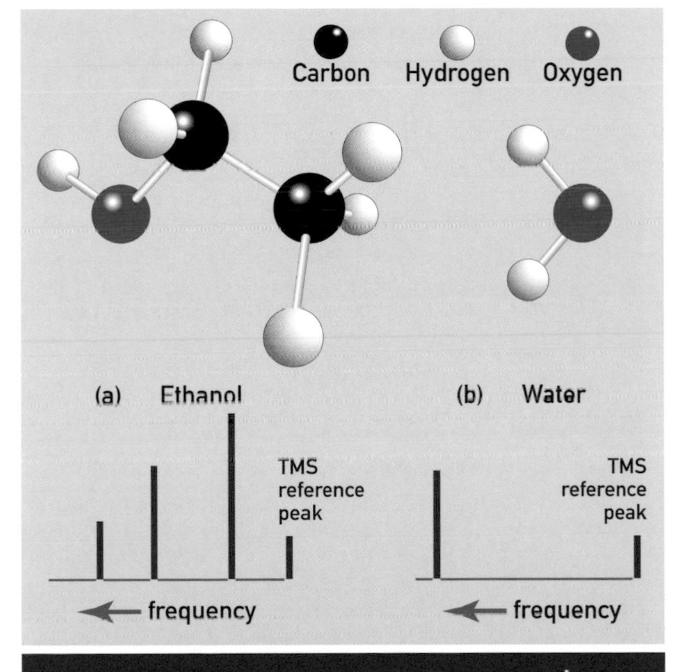

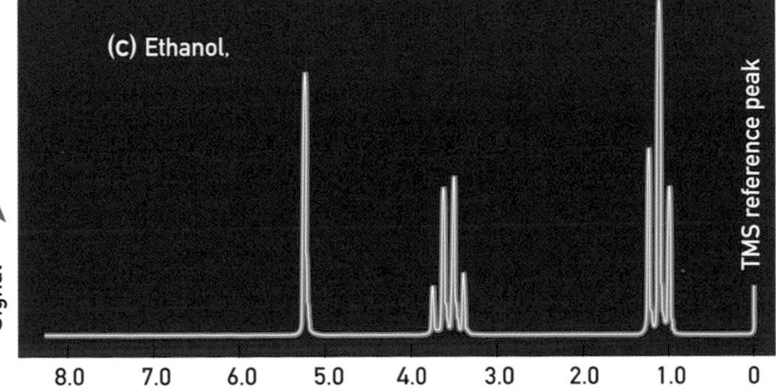

Figure 5. The proton NMR spectra for ethanol (CH_3CH_2OH) and water show up the different environments of hydrogen nuclei in the two compounds. (c) The bottom spectrum shows a higher resolution proton NMR spectrum of ethanol, illustrating how the lines are split by coupling between two sets of hydrogen nuclei.

Typical proton chemical shift values (δ)

Type of proton	Chemical shift /ppm
R–CH_3	0.9
R–CH_2–R	1.3
R–C(=O)–CH_3	2.1
R–CH_2–Hal	3.2–3.7
R–O–CH_3	3.8
R–O–H	4.5
RHC=CH_2	4.9
RHC=CH_2	5.9
C_6H_5–H (benzene ring–H)	7.3
R–C(=O)–H	9.7
R–C(=O)–O–H	11.5

Table 2. Typical proton chemical shift values (δ).

Spin-spin coupling

If we look more closely at the ethanol spectrum (a so-called high-resolution spectrum), Figure 5(c), we see that the -CH_3 and -CH_2- signals are, in fact, not single lines, but split into three and four components respectively. This is because the spins, or 'nuclear bar magnets', of the hydrogen nuclei 'couple' with each other. The effective magnetic field 'felt' by one proton depends on whether the spin of an adjacent proton is lying parallel or antiparallel to the applied field. A proton with spin lying parallel to the applied field strengthens the field slightly and one with spin lying antiparallel weakens it slightly. This gives rise to two lines, one either side of where the uncoupled peak would have been. This is called spin-spin coupling.

If two or more protons are equivalent, that is, with the same chemical shift, such as the three protons in a -CH_3 group, they do not split each other's signal. But in the case of two or more adjacent groups of protons that are not equivalent (such as those in the -CH_3 group and those in the -CH_2- group in ethanol), the spin-spin coupling is additive, producing clusters of lines depending on the number of ways the nearby protons can affect the applied field. A single proton splits the line into two, two protons split it into three and so on – the so-called n+1 rule.

This is the scenario for the methyl (-CH_3) hydrogens in ethanol. Each is coupled to the two -CH_2- hydrogens. Each methyl hydrogen 'feels' three possible net fields, one where the spins of both -CH_2- hydrogens are parallel to the external field, one where they are both antiparallel, and one where one -CH_2- hydrogen's spin is parallel and the other antiparallel. This third arrangement can occur in two possible ways, making it twice as likely as the other two, so the middle of the three lines has twice the intensity of the outer ones, Figure 6.

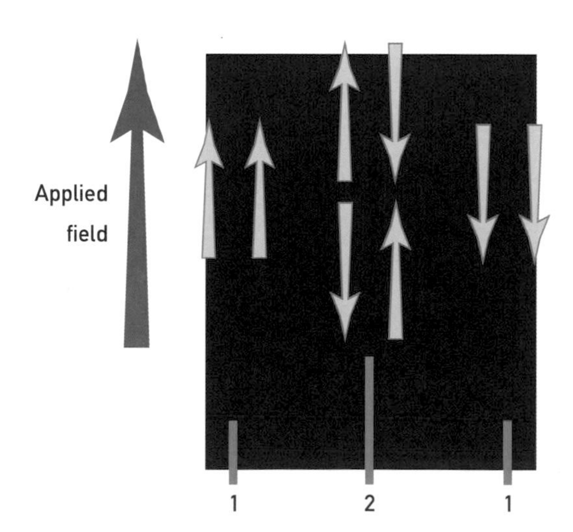

Figure 6. The possible orientations of the spins of the -CH_2- protons as 'felt' by the -CH_3 protons in ethanol. They cause the CH_3 peak to be split into three peaks of height ratios 1:2:1.

11. In ethanol, the signal from the -CH_2- protons is split by the -CH_3 protons into four peaks with intensities in the ratio 1:3:3:1. Draw a diagram similar to that of Figure 6 to show how the three protons of the -CH_3 group can line up to produce these intensities.

Once the chemical shift had been discovered, chemists were very quick to realise the potential of NMR. However, the pioneering chemists could not have realised the eventual power of the technique. The first real applications of NMR to problem-solving in structural chemistry began in 1953. The information NMR provided was particularly useful in deciding between different structures having the same chemical formula (isomers). A simple example of how low resolution NMR can distinguish between isomers is shown in Figure 7.

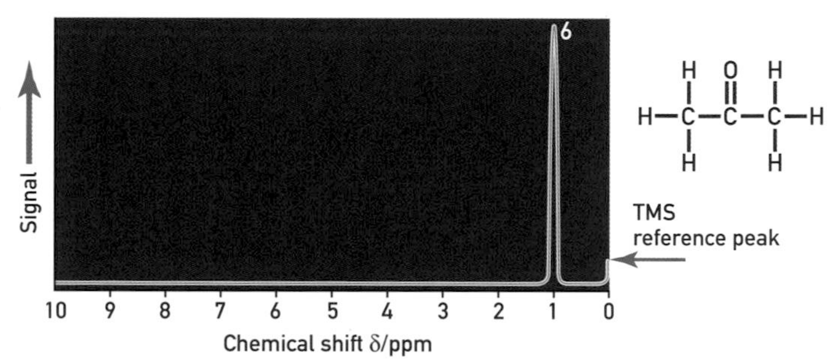

Propanone shows just one peak as all six protons are identical.

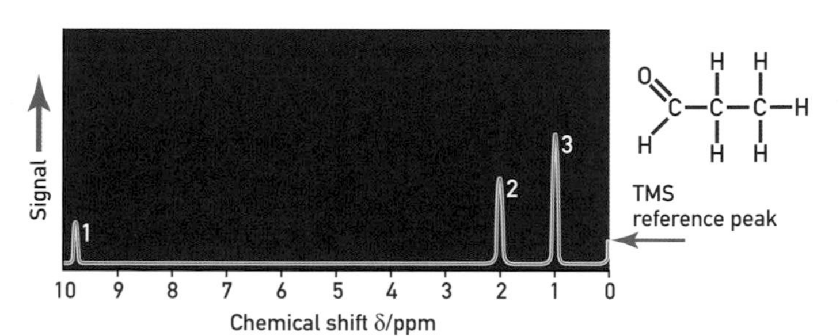

Propanal has three peaks in the ratios 3:2:1. These represent the -CH_3 the -CH_2- and the -CHO protons respectively.

Figure 7. Propanone and propanal are isomers. However, they have very different proton NMR spectra – even at low resolution where spin-spin coupling is not seen.

12. Ethanol (CH_3CH_2OH) has an isomer methoxymethane (CH_3OCH_3). Draw the displayed formula of methoxymethane and predict how many lines there would be in its proton NMR spectrum. Would there be any spin-spin coupling between the protons?

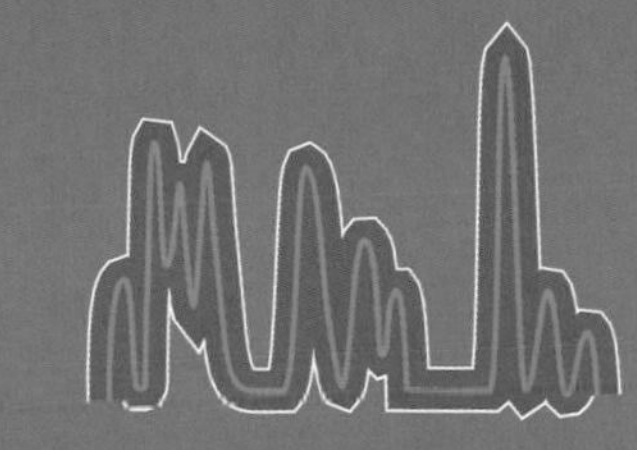

The processing of the experimental data from an NMR spectrometer into a frequency spectrum is performed by today's modern computers in the twinkling of an eye. However, back in the 1960s, computers were somewhat slower. The digital data from the experiment came out on paper tape and then had to be parcelled up and posted to IBM to be punched onto computer cards, then put onto magnetic tape, and only then fed into the most powerful computer then available. The results were posted back a week later, but if the computer operator tore the tape or dropped the cards, months of work were lost! Developments in computers have been essential to the advancement of NMR.

Magnet technology also had to develop before many of today's applications of NMR could be realised. Even simple molecules may yield NMR spectra with a forest of lines, with neighbouring lines overlapping extensively. The separation of these lines in the spectrum increases with the strength of the magnetic field. Not only this, but the sensitivity of the experiment increases rapidly as well. Consequently, there has been a drive towards ever stronger and better quality magnets since the beginning of NMR.

One of the most important developments here has been superconducting magnets which operate at the temperature of liquid helium. Superconducting materials pass an electric current without any resistance and the electromagnets made from these materials produce extremely strong fields.

The first NMR experiments used a magnetic field strength of 0.2 tesla produced by a cast-off electromagnet. Today, superconducting magnets producing field strengths of 18.8 tesla are available, with a 23 tesla magnet on the drawing board.

These magnets come at a price of course – around £2 million for an 18.8 tesla superconducting magnet – and they are also somewhat larger than the magnets used by the early researchers in NMR: an 18.8 tesla magnet stands some 4 metres high, the height of a small house!

It is not only the attainment of higher magnetic field strengths which improves the quality of NMR data. The design of the radiofrequency probes and hardware is continually being upgraded, with a new era of ultra-low noise, cooled 'cryoprobes' and amplifiers on the horizon.

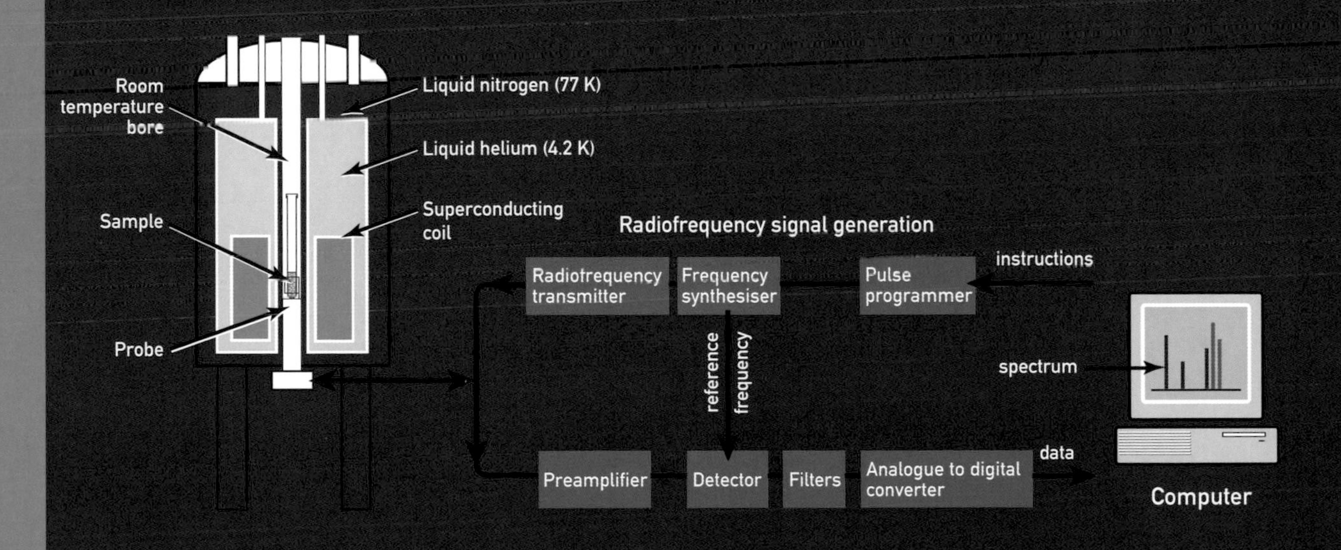

The layout of a modern spectrometer.

Columbianetin

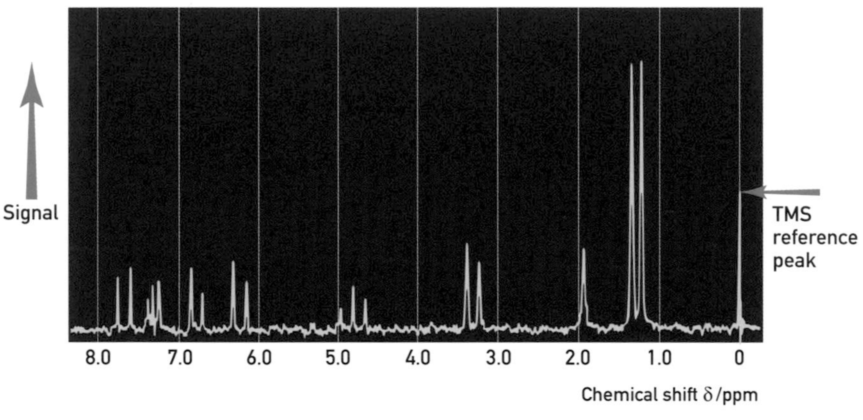

Figure 8. The NMR spectrum of a complex molecule can be difficult to interpret. Which peak belongs to which proton?

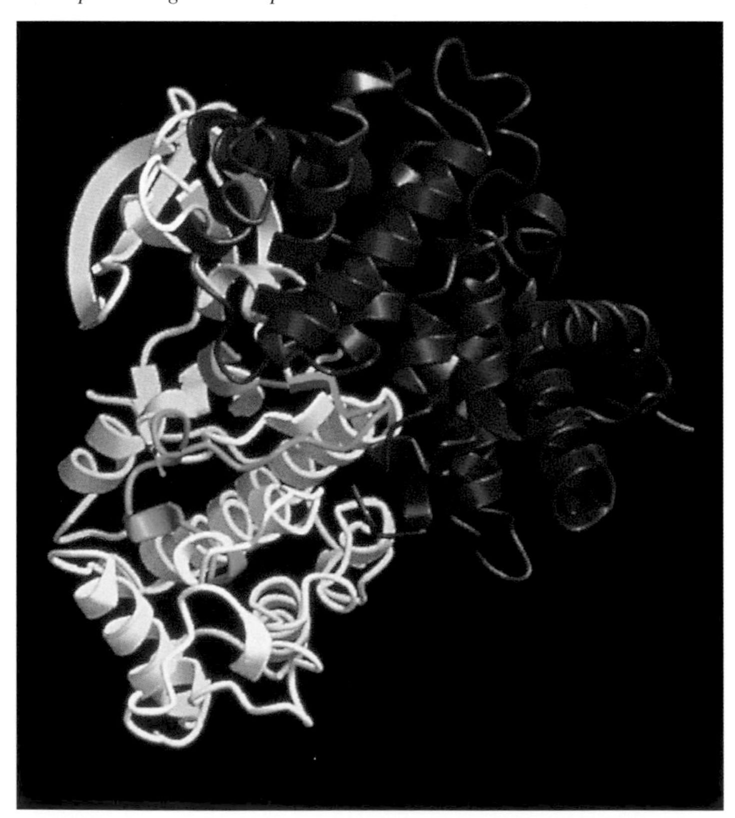

Determining the structures of biological molecules by NMR is a highly competitive field.
Mol. Biophysics, Oxford Univ. Wellcome Trust Medical Photographic Library.

More complex molecules

Interpreting the information provided by the NMR spectrum of a complex molecule such as columbianetin, Figure 8, relies on being able to identify peaks as belonging to certain nuclei on the basis of their chemical shifts alone. Unfortunately, our current ability to use the chemical shift is such that it is useful only for pigeonholing certain types of nuclei, as, for example, the protons in hydroxyl (-OH) or methyl (-CH_3) groups in the spectrum of ethanol shown earlier. What happens if you have more than one methyl group in your molecule and, therefore, more than one methyl resonance in the proton NMR spectrum?

Clearly, the more peaks there are in the spectrum, the harder it is to assign them to a particular nucleus. The information is there, but how do we extract it?

There is another problem encountered with larger molecules because there is a limit to the number of peaks that can physically fit into a particular region of the spectrum before they start overlapping. In severe cases, the peaks merge to a broad envelope in the crowded regions. How do we disentangle the spectrum?

Both of these problems have been overcome, and solving each of them represented a huge leap forward in the usefulness of NMR. The details of the techniques are complex and we will just hint at the possibilities in the examples that follow.

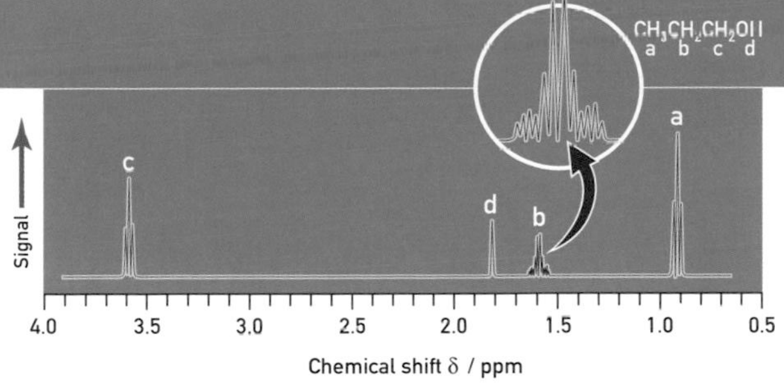

Figure 9(a). The proton NMR spectrum of propan-1-ol. $\underset{a}{CH_3}\underset{b}{CH_2}\underset{c}{CH_2}\underset{d}{OH}$

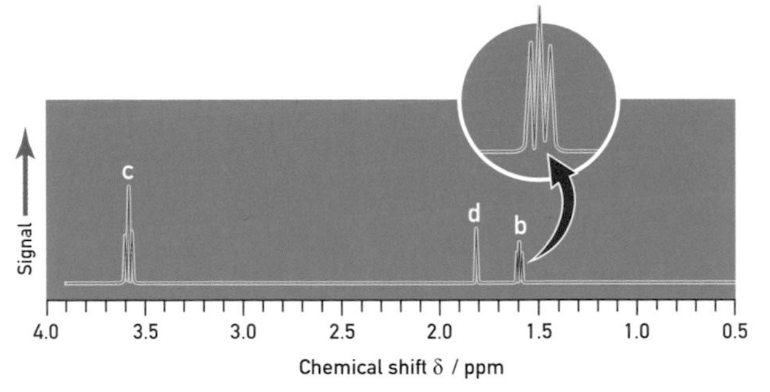

Figure 9(b). The proton NMR spectrum of propan-1-ol 'decoupled' by flooding the sample with radiation of the frequency absorbed by the H_a *protons. The complex peak representing* H_b *has reduced from 12 sub-peaks to 3. The* H_a *peak also disappears.*

Decoupling

Sometimes a spectrum can be simplified by removing some of the spin-spin coupling. For example in the NMR spectrum of propan-1-ol, Figure 9(a), the protons labelled H_b are coupled with those labelled H_a *and* those labelled H_c. This means that the H_b peak is split into four peaks by the three H_as. Each of these four sub peaks is itself split into three by the two H_cs, giving a total of 12 peaks and making the spectrum very complicated – see inset Figure 9(a).

It is possible to remove spin-spin couplings by saturating the sample, while the spectrum is being recorded, with radiation of the frequency at which one of the groups of protons resonates. The spins of these protons then 'flip' rapidly and are unable to line up with or against the magnetic field and their coupling disappears. So if we flood propan-1-ol with radiation of the frequency absorbed by the H_a protons, the coupling with H_b disappears and the 12 peaks 'collapse' into three, Figure 9(b). This simplifies the spectrum and at the same time makes it clear which protons are causing which coupling.

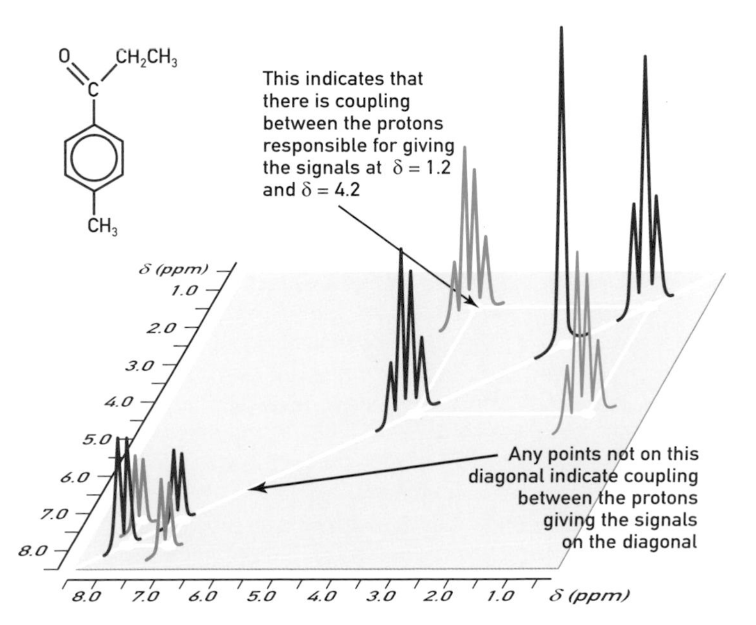

Figure 10. The 2-D proton NMR spectrum of ethyl 4-methylbenzenecarboxylate (ethyl 4-methylbenzoate).

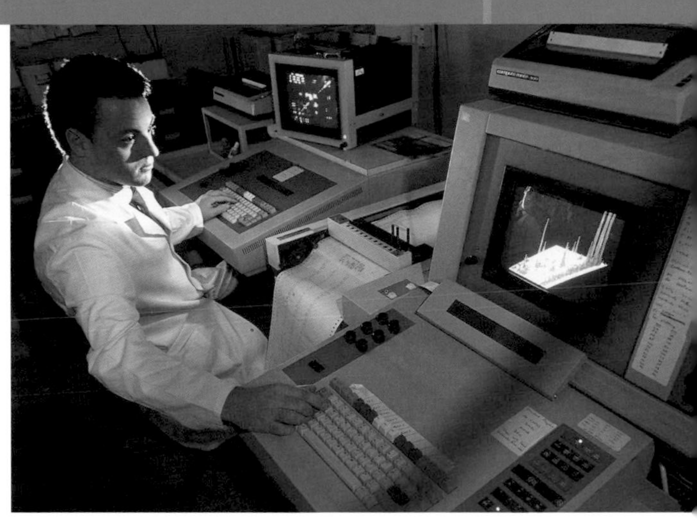

An NMR instrument with a 2D spectrum on screen.
Geoff Tompkinson/Science Photo Library.

? 13. What happens to the H_b proton peaks if we saturate the sample with radiation of the frequency at which H_c protons resonate?

A development of this technique, which provides the same information much more quickly, is to present the NMR spectrum in two-dimensions. The conventional spectrum is seen along the diagonal and peaks off the diagonal show which protons are coupling with which, Figure 10.

Techniques such as these are made possible by the Fourier transform technique in which all the information in the spectrum is recorded from a series of pulses of radiation and the spectrum is constructed by mathematical manipulation using a computer.

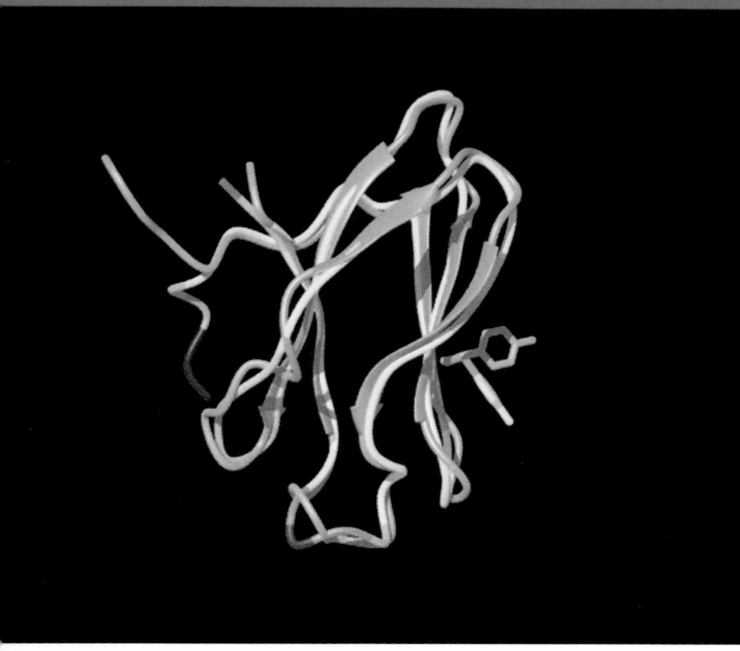

Structures of Tendamistat obtained by NMR and X-ray superimposed to show similarity.

NMR in biochemistry

The 2-D technique is helpful in working out the structures of complex biomolecules such as proteins. The first such structure to be solved by NMR was that of BUSI (bull seminal plasma inhibitor) by a Swiss team in 1984. However other chemists were sceptical, as the stuctures of similar proteins were already known. However the stucture of a completely new protein, Tendamistat, has since been worked on simultaneously by both NMR and X-ray crystallography, and virtually identical stuctures were found by both methods. Currently proteins of relative molecular masses up to 40 000 are being worked on and hundreds of structures of different proteins have been solved by NMR.

As well as ^{1}H-NMR, ^{13}C- and ^{15}N-NMR are useful with proteins (^{12}C has zero spin, and it is difficult to do NMR on ^{14}N). However, these nuclei are not very abundant naturally – ^{13}C's abundance is about 1% and ^{15}N's is about 0.4%.

Proteins are made in the laboratory by inserting the gene that makes the protein into the DNA of an organism which can be grown quickly and easily in large numbers in laboratory conditions, for example, the bacterium *E. coli*. To make protein enriched in isotopes ^{13}C or ^{15}N, the bacteria are grown in media in which the starting materials for making the protein, glucose and ammonium chloride, are enriched with these isotopes.

Aside from structure determination, NMR is an excellent source of information on the dynamic properties of biomolecules (as it is for other chemical compounds), and is a useful probe of transient interactions, (processes where atoms such as hydrogens are exchanged), and the mechanisms of biochemical reactions. Using NMR, it is possible to gain an understanding of how biological molecules interact in nature, for example, how proteins bind to DNA and how enzymes recognise their substrates. These interactions are not only interesting for scientific curiosity: with this information, we can understand diseases at the molecular level, and so better drugs can be designed. Another useful feature of NMR is that it can be done in 'real time' enabling, for example, studies of the complex way in which proteins fold up. All these subjects are currently active and competitive areas of research.

NMR of solids

Although the first NMR experiments were done on solids, NMR developed in the 1950s into a technique used solely on solutions. This is because in solids, the nuclear 'bar magnets' interact strongly and give spectra with very broad lines which overlap and prevent the spectrum being analysed. This does not occur in solution because the rapid molecular motion averages out these interactions. A similar effect can be brought about in solids by rapidly spinning the sample at a particular angle to the magnetic field and this technique has been developed recently. This so-called magic angle spinning has enabled solid-state NMR to develop. So polymers can be studied and in particular the degree of crystallinity and how rotation of molecules in the solid state enables materials to absorb mechanical energy, making them tough and flexible.

Glossary

Non-invasive: A medical technique which does not involve cutting open the patient.

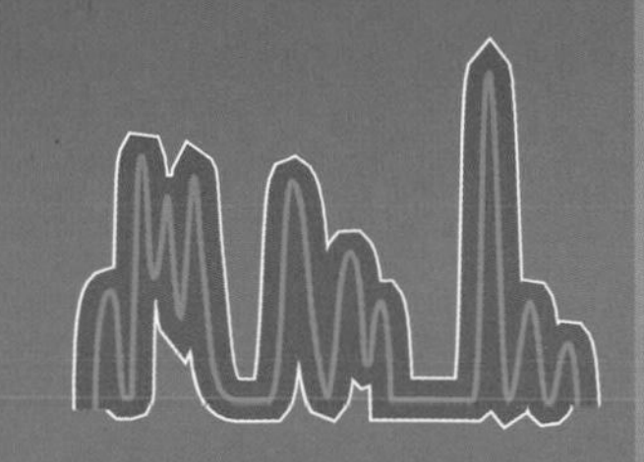

Magnetic resonance imaging

In recent years, NMR has been applied to medical scanning where it is usually called magnetic resonance imaging, MRI, (the 'nuclear' in the name has been dropped because some of the public might associate it with radioactivity).

Soon after the first successful NMR experiments, one of the pioneers of NMR, Felix Bloch, obtained a strong proton NMR signal when he positioned his finger in his spectrometer. Most of the signal came from water, which makes up 70% of the human body. However, water in the body is in a very different environment to free water, and this difference can be measured by NMR.

In 1971, Raymond Damadian a doctor at the State University of New York found that he could discriminate between normal cells and cancer cells in rat tissue by measuring the NMR signal from the water protons within the cells. It was the observation of similar experiments which inspired the chemist Paul Lauterbur to invent MRI in 1972. He realised that NMR information could be obtained directly from intact living organisms and wondered if there was any way to locate where an NMR signal came from within a three-dimensional object, without opening it up.

The answer was to use a magnetic field which varied across the object being examined. The resonance frequency of the protons in, say, water depends on their chemical shift and also the magnetic field. So with a variable magnetic field, the frequency of the signal from water depends on how deep the water is inside the object.

Magnetic resonance imaging is now one of the most important tools available in biology and medicine. The key is that the method is **non-invasive**, and unlike the medical use of X-rays, is extremely sensitive to differences in parts of the body with a high water content. Thus, although broken bones remain the province of X-rays, MRI is used for the study of abnormalities in tissues, muscles, and other soft parts of the body, and is also increasingly useful for studies of blood flow. As a relatively young technique, it has a rich future.

Magnetic resonance imaging also has nonmedical applications including imaging of materials, and even real-time changes to these materials, for example, the processes of drying cement.

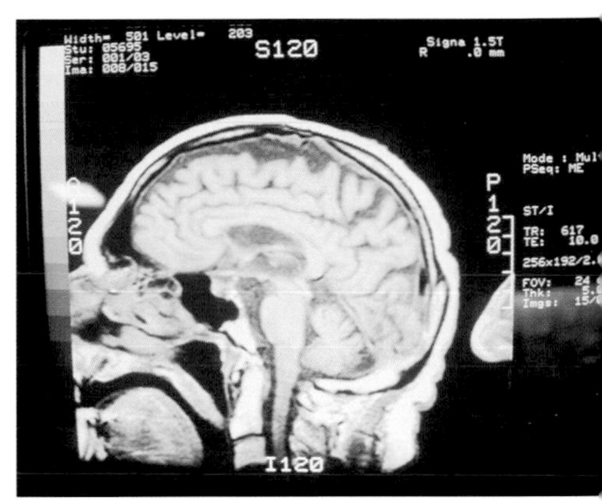

MRI scan of brain.

MRI scan in progress.

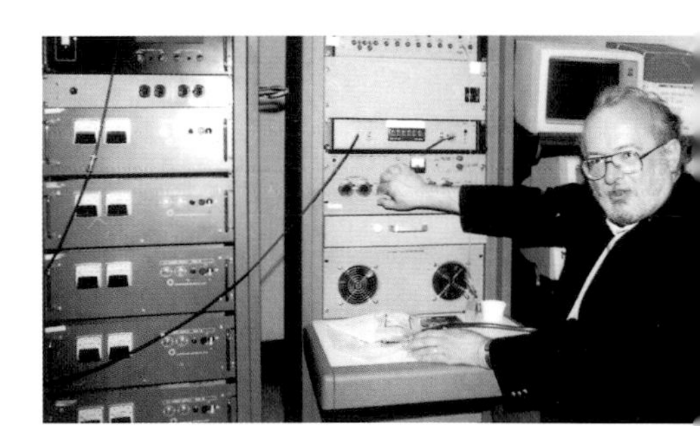

Paul Lauterbur, who helped develop MRI.
University of Illinois, Chicago.

The future

Technological advances are likely to increase further the range of materials suitable for study by NMR spectroscopy and MRI. Recently, a research group in New Zealand performed NMR experiments on one metre long samples of ice in Antarctica using the Earth's own magnetic field as the magnetic field in the experiment; further studies are planned for the future. This may well be the first of many future experiments where large samples are able to be examined by NMR.

At the other end of the size scale, NMR is being developed into a tool for producing high resolution images of the interiors of very small objects.

Answers

1. $x = 8$.

2. (a) 2, (b) 3.

3. $2Na_2S_2O_3(aq) + I_2(aq) \rightarrow 2NaI(aq) + Na_2S_4O_6(aq)$
The end point is when the yellow-brown colour of the iodine just disappears. This is made easier to see by adding starch when the colour of the solution becomes that of pale straw. The starch forms a blue-black complex with the iodine which then disappears.

4. Water
$2YBa_2Cu_3O_8(s) + 9H_2(g) \rightarrow Y_2O_3(s) + 4BaO(s) + 6Cu(s) + 9H_2O(g)$

5. Carbon dioxide, nitrogen oxides, water and sulfur dioxide. (In fact, in most combustion analysis instruments the nitrogen oxides initially produced are then reduced to elemental nitrogen to measure the amount of nitrogen.)

6. They have only one electron.

7. Deuterium has two nucleons (one proton and one neutron) – twice as many as hydrogen.

8. (a), (i) H_2O, (iii) CO and (iv) NH_3. Both the atoms in O_2 have the same electronegativity so there is no asymmetry in the charge distribution. In each of the other molecules, there is asymmetry in the charge distribution in each bond because of the electonegativity difference between the atoms, and the shapes of the molecules are such that these do not cancel out.

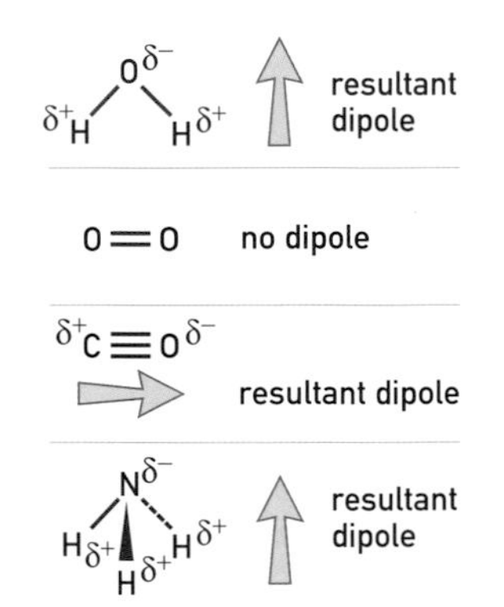

(b) Both the S=O and C=O bonds have a dipole; carbon dioxide is linear, so the bond dipoles cancel, while sulfur dioxide is angular so the bond dipoles do not cancel.

9. Hydrogen is surrounded by more electron density in methane. In water, the electronegative oxygen atom pulls electrons away from the hydrogen atoms.

10. $Si(CH_3)_4$ (structural formula: central Si bonded to four CH_3 groups)

They all have the same environment.

11.

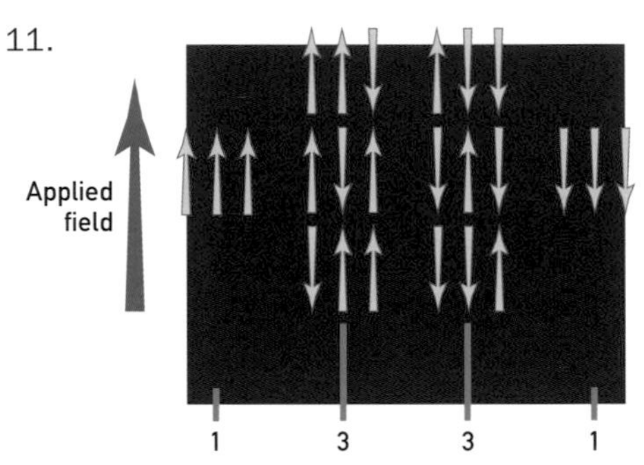

12. $H_3C{-}O{-}CH_3$ (structural formula: H–C(H)(H)–O–C(H)(H)–H)

One line only with no coupling as all the protons are identical.

13. They will 'collapse' into four peaks.

Chemical marriage brokers

The Chinese and Japanese languages use the same pictorial character for the word 'catalyst' as for 'marriage broker' or arranger of marriages – this seems an appropriate description of what catalysts do.

An example of acid production early in the 20th century.
Reproduced courtesy of Public Record Office MUN 5/297/626.

Jöns Jacobs Berzelius, at the age of 47.
Reproduced courtesy of the Library and Information Centre, Royal Society of Chemistry.

The majority of industrial chemical processes which make the useful materials that society relies on depend crucially upon catalysts. These speed up reactions, which would otherwise be too slow to be used commercially, without themselves being used up. Similarly, most molecular processes in living organisms involve biological catalysts called enzymes. Catalysis is thus vitally important to our everyday lives. It is an extremely active area of research, as chemists search for more efficient catalysts to create energy-saving and thus cheaper chemical processes which may also be less harmful to the environment.

Modern catalysis plays a huge part in our lives. Take the car, for example: its fuel – a carefully designed mixture of hydrocarbons of specified chain length and degree of chain branching – is produced by catalytic processing of crude oil; the plastics which are increasingly used in vehicle components rely on catalysed polymerisation processes; and the antifreeze in the radiator is made by a catalytic oxidation process. Even the exhaust gases from the engine are passed over a catalyst to convert them into less harmful molecules before being released into the atmosphere.

The history of catalysis

The first person to formulate the idea of catalysis was the Swedish chemist Jöns Jacob Berzelius. In 1835, he described catalysts as substances: '.....able to awaken affinities which are asleep at this temperature by their mere presence and not by their own affinity'. Substitute the word 'reactivity' for the word 'affinity' and you arrive at something close to our current understanding of a catalyst as: 'A substance that participates in a chemical reaction and increases its rate without an overall change in the amount of the catalyst in the system'. In particular, Berzelius had noticed that catalysts help reactions to occur at lower temperatures than normal.

Catalytic converters are now in widespread use throughout the motor industry, reducing exhaust emissions and giving a cleaner environment.
Courtesy of Rover Group.

The rise of the use of catalysts in industry

The use of catalysis on a major industrial scale began late in the 19th century with the production of sulfuric acid – then used mainly in the dye industry – over a platinum catalyst. Sulfuric acid is still made using this method, called the Contact process, though a catalyst of vanadium(V) oxide, V_2O_5, has replaced the platinum one. Sulfuric acid is such a vital component of so many industrial processes that the amount of this acid used by a nation was once used as a measure of its technological development!

1. Among the main uses of sulfuric acid are:

- **the treatment of metals;**
- **the manufacture of paints; and**
- **the manufacture of fertilisers.**

Suggest why the statement made by Justus von Liebig in the early 19th century that 'we may fairly judge the commercial prosperity of a country from the amount of sulfuric acid it consumes' is less true now than it was then.

Catalysis is still one of the most active areas of research in chemistry, though its aims have changed significantly. In these environmentally-conscious days, the ability of a catalyst to promote a reaction with great **selectivity** is just as important as being able to drive the reaction with lower energy costs.

Glossary

Catalytic cycle: A sequence of chemical reactions that can be linked to form a loop. The catalyst cycles round the loop converting reactants into products.

Selectivity: The ability of a catalyst to produce the desired product with the minimum of unwanted by-products.

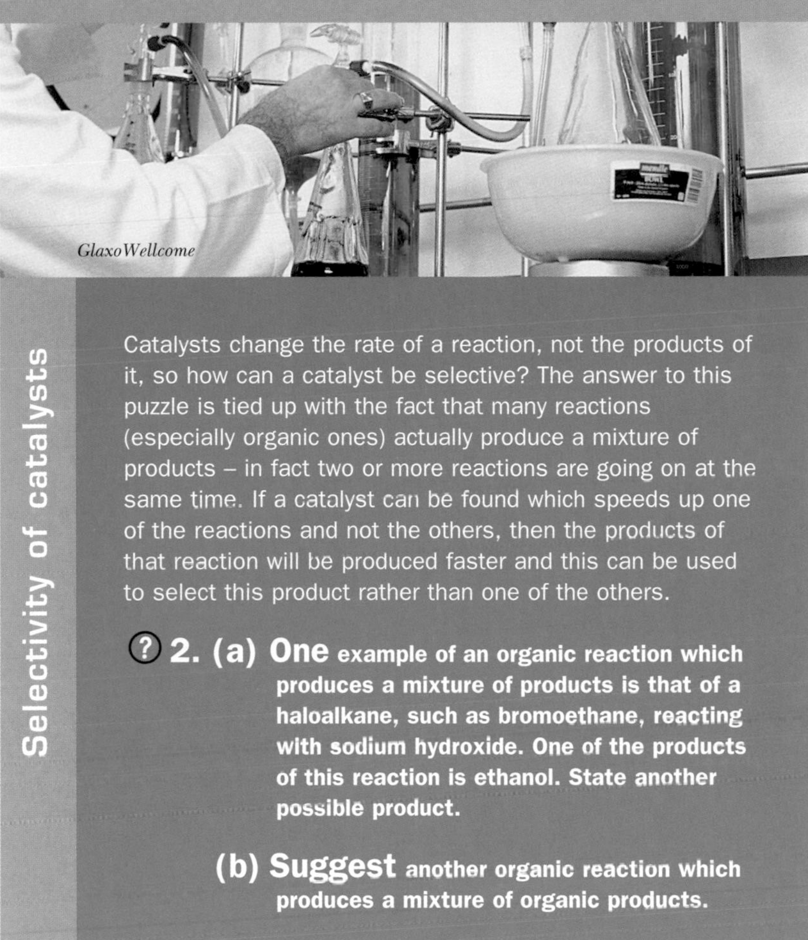

GlaxoWellcome

Selectivity of catalysts

Catalysts change the rate of a reaction, not the products of it, so how can a catalyst be selective? The answer to this puzzle is tied up with the fact that many reactions (especially organic ones) actually produce a mixture of products – in fact two or more reactions are going on at the same time. If a catalyst can be found which speeds up one of the reactions and not the others, then the products of that reaction will be produced faster and this can be used to select this product rather than one of the others.

? 2. (a) One example of an organic reaction which produces a mixture of products is that of a haloalkane, such as bromoethane, reacting with sodium hydroxide. One of the products of this reaction is ethanol. State another possible product.

(b) Suggest another organic reaction which produces a mixture of organic products.

How do catalysts work?

There is no single answer to this question, as different types of catalyst have different mechanisms of action, some of which are dealt with later. What they have in common, however, is that they all provide an alternative reaction pathway which has a lower activation energy barrier than the uncatalysed reaction, see Box – *Activation energy*.

The alternative route for the reaction usually involves the formation of a short-lived intermediate compound of the reacting molecules and the catalyst. Once the product molecule forms, the catalyst is regenerated so that it can go on to catalyse reactions between further molecules and so on in a continuous **catalytic cycle**.

Below is an energy profile (strictly speaking, an **enthalpy** profile) of a typical **exothermic** reaction. The products have less energy than the reactants, so the energy difference is given out as the reaction takes place. However, for the reaction to start, bonds in the reactants have to start to break, and energy has to be put in for this to occur. There is therefore a hump in the reaction profile and this acts as a barrier to the reaction taking place. The height of this barrier, called the activation energy, is the main factor controlling the rate of the reaction. Only molecules which collide with energy greater than the activation energy are able to react.

The species present at the highest point of the energy profile for the reaction is called the transition state or activated complex. It is a short-lived species in which bonds in the reactants are in the process of breaking and new bonds in the products are in the process of forming.

Catalysts enable reactions to go via different pathways, or follow different mechanisms, which have transition states of lower energy. The activation energy is therefore less and this allows more pairs of colliding molecules to have sufficient energy to react at a particular temperature. Notice that the catalyst does not affect the energies of the reactants or the products.

The shape of the graph of the number of collisions with a given energy against energy, below, shows that a small decrease in the activation energy brings about a large increase in the number of collisions that can lead to reaction and therefore a large increase in the reaction rate.

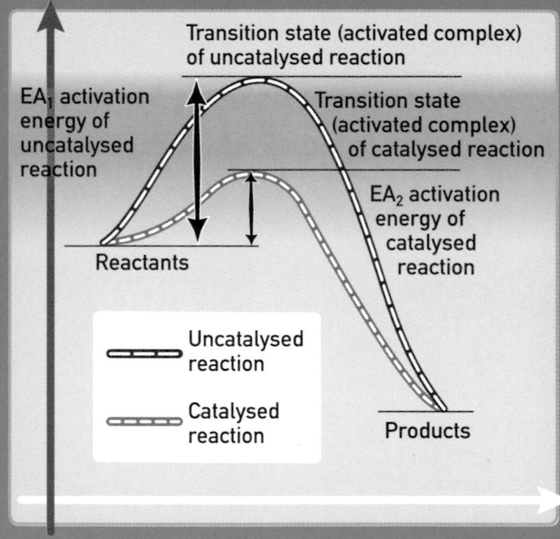

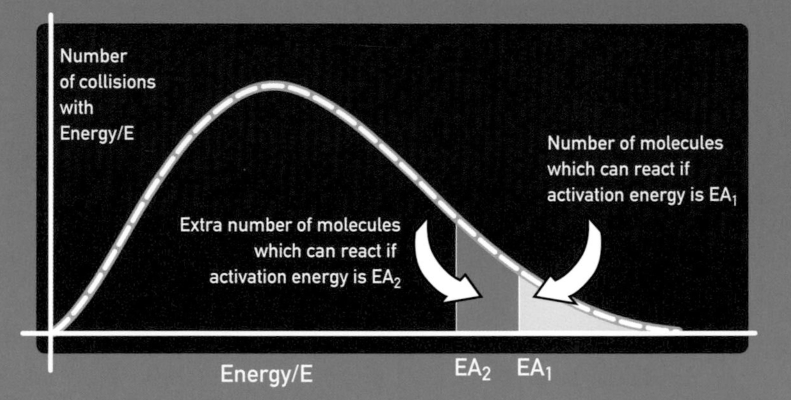

Glossary

Enthalpy: Energy measured under specified conditions *ie* a temperature of 298 K and a pressure of 100 kPa.
Exothermic: Describes a reaction in which heat energy (enthalpy) is given out.

Ammonia-based fertilisers increase crop yields.

The manufacture of ammonia – the Haber process

The synthesis of ammonia (NH_3) from nitrogen (N_2) and hydrogen (H_2) by the reaction

$$N_2(g) + 3H_2(g) \rightleftharpoons 2NH_3(g)$$

over an iron catalyst is a classic example of catalysis. The nitrogen molecule is held together with a strong triple bond that requires about 950 kJ mol^{-1} to break it. Without a catalyst, we would expect the activation energy of the reaction to be about this size because this bond must break in the reaction. In the gas phase, the only feasible way of breaking this bond is with an electric discharge (a spark) and, except where electricity is cheap, that is a very expensive procedure. By 1898, more than 300 000 tons of nitrogen derived from deposits of nitrate salts were being used to produce ammonia to make fertiliser every year. It became clear that unless a cheap way could be found to 'fix' nitrogen from the air, all other known sources of the element would be exhausted by the mid-20th century. Chemists around the world worked on the problem until 1905, when a German chemist, Fritz Haber, made a breakthrough: he discovered that iron acted as a catalyst for the reaction.

His discovery, which earned him a Nobel prize, was developed into a full-scale industrial process (still used today) for the Badische Anilin und Sodafabrik Company (BASF) by Carl Bosch in 1914. It is often claimed that the Haber-Bosch process actually saved the world from starvation by making the ammonia for fertiliser manufacture. It is also possible that the discovery of this process lengthened World War I as, without ammonia to make fertilisers (to feed its population) and explosives, Germany would have been unable to fight for so long.

The role of the iron catalyst is to provide an easy way for the strong nitrogen-nitrogen triple bond ($N \equiv N$) to break. It does this by forming new bonds to the nitrogen atoms. The nitrogen atoms are then able to combine with hydrogen atoms at the catalyst surface one at a time. The ammonia readily departs from the metal, regenerating the clean iron surface. The activation energy for the catalysed reaction is about 160 kJ mol^{-1}.

What makes a good catalyst?

Catalysts are often divided into two groups:

- **heterogeneous catalysts where the catalyst is in a different phase from the reactants and products, Figure 1;**

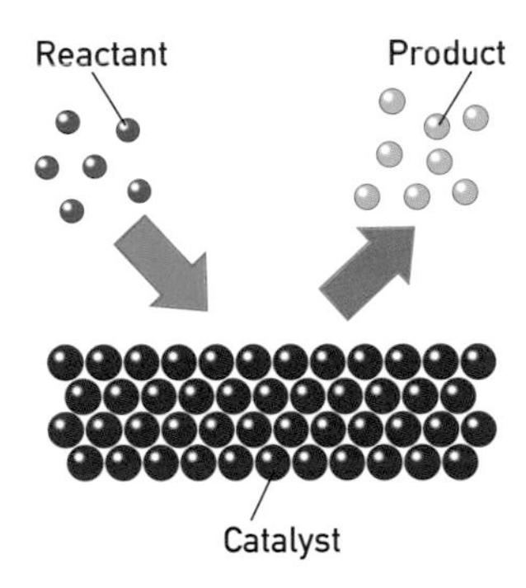

Figure 1. Heterogeneous catalysis.

- **homogeneous catalysts where the catalyst is in the same phase as the reactants and products, Figure 2.**

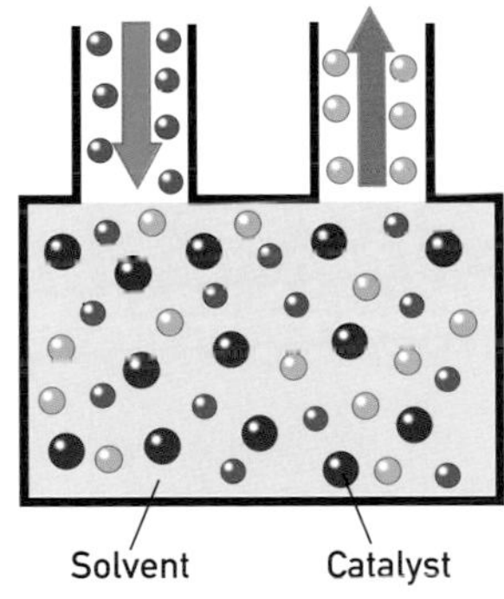

Figure 2. Homogeneous catalysis.

The solid iron catalyst in the Haber process is an example of a heterogeneous catalyst. This is because the reactants and products are all gases and the catalyst is a solid.

The acid catalyst in the reaction of ethanol and ethanoic acid to form ethyl ethanoate

$$C_2H_5OH(aq) + CH_3CO_2H(aq) \xrightarrow{H^+} CH_3COC_2H_5(aq) + H_2O(l)$$

is an example of a homogeneous catalyst because reactants, products and catalyst are all in aqueous solution.

3. Classify the following examples of catalysis as homogeneous or heterogeneous.

(a) The vanadium(V) oxide catalyst for converting sulfur dioxide into sulfur trioxide in the Contact process for manufacturing sulfuric acid.

(b) The enzymes used in fermenting sugars into ethanol in the brewing process.

Many catalysts are metals or metal oxides, particularly those of the transition elements (such as platinum, iron, cobalt and rhodium). Transition elements have electronic structures which allow the formation of variable numbers of differing types of bonds with many kinds of atoms and molecules. It is this ability that allows them to offer alternative low-energy pathways for reacting molecules. In some instances the transition metal may be bonded with other atomic or molecular units called **ligands** to form a **complex**.

Other materials also behave as catalysts. Acids in solution catalyse a wide variety of reactions, and some of the new solid catalysts used in industry (such as those used for cracking petroleum derivatives) are also acid catalysts. Another important group of catalysts is the enzymes, which are large protein molecules found in all living things.

In all cases, for catalysts to work efficiently, the active site – the location on the catalyst at which the reaction takes place – must fulfil a number of conditions.

- **It must have the right electronic characteristics to form some kind of bond with at least one of the reactants. These bonds must break readily after the reaction.**
- **It must have the right shape to give the reactants access to it. Indeed the configuration around the active site often determines the catalyst's selectivity.**
- **There needs to be a large number of active sites which the molecules involved in the reaction can readily reach.**

In heterogeneous catalysis, the number of active sites on a solid surface is limited (only about 1 in 10^8 atoms of a typical solid is at the surface). Because so little of the catalyst is accessible to the reactants, the catalyst must reduce the activation energy considerably to be effective.

Heterogeneous catalysts do, however, have one outstanding advantage – being in a different phase they are easily filtered off from the reaction mixture. Because heterogeneous catalysts are easier to handle, they are used in 85% of all catalysed chemical processes.

In contrast, since homogeneous catalysts are dissolved in solution along with the reactants, every catalyst atom can be an active site, so that a quite small reduction in activation energy can speed up the reaction significantly. On the other hand, homogeneous catalysis requires expensive separation techniques to recover the catalyst from the reaction mixture.

Heterogeneous catalysis and fuels

The economic role of catalysts is revealed clearly in one particular sector, that of providing the world's fuel.

Many crude oil deposits are found under the sea bed.

Today, most of the energy needed to run vehicles comes from crude oil, a complex mixture of hydrocarbons which includes long-chain alkanes, branched alkanes, alkenes (which contain carbon-carbon double bonds) and aromatic hydrocarbons. To make crude oil suitable for use it must be chemically changed by several processes after the initial separation by distillation. The two major processes are reforming, in which linear alkanes are converted into branched or cyclic versions, and cracking, which is breaking up the longer chain hydrocarbons into shorter ones, some of which contain double bonds. Both processes rely on heterogeneous catalysis.

4. Draw the displayed formula of an unbranched alkane, a branched alkane, an alkene, a cycloalkane (an alkane with a ring) and an aromatic hydrocarbon each of which contains six carbon atoms.

Glossary

Complex: A molecule or ion formed when ligands bond to a metal atom or ion.

Isomerisation: The conversion of a molecule into another molecule which has the same molecular formula but a different arrangement of the atoms in space.

Ligand: An atom or group of atoms bonded to a metal by donating a lone pair of electrons to partly-filled d-orbitals on the metal.

'Never in the field of human conflict has so much been owed by so many to so few.'

Perhaps the honoured 'few' of Winston Churchill's famous tribute to the RAF's fighter pilots in the Battle of Britain, should also include a group of scientists whose feet never left the ground. These were the chemists who, shortly before the battle, developed a new reforming catalyst capable of producing 100-octane fuel from the distillate produced in the oil refinery. The new fuel helped British planes perform so much better that they were able to defeat an enemy who had beaten them in the same aircraft just a few months previously in the skies over France.

Reforming is essential because linear alkanes explode prematurely rather than burning in petrol engines causing the effect called knocking which makes the engine run roughly. Branched chain alkanes and cycloalkanes or aromatic compounds such as benzene and methylbenzene burn more smoothly. Branched chains are produced by reorganising the bonds in the hydrocarbon chains (a process called **isomerisation**) and aromatic compounds are produced by removing some of the hydrogen and forming rings. The success of the processes is indicated by the octane number of the resulting fuel. This is a measure of its anti-knocking properties – the higher the number the better the fuel. The Table below gives some octane numbers of typical hydrocarbons.

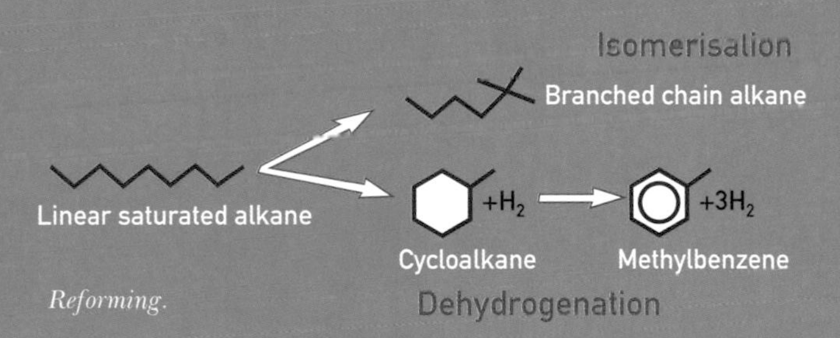

Reforming.

Hydrocarbon Name:	Structure:	Octane No.
n-Hexane		19
n-Heptane		0
n-Octane		-19
2,2-Dimethylpentane		89
2,2,3-Trimethylbutane		113
Cyclohexane		110
Benzene		99
Methylbenzene		124

Octane numbers for typical hydrocarbons found in petroleum. The formulae are given in skeletal notation where '—' represents a carbon-carbon bond.

The natural mineral, stilbite, discovered by Axel Fredric Cronstedt while walking his dog. The story that Cronstedt named the mineral after his dog is probably wrong. The sample shown was obtained from Lonaula, Pune, Maharashtra, India and is part of the collection in the National Museum and Galleries of Wales.
Photograph reprinted with permission from the National Museum and Galleries of Wales. ©NMGW. NMW 94.23G.M.2.

Oil companies are highly secretive about the catalysts they use because their business is so competitive. However, many of the catalysts involve small platinum particles, 2 nm across, supported on a base of aluminum oxide. This is a typical form for a heterogeneous catalyst. The support is there to provide a high surface area to maximise the number of active sites. It is usually inert, although in the case of reforming catalysts the oxide support also plays an active role. It catalyses the isomerisation reactions while the platinum is responsible for hydrogenation and dehydrogenation steps.

Zeolites – cracking good catalysts

The larger molecules in crude oil burn too slowly to be used in a petrol engine. So, to get the maximum amount of petrol from crude oil, these heavier hydrocarbons are split into smaller ones. This involves breaking relatively strong carbon-carbon bonds and so requires high temperatures. Catalysts such as platinum on aluminium oxide could not withstand such temperatures. Instead, the cracking catalysts which are chosen are zeolites. These natural minerals were discovered by the Swede, Axel Fredric Cronstedt, in 1756.

The name zeolite derives from the Greek words *zein*, meaning 'to boil', and *lithos*, meaning 'stone'. It was chosen by Cronstedt because he found stilbite – the first zeolite he discovered – released large amounts of water when heated, giving the impression that the substance was boiling. In fact, this is a clue to why zeolites are such valuable catalysts. They are riddled with tiny interconnecting channels, or pores, which provide a huge surface area of anything up to 500 square metres per gram. That's about the area of two tennis courts in an amount of catalyst you can pick up with a teaspoon! The channels, which are between 0.3 and 1 nm wide, are constructed such that almost every atom of the catalyst's crystal lattice is at a surface and therefore available as an active site.

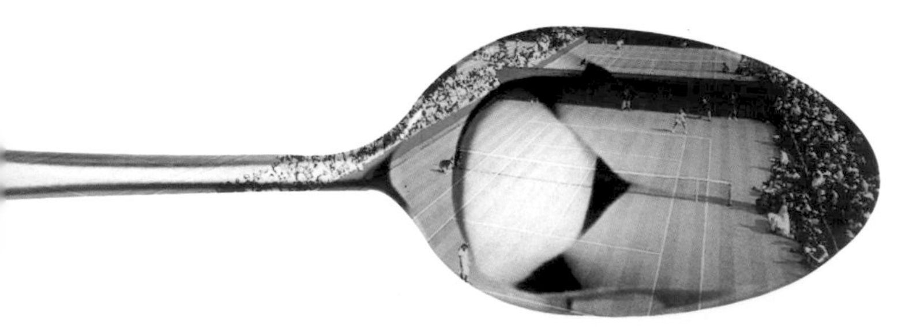

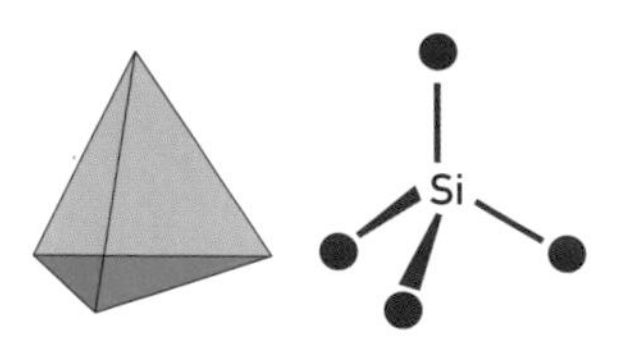

Figure 3(a). The building blocks of zeolites. Zeolites are constructed from tetrahedral units such as the SiO_4 unit.

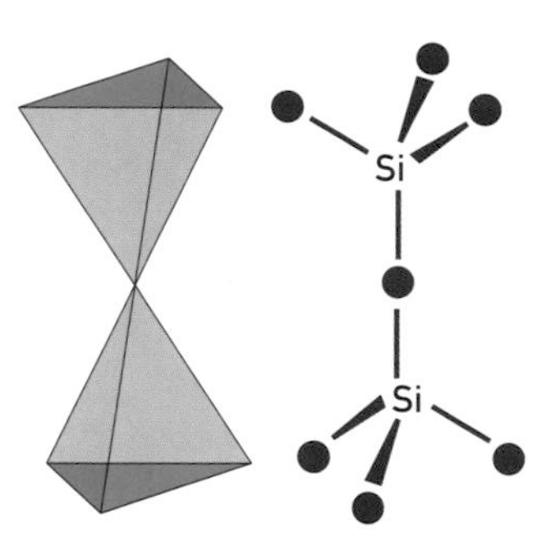

Figure 3(b). The tetrahedral units link together through the oxygen atoms.

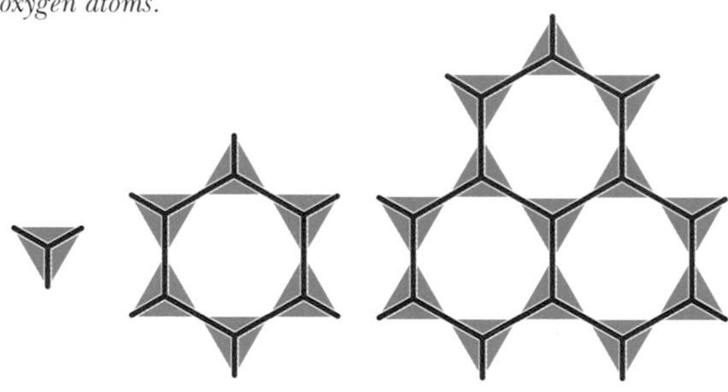

Figure 3(c). The linked tetrahedra can form rings which stack together to form planes.

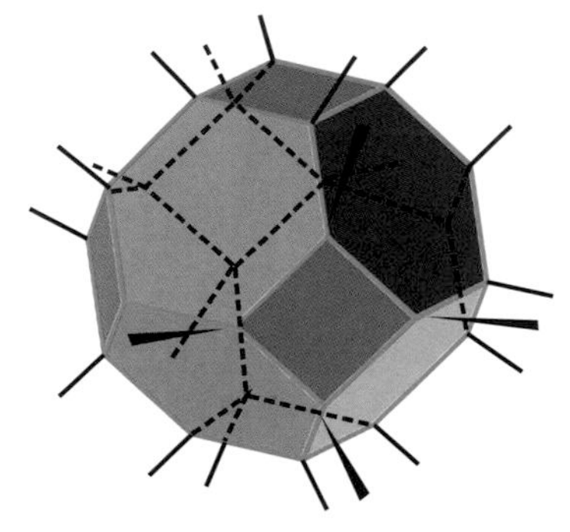

Figure 3(d). The linked tetrahedra can form rings which fold to form cages.

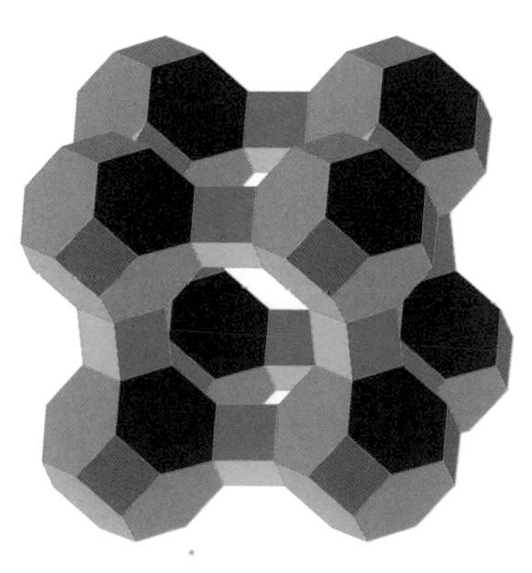

Figure 3(e). The cages can stack together giving linked channels.

Zeolites are thus sometimes referred to as three-dimensional surfaces. The channels are formed by stacking rings made up from a lattice of SiO_4 units in which silicon forms the centre and the oxygens form the corners of a tetrahedron. These tetrahedra link together at the corners. The rings can involve 4, 6, 8, 10 or 12 tetrahedra, Figure 3.

There are about 30 naturally-occurring zeolites which are made up primarily of SiO_4 units with some replaced by AlO_4 ones. AlO_4 units require an extra positive ion such as sodium or calcium, because essentially an Al^{3+} ion has replaced an Si^{4+} ion. Since Cronstedt's discovery, chemists have also made more than 150 synthetic zeolites, incorporating, for example, zinc, germanium and phosphorus.

5. Explain why aluminium tends to form Al^{3+} ions while silicon forms Si^{4+} ions.

In the 1960s it was found that the new synthetic zeolites also had very good catalytic properties. Zeolites have two characteristics which caused particular excitement.

The first is their high, but controllable acidity – the ability to donate hydrogen ions (protons). This is important because it is the acidic sites that are catalytically active. They are formed where an aluminium atom replaces a silicon atom. Aluminium has only three electrons in its outer shell whereas silicon has four. This leaves one of the oxygen atoms in the lattice one electron short. This oxygen atom picks up a hydrogen atom to form a hydroxyl (OH) group, Figure 4.

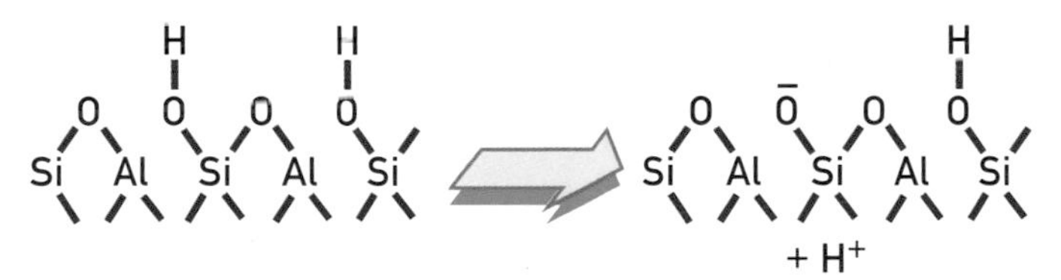

Figure 4. The acidity of a zeolite containing aluminium instead of silicon.

This OH group is acidic because it can later donate the hydrogen as a H^+ ion. The total number of acid sites therefore depends on the aluminium concentration in the zeolite, and controlling the zeolite's acidity allows chemists to decide which reactions occur.

The second useful aspect of zeolite structure is that you can control, even design, the precise shape of the pores. The size and shape of the channels affect how easily particular molecules diffuse through them and thus what products form. For example, the cracking of large hydrocarbon molecules in some zeolites produces almost exclusively 1,4-dimethylbenzene. 1,2-Dimethylbenzene and 1,3-dimethylbenzene are too wide to escape from the catalyst pores, while the 1,4-dimethylbenzene can squeeze out quite easily, Figure 5. The 1,2- and 1,3-products cannot escape unless they rearrange themselves into the 1,4- isomer. In this way zeolites can select the products of the reaction.

A computer graphic of molecules in the pores of a zeolite.
Reproduced from Faraday Discussion 105 with permission of the authors.

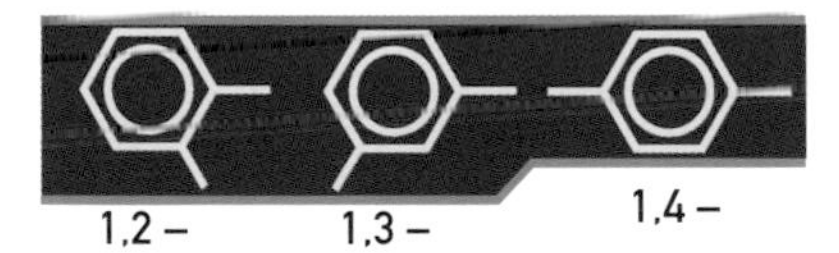

Figure 5. 1,4-Dimethylbenzene can squeeze through a narrower channel than can either 1,3- or 1,2-isomers.

Tailored catalysts

In recent years, chemists have been designing and making zeolite catalysts with channels of precise dimensions. They do this by using organic molecules as templates around which a zeolite crystal can grow from solution.

A parallel approach to 'designer catalysts' is to make zeolites with much wider pores (so that large molecules can easily diffuse in and out of them) and graft on to them other molecules with catalytic activity. This might prove to have the advantages of both homogeneous and heterogeneous catalysts. The catalyst is attached to the zeolite and can be easily filtered off but the vast surface area of the zeolite makes the active sites as accessible as homogeneous catalysts.

Fuels for the future

Ultimately the world's oil reserves will run out. Whether it takes 50 or 100 years, we will eventually need alternative sources of fuel. One possibility is to make hydrocarbons from coal by the Fischer-Tropsch reaction in which coal is first broken down into a mixture of hydrogen and carbon monoxide called synthesis gas. These gases in the mixture are then reacted together over a heterogeneous metal catalyst to form long chain hydrocarbons similar to those from crude oil.

This reaction has been used to make hydrocarbons in countries with no access to crude oil, such as Germany during World War II and South Africa during its years of political isolation. It is attracting interest again as the possibility of crude oil running out becomes more real. So chemists are trying to find out more about how the reaction works to try and develop better catalysts.

There seem to be two possible reaction mechanisms –

- **either the carbon monoxide is adsorbed on the surface of the metal catalyst (forms weak chemical bonds with the surface) and remains intact, reacting with the hydrogen, Figure 6;**

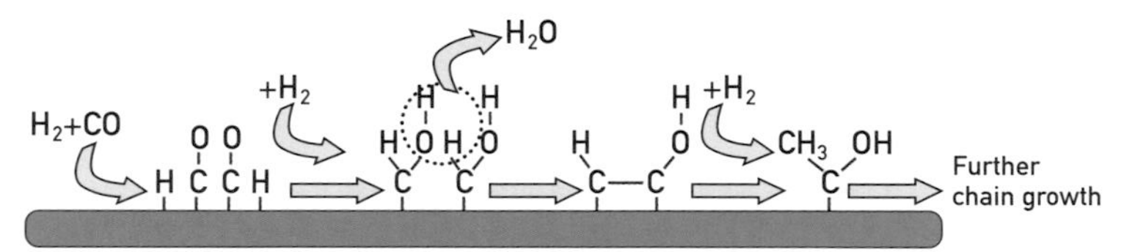

Figure 6. A possible mechanism for the Fischer-Tropsch reaction in which carbon monoxide remains intact on the catalyst surface.

- **or the carbon monoxide breaks up on the catalyst surface and the resulting carbon atom reacts with the hydrogen, Figure 7.**

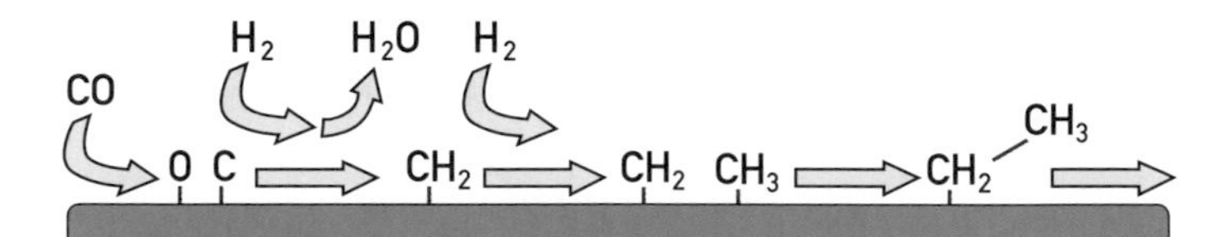

Figure 7. An alternative mechanism for the Fischer-Tropsch reaction in which the carbon monoxide breaks down on the catalyst surface.

To decide between these two mechanisms, chemists used a technique called photoelectron spectroscopy to investigate the molecules adsorbed on the surface of the catalyst. This method uses light to eject electrons from molecules at the surface, and the energies of the resulting electrons give information about the adsorbed molecules. The results revealed that no carbon monoxide was present on good catalysts but was present on poor ones, suggesting that the second mechanism (in which the carbon monoxide breaks up) was the correct one.

With the possibility of crude oil running out, chemists are searching for alternative sources of fuel.

Methanol – a fuel for the future

Methanol, CH_3OH, is also a candidate to replace fuels based on crude oil. It is already made in large quantities from a type of synthesis gas (one part carbon monoxide, one part carbon dioxide, eight parts hydrogen) using a copper/zinc oxide catalyst, but no one is sure how the catalyst works. Radioactive labelling experiments show that all the methanol comes from the carbon dioxide rather than the carbon monoxide. It is known that the reaction occurs on the copper rather than the zinc oxide and, from infrared spectroscopic measurements, that the methanoate group (HCO_2) is involved. One puzzle is that the rate of reaction of carbon dioxide with copper is slower than the rate at which methanol is produced in the reaction.

6. (a) What is radioactive labelling? State briefly how you could use a supply of $^{14}CO_2$ (^{14}C is radioactive) to show that this was the source of the methanol.

(b) Suggest how infrared spectroscopy would help to show that the methanoate group (HCO_2) is involved.

The reaction is difficult to study directly because of the high pressures (50 – 100 atmospheres, 5000 – 10 000 kPa) and temperatures (200 – 300 °C, 473 – 573 K)) involved. So chemists have looked at the reactions of different components and, surprisingly, the clue has come from studying oxygen, which is also present in the gas mixture as a result of other reactions catalysed by copper.

Chemists at Cardiff University have used a family of techniques called scanning probe microscopy, SPM, (see Box – *Scanning probe microscopy*) to 'see' oxygen on copper surfaces.

Scanning probe microscopy

We cannot 'see' atoms by reflection of visible light because its wavelength is too long. In scanning probe microscopy, a metal probe with a tip sharpened to a single atom is positioned very close to a surface and then scanned across it. A voltage is applied between the probe and the surface and this draws a current of electrons from the surface – an effect called 'tunnelling'. The smaller the distance between the probe and the surface, the bigger the current. An electronic control system moves the probe up and down above the surface so the current is kept constant, *ie* the gap between probe and surface remains constant. So the probe rises over atoms adsorbed on the surface and falls into the 'valleys' between them. A computer generates an image of the surface from the probe's movement.

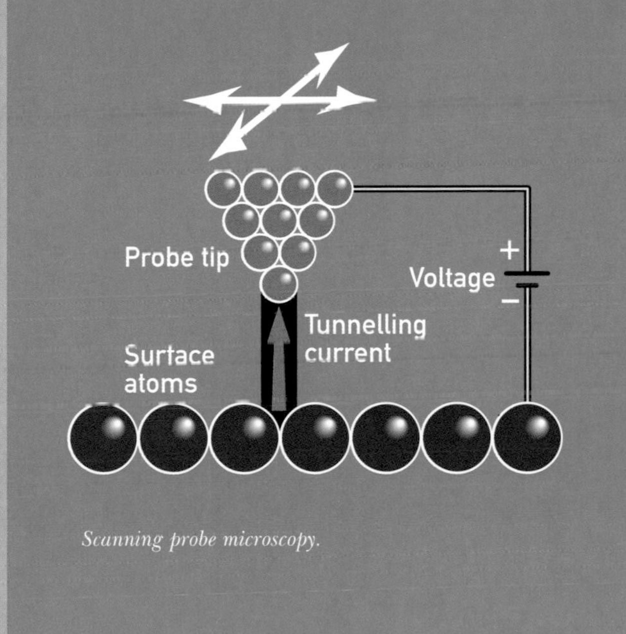

Scanning probe microscopy.

The Cardiff chemists have found that oxygen molecules are adsorbed on the surface and first dissociate to form highly reactive O^- ions before forming unreactive oxide ions, O^{2-}. The O^- ions have a lifetime of about 10^{-7} seconds during which they travel over the surface and react with carbon dioxide to form carbonate which then reacts with hydrogen to produce methanoate and eventually methanol. Figure 8 is an SPM picture of oxygen on a copper surface showing an undissociated oxygen molecule and separate O^- ions.

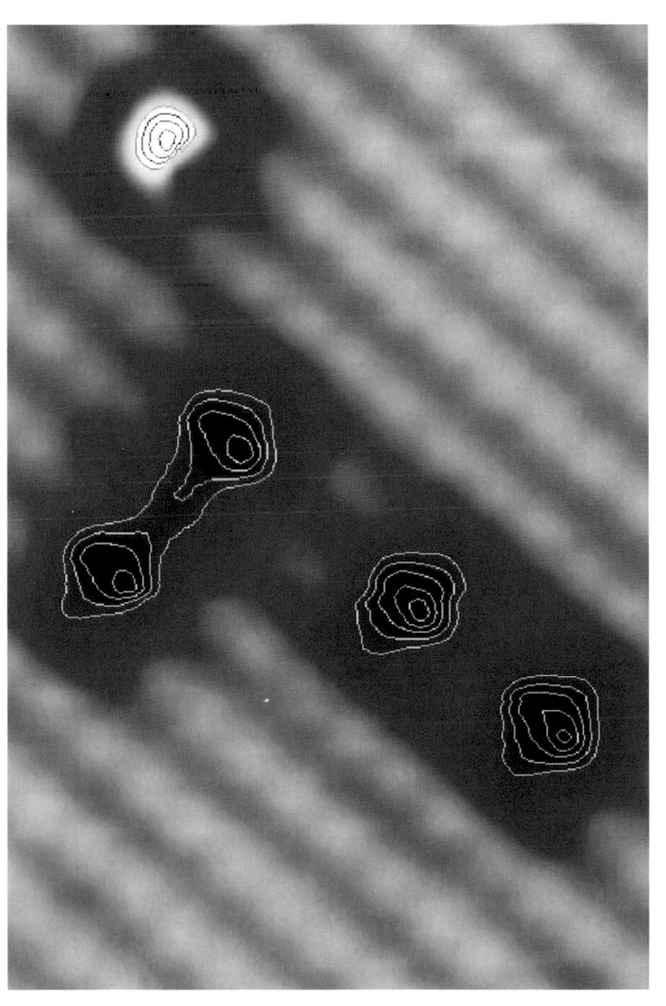

Figure 8. A scanning probe microscope (SPM) picture of oxygen on a copper surface. The orange coloured diagonal array is the copper atoms. The object at the top is an undissociated oxygen molecule. It is bright because of its high concentration of electrons. The other four objects are reactive O^- ions. Their distance apart, about 0.80 nm, shows that they are not bonded together (the oxygen-oxygen distance in an O_2 molecule is 0.12 nm) despite the electron density contours which appear to link the upper two.

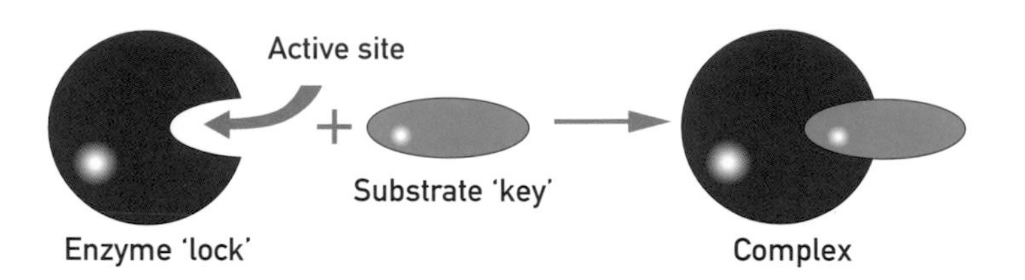

Figure 9. The enzyme has an active site, or 'lock', which recognises and binds the substrate 'key'. The chemical reaction occurs within the resulting complex.

Homogeneous catalysts

These are catalysts dissolved in a solution along with the reactants. This leads to more effective collisions between catalyst and reactant than with a heterogeneous catalyst. Homogeneous catalysts therefore tend to work under milder conditions. Two common types are enzymes and transition metal complexes.

Enzymes are proteins, large molecules which are mostly soluble in an aqueous environment. Their complex shapes contain active sites – 'clefts' of precise geometrical shape designed to 'recognise' and hold in place a particular molecule while it reacts. A simple way to imagine the way in which an enzyme works is the 'lock and key' model, Figure 9.

In some enzymes, a metal ion lies at the heart of the active site where it plays a key role in the catalytic activity. Carbonic anhydrase, for example, Figure 10, is a zinc-containing enzyme which catalyses the reaction of water with carbon dioxide in living organisms. The metalloenzyme speeds up the rate of the reaction 1 billion (1×10^9) times, but if the zinc ion is stripped away, the enzyme is no longer active.

Enzymes are used in industry for specific reactions (see Box – *Enzymes in industry*). They are highly selective and work under very mild conditions. In some ways they are too selective, usually acting only on one particular molecule, or substrate. Many chemists would like to devise catalysts based on transition metal complexes that work as effectively as enzymes but which act on a wide variety of substrates.

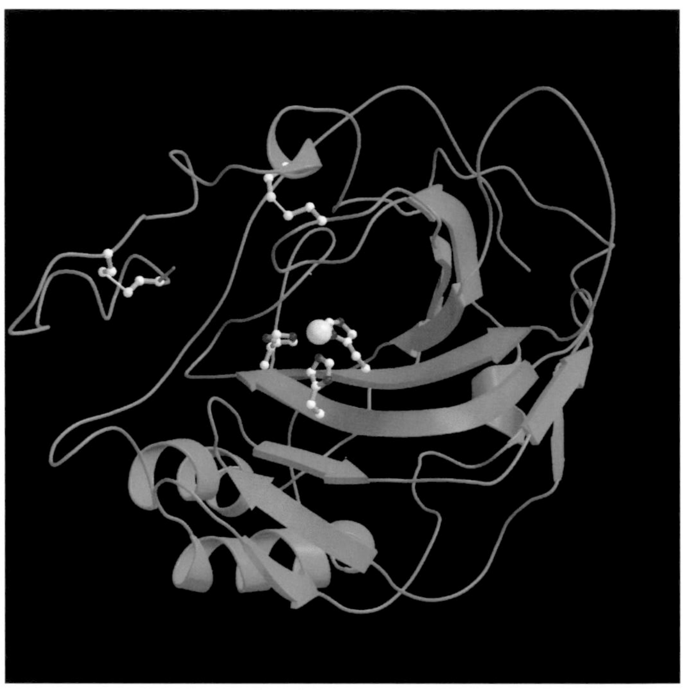

Figure 10. A computer graphic of the enzyme carbonic anhydrase.

X-ray picture of an enzyme clearly showing complex, partially helical structure. Courtesy of Professor Stan Roberts.

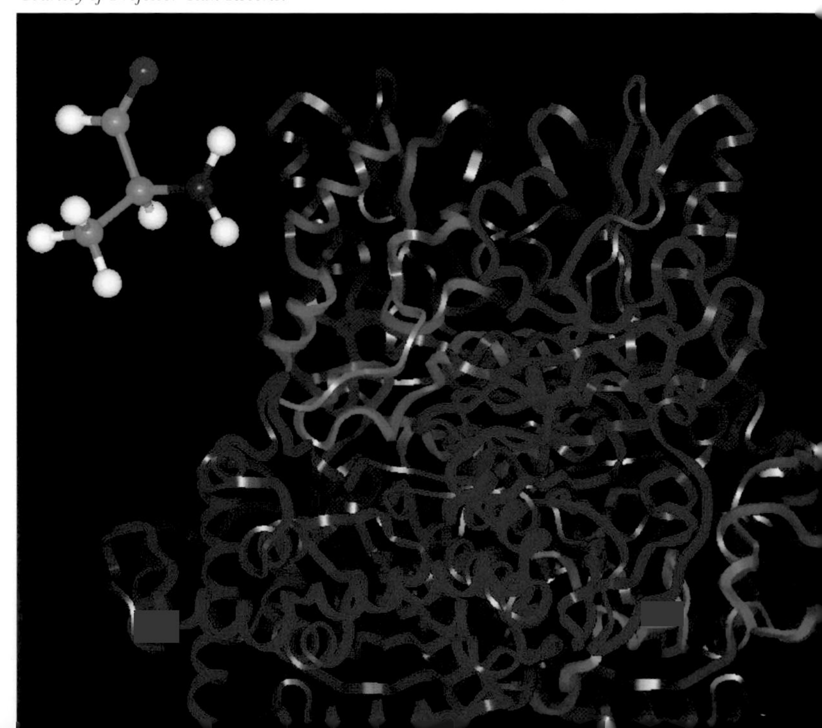

Chemical marriage brokers

Glossary

Amide: An amide is made from an amine and a carboxylic acid and has the functional group R-CONHR[I].

Enantiomer: One of a pair of non-identical mirror image isomers.

Ester: An ester is made from an alcohol and a carboxylic acid and has the functional group R-COOR[I].

Stereoselectivity: Describes a reaction in which a particular isomer of the product is produced in preference to any other.

Enzymes, the catalysts of nature, are present in all living matter, and promote a wide variety of chemical transformations. Typical reactions catalysed by enzymes include the hydrolysis of **esters** and **amides** – in other words their breakdown by water into the organic acids and alcohols or amines from which they were derived. Enzymes promote reactions extremely efficiently under very mild conditions, for example at temperatures just above room temperature, a pH of near 7.0 and at atmospheric pressure. This is very appealing to the chemical industry.

The natural catalysts are extremely selective in their actions; thus enzymes that hydrolyse esters (called esterases) do not affect amides and amidases hydrolyse amides but not esters. Enzymes are also stereoselective. This means that they distinguish between compounds that have the same formula but whose atoms are arranged differently in three dimensional space. These include chiral (handed) compounds which exist as pairs of non identical mirror images of each other, called **enantiomers** – see Box – *Chirality* in Chapter 1 – *Make me a molecule.*

? 7. What is the general name for compounds that have the same structure but whose atoms are arranged differently in three-dimensional space?

For example an esterase, when given the opportunity to hydrolyse a mixture of chiral esters ($R^IR^{II}CHCOOCH_3$), to their parent acids and alcohols, will often hydrolyse one enantiomer more rapidly than its mirror image. This leaves behind the other chiral ester and also produces a chiral acid. This process is important in the production of non-steroidal anti-inflammatory medicines, such as ibuprofen, for example.

? 8. Write the equation for the reaction of an ester such as methyl ethanoate with water. Use your answer to help you explain the term 'hydrolysis'.

Other enzymes catalyse reactions other than hydrolysis. Reductase enzymes catalyse reduction reactions, for example the reduction of a ketone ($R^IR^{II}CO$) to a secondary alcohol ($R^IR^{II}CHOH$). Once again, if the product is chiral, one enantiomer is formed preferentially in many cases. The reductase enzymes are more complicated than the hydrolase enzymes. This is because the hydrogen atoms in the secondary alcohol $R^IR^{II}CHOH$ are not derived from the enzyme but from an associated cofactor and the solvent. The cofactor is another molecule that the enzyme needs in order to work. Because of the need to have an enzyme and cofactor together to bring about the reduction, reductase enzymes are often used in their natural habitat, in other words in a whole cell such as that of a bacterium or fungus. Baker's yeast (a fungus) has a set of reductases which, with their cofactor(s), reduce many ketones stereoselectively to the corresponding alcohols, which are important components in many pharmaceutical products.

? 9. Write an equation for the reduction of butanone to the corresponding alcohol (use [H] to represent the hydrogen being added). Name, and draw the displayed formula of the alcohol formed. Mark the chiral centre. Would the reduction of propanone produce a chiral alcohol? Explain your answer.

It is not uncommon to use whole cells in biochemical reactions. Not only are the necessary cofactors in place but unstable enzymes can be used without being damaged by the extraction procedures. This is particularly true in the area of oxidations. Interesting enzyme-catalysed oxidations include reactions that add a hydroxy (OH) group to a molecule. Some of these hydroxylations are particularly important in the pharmaceutical industry. The whole-cell catalysed conversion of progesterone into 11-hydroxyprogesterone opened the way to the simple preparation of a wide range of anti-inflammatory steroids.

? 10. What do you think is meant by 'damage' to an enzyme molecule? Give two conditions in an extraction process which might damage an enzyme molecule. How does this damage come about?

It has long been recognised that enzymes catalyse reactions of natural starting materials. In recent years it has become increasingly apparent that many enzymes will equally well catalyse their particular reactions on non-natural substrates, while retaining chemical and **stereoselectivity**. Even if the enzyme is not exactly right for a particular transformation, its stucture can be altered by genetic engineering to confer better properties. For example, soap powders contain an amidase (subtilisin) which breaks down protein. The enzyme has been adapted by controlled mutation to be effective at different temperatures, and different forms are added to washing powders depending on whether they are going to be used in a 40 °C, 50 °C or 60 °C wash.

? 11. Suggest a type of stain that might be broken down by an amidase. Suggest why it is thought necessary to produce enzymes which are effective in higher temperature washes rather than instruct users to do their washing at lower temperatures.

The first sandwich compound to be discovered was ferrocene, $Fe(C_5H_5)_2$, which forms orange crystals. It contains an iron atom sandwiched between two flat rings which consist of five carbon atoms, each bonded to a hydrogen atom. These C_5H_5 rings are represented by blue discs in the diagrams below. The chemical reactivity of the rings is similar to that of benzene (C_6H_6) because they have delocalised electrons.

We now know of a whole class of sandwich compounds which contain various transition metals as the 'filling'. The 'bread' of the sandwich can also come in various 'flavours', where the hydrogen atoms of the C_5H_5 rings are replaced by more complex chemical structures. Chemists have recently become very interested in these so-called metallocenes, primarily because of the discovery that they can be used as catalysts for making valuable polymers such as poly(ethene) and poly(propene), which are manufactured on a huge scale worldwide. Polymers of this type have long been made using heterogeneous catalysts based on titanium and aluminium. These were discovered by Karl Ziegler and Giulio Natta, who were awarded the Nobel Prize in 1963. Their work is described in Chapter 7 – *The age of plastics*.

In the late 1970s Walter Kaminsky's research group at the University of Hamburg found that metallocenes containing the metals titanium, zirconium or hafnium were extremely good polymerisation catalysts when they added certain aluminium compounds to the reaction mixture as promoters. Later work, notably by Hans Brintzinger at the University of Constance has helped us to understand how these catalysts work.

In metallocene polymerisation catalysts, the rings tilt to give a bent 'sandwich'. This creates an active site on the metal atom where a polymer chain can grow by incorporating monomer units, one by one. In the example shown, the monomer is ethene, C_2H_4, and the resulting polymer is poly(ethene) (polythene). A range of other polymers can be made using similar catalysts. For example, in poly(propene), one of the red hydrogens on each monomer unit is replaced by a methyl (CH_3) group. The way these side groups are positioned along the polymer chain affects the physical properties of the resulting plastic. If all the methyl groups are attached to one side of the polymer backbone, the plastic is more rigid, whereas an alternate up-down arrangement gives a more transparent material. There are more details in Chapter 7 – *The age of plastics*. Which form is produced can be controlled by making subtle changes to the rings attached to the metal, which influence the size and shape of the active site. The metallocene catalysts shown here each produce a polymer with different properties. Such control is less easily achieved for heterogeneous catalysts in which the active sites are less well defined.

These homogeneous metallocene catalysts have been marketed by a number of companies and are beginning to compete with the established Ziegler-Natta catalysts in certain areas of the plastics market.

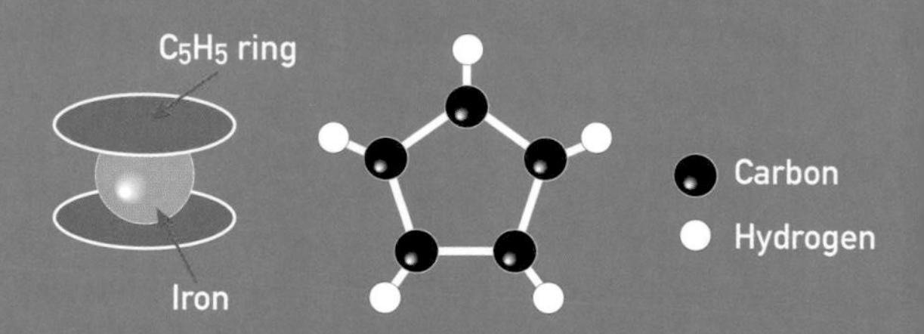

The structure of ferrocene. An iron atom is sandwiched between two C_5H_5 hydrocarbon rings, represented by the blue discs.

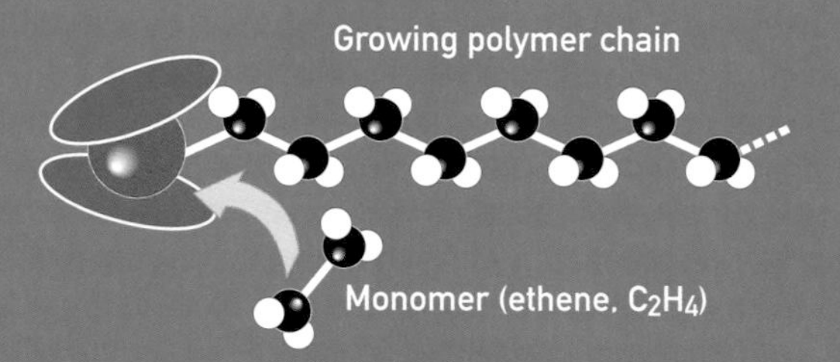

The mechanism for polymerisation of ethene by a metallocene catalyst. The polymer chain grows by incorporating ethene molecules one at a time.

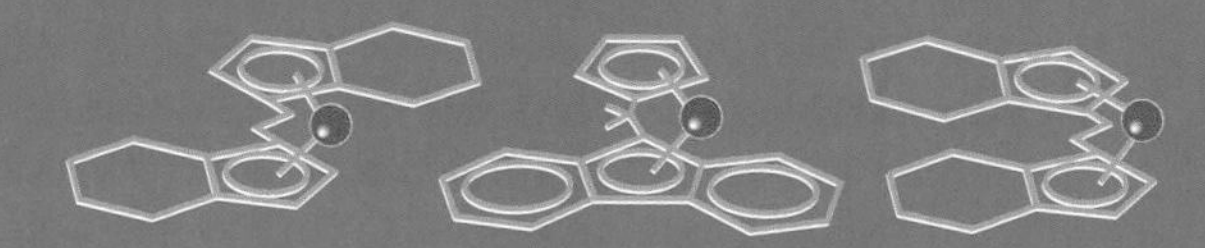

Metallocene polymerisation catalysts. Chemists can change the shape of the active site by altering the structures of the organic rings sandwiching the metal atom.

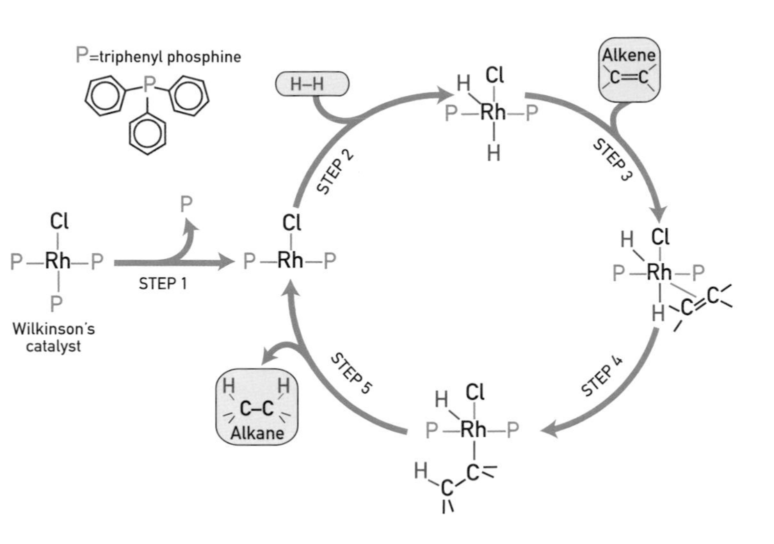

Figure 14. Catalytic cycle for hydrogenation of a carbon-carbon double bond using Wilkinson's catalyst.

Aspartame is widely used in soft drinks.

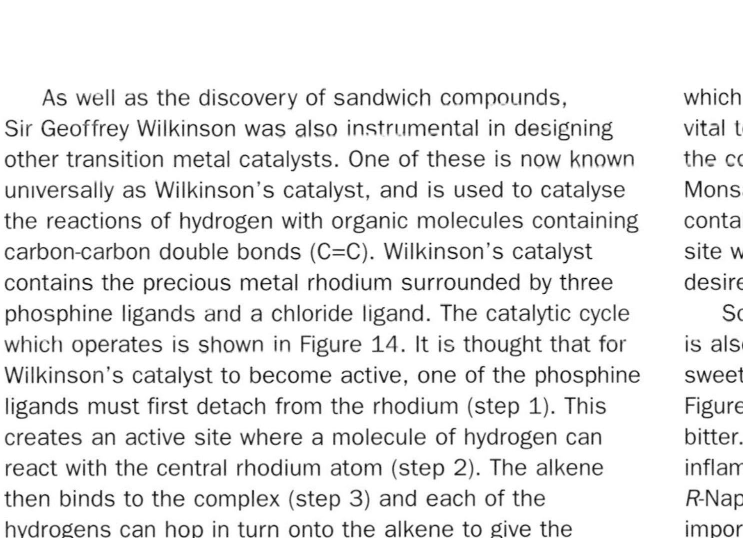

As well as the discovery of sandwich compounds, Sir Geoffrey Wilkinson was also instrumental in designing other transition metal catalysts. One of these is now known universally as Wilkinson's catalyst, and is used to catalyse the reactions of hydrogen with organic molecules containing carbon-carbon double bonds (C=C). Wilkinson's catalyst contains the precious metal rhodium surrounded by three phosphine ligands and a chloride ligand. The catalytic cycle which operates is shown in Figure 14. It is thought that for Wilkinson's catalyst to become active, one of the phosphine ligands must first detach from the rhodium (step 1). This creates an active site where a molecule of hydrogen can react with the central rhodium atom (step 2). The alkene then binds to the complex (step 3) and each of the hydrogens can hop in turn onto the alkene to give the product alkane (steps 4 and 5). The whole cycle can then repeat.

A whole range of catalysts, many also containing rhodium, have been developed to catalyse these kinds of reactions. A commercially important example is the hydrogenation of vegetable oils to make butter substitutes. Another application is in the manufacture of pharmaceuticals such as L-DOPA, a drug used in the treatment of Parkinson's disease. (It is said that the production of L-DOPA ceased in 1986 because the catalytic synthesis was so efficient that it led to stockpiles of the drug that could meet demand for several years!) The drug can exist in two chiral, or mirror-image forms, only one of which is active for the treatment. It is therefore vital to find a way of making the drug which gives the correct mirror image. The US company, Monsanto, developed a rhodium catalyst containing ligands designed to create an active site with just the right shape to produce the desired version of the drug.

So-called asymmetric catalytic hydrogenation is also used in the synthesis of the artificial sweetener, aspartame (trade name Nutrasweet®, Figure 15) whose mirror image molecule tastes bitter. Another example is *S*-Naproxen, an anti-inflammatory drug, whose mirror-image molecule, *R*-Naproxen, is a liver toxin, showing just how important it is to obtain the drug in the correct form.

14. Aspartame has two chiral centres. Mark them with a * on a copy of Figure 15.

Figure 15. The artificial sweetener aspartame (Nutrasweet®).

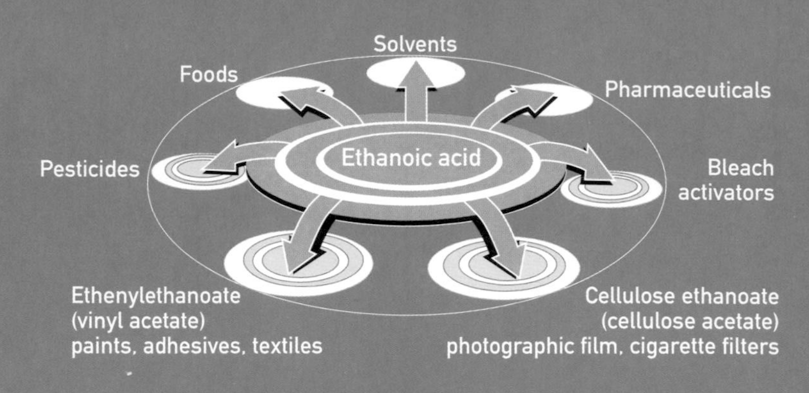

Major uses of ethanoic acid.

If you look at the bottles of vinegar in the supermarket, you may well see that some cheaper brands are labelled 'non-brewed condiment'. Traditionally, vinegar is made by oxidation of alcohol – itself produced by fermentation – to ethanoic (acetic) acid. By using a metal catalyst, ethanoic acid can now be made in a much more efficient way. It is made on a very large scale as a bulk chemical which has a whole range of uses, apart from cheaper vinegar.

The global demand for ethanoic acid has risen rapidly during the 20th century. In response, the chemical industry has evolved a series of processes to manufacture the product, each one more efficient than the last. The older processes used ethene (C_2H_4) or ethyne (C_2H_2) as their hydrocarbon feedstocks. The figure opposite shows a number of routes to ethanoic acid, all involving the use of catalysts.

After World War II, butane (C_4H_{10}), from crude oil became the preferred feedstock, and it was oxidised using a homogeneous cobalt-based catalyst. By 1973 this technology accounted for 40% of ethanoic acid production.

However, oxidation reactions suffered from a lack of selectivity, so a substantial price had to be paid to separate ethanoic acid from the various by-products. A new approach, using a more controlled catalytic reaction overcame these problems, and it is now regarded as one of the foremost successes of homogeneous catalysis.

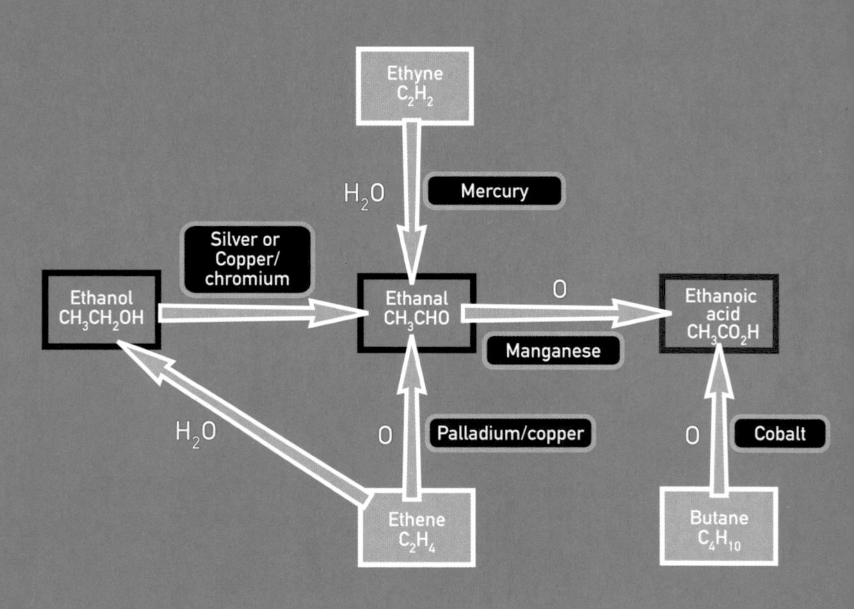

Industrial routes to ethanoic acid involving oxidation of hydrocarbons.

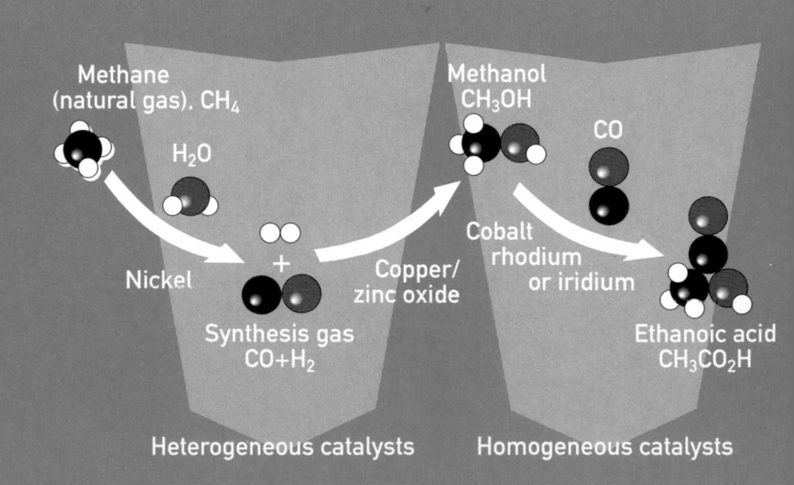

The route to ethanoic acid from natural gas, using both heterogeneous and homogeneous catalysts.

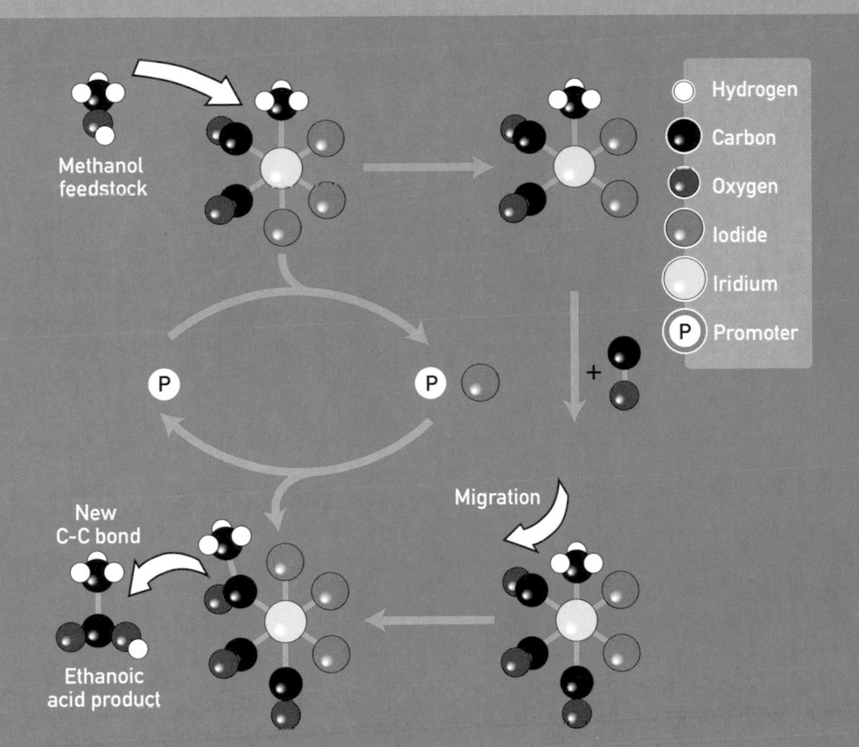

©BP AMOCO p.l.c. (2000).

The story began in the early 1960s when the BASF company developed a new process for making ethanoic acid by reacting methanol with carbon monoxide. The reaction is termed an insertion, because a molecule of carbon monoxide is inserted into the carbon-oxygen bond of methanol (CH_3OH). The homogeneous catalyst used was cobalt-based with an iodide promoter, which improved the catalytic activity. One of the advantages of this process was that both starting materials, methanol and carbon monoxide, can be made from natural gas, or alternatively, coal. The economics of the process therefore no longer depended on the price of oil which was subject to large fluctuations.

Despite this advance, the BASF process was soon improved upon by researchers at Monsanto. They developed a similar catalyst based on the metal rhodium (which lies directly beneath cobalt in the Periodic Table). In 1968, Monsanto announced its new process, which operated at lower temperature and pressure than the cobalt system. The rhodium catalyst converted methanol into ethanoic acid with less than 1% of unwanted byproducts. Many plants based on this technology were built and in 1993, 60% (more than 3 million tonnes annually) of the world ethanoic acid capacity used the rhodium catalyst system.

Much research has gone into understanding how the Monsanto process works. Chemists identified the chemical structures of a number of the rhodium compounds involved in the catalytic cycle, similar to those shown in Figure 14 for hydrogenation via Wilkinson's catalyst. They were able to measure the infrared spectrum of a real catalytic reaction at high temperature and pressure. The spectrum of the rhodium compound which they observed under the extreme conditions of this experiment was identical to one they could isolate in the laboratory. By studying the chemistry of this compound, they could build up a series of reaction steps which formed the catalytic cycle.

Further studies have continued to this day. In the early 1990s, chemists at the University of Sheffield proved the existence of one of the molecules which had eluded Monsanto. This helped to explain the extremely high selectivity of the catalytic process. The story is not quite finished. In 1996, a new catalyst was found that works even better than rhodium. The journey down the Periodic Table has continued to iridium. This element was originally tested by Monsanto, but under the conditions that they used, it did not work as well as rhodium. BP Chemicals (who bought the technology from Monsanto) has now discovered additives that promote the iridium catalyst, and can improve its activity so that it out-performs rhodium. The new catalyst is already installed on plants in Hull and in the US.

Improvement of the iridium catalyst depended on finding promoters that could speed up the slowest reaction in the catalytic cycle to give a faster output of product. Chemists at BP and the University of Sheffield think that they have unravelled how the promoters work – by encouraging a methyl (CH_3) group to combine with a molecule of carbon monoxide (CO) while both are bonded to the iridium catalyst. To do this it appears to be necessary to remove one of the three iodide ligands from the central iridium atom. The job of the promoter is to keep a hold of this iodide until the bond between the CH_3 and CO groups has formed. A carbon-carbon bond is formed in this sequence of steps, giving an ethanoyl (CH_3CO) group which is a fundamental unit of the final product, ethanoic acid (CH_3COOH).

Future challenges

The search continues for more active and selective catalysts in all parts of the chemical industry. One long-term aim is the selective conversion of hydrocarbons into more valuable chemicals, for example, via selective oxidation. The biggest advances are expected to involve completely new processes, particularly replacements for current non-catalytic reactions such as the conversion of propene ($CH_3CH{=}CH_2$) to propene oxide (CH_3CHOCH_2). Catalytic routes for these processes would be cheaper and do less harm to the environment as they would use less fuel to heat the reactants.

Building organic compounds

One of the world's most abundant resources is natural gas, the main constituent of which is methane (CH_4). As well as being burnt to provide energy, methane is a source of carbon which can be used as a building-block for larger organic chemicals. Currently, this is achieved by rather circuitous routes, via synthesis gas rather like those described above. Chemists are exploring a number of avenues by which methane and similar hydrocarbons could be converted directly into other compounds like methanol.

Mirror image catalysis

Another key goal, mentioned earlier in this chapter, is so-called asymmetric catalysis, where a chiral product is made in preference to its mirror image partner. Chiral compounds are of vital importance for pharmaceuticals and other biologically-related products. The processes used at the moment are often not catalytic, and if so they are usually homogeneous. The chemical industry prefers heterogeneous catalysts, so researchers are currently trying to develop chiral catalysts that work heterogeneously. One approach is to tether chiral catalysts onto a substrate or to the internal surface of a zeolite, as described earlier. Another method is to modify the surface of a catalyst with adsorbed organic molecules which act as a template, controlling the configurations of the intermediates to produce only one chiral molecule.

More generally, the chemical industry requires highly selective, energy-efficient processes with minimal adverse environmental consequences. The role of chemists is to explore and understand the behaviour of catalytic systems and to use this knowledge to help design the next generation of catalysts.

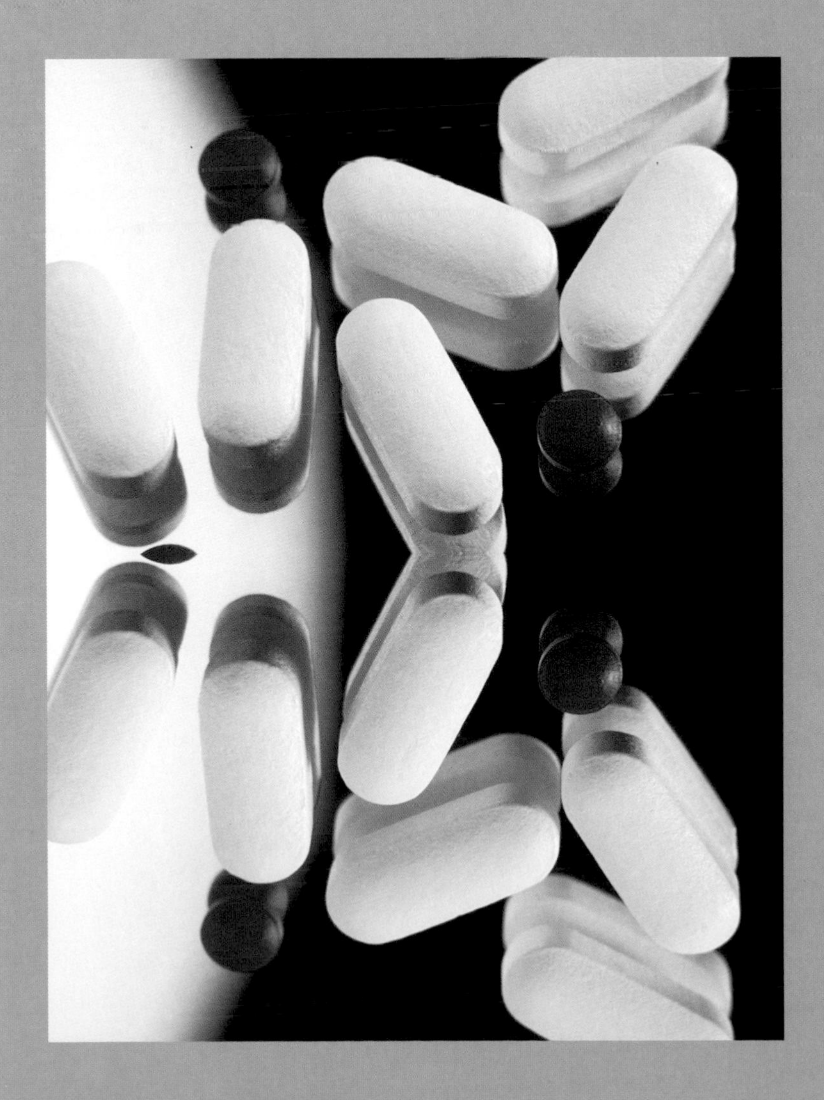

Answers

1. Modern Western economies rely more on service industries (which do not manufacture goods) than they once did. Many 'high tech' goods such as electronic devices rely less on metals and paint. Economies are less geared to agriculture.

2. (a) Ethene.
   (b) There are many possibilities. Oxidation of a primary alcohol to an aldehyde and a carboxylic acid is one.

3. (a) Heterogeneous,
   (b) Homogeneous.

4. Other answers are possible.

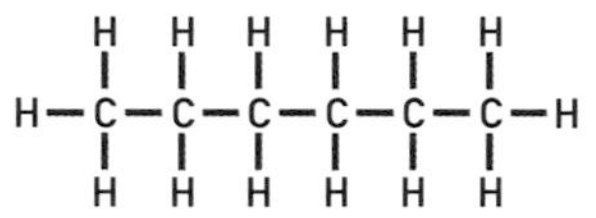

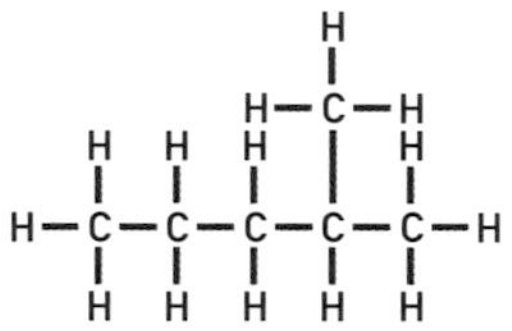

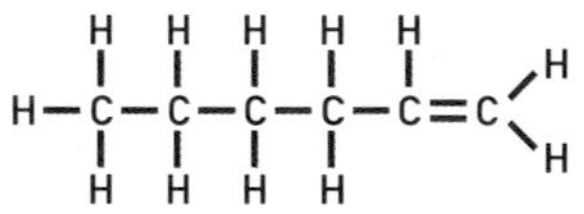

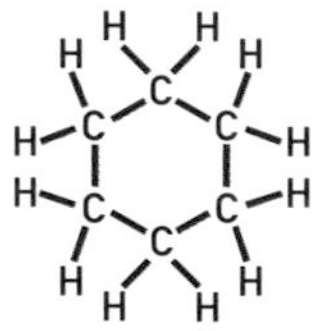

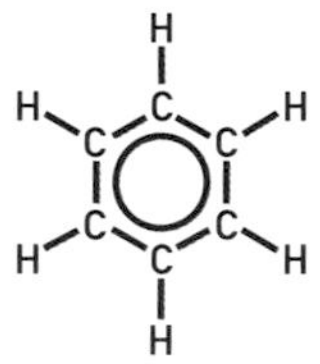

```
    H  H  H  H  H  H
    |  |  |  |  |  |
 H—C—C—C—C—C—C—H
    |  |  |  |  |  |
    H  H  H  H  H  H

                H
                |
             H—C—H
    H  H  H     |  H
    |  |  |     |  |
 H—C—C—C—C—C—H
    |  |  |  |  |
    H  H  H  H  H

    H  H  H  H  H
    |  |  |  |  |      H
 H—C—C—C—C—C=C
    |  |  |  |         H
    H  H  H  H

      H   H
    H  \ /  H
     \  C  /
  H—C     C—H
     |     |
  H—C     C—H
     /  C  \
    H  / \  H
      H   H

        H
        |
  H    C     H
    C     C
    |  O  |
    C     C
  H    C     H
        |
        H
```

5. Aluminium has three electrons in its outer shell, silicon has four.

6. (a) Radioactive labelling is the use of radioisotopes to trace the fate of particular atoms in chemical reactions by measuring where the radioactivity ends up.
   Use $^{14}CO_2$ and unlabelled CO in the reaction. Look for radioactivity in the product. Repeat using ^{14}CO and unlabelled CO_2.
   (b) Infrared spectroscopy looks at vibrations of parts of molecules and would identify stretching frequencies of the carbon-hydrogen and carbon-oxygen bonds in the methanoate group.

7. Stereoisomers.

8. $CH_3COOCH_3(aq) + H_2O(l) \rightarrow CH_3COOH(aq) + CH_3OH(aq)$
   Hydrolysis means 'breaking up by reaction with water'.

9. $C_2H_5COCH_3 + [2H] \rightarrow C_2H_5CH(OH)CH_3$
   The product is butan-2-ol.

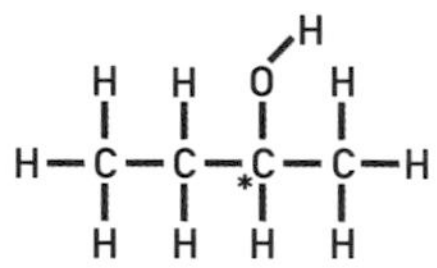

```
              H
             /
    H  H  O  H
    |  |  |  |
 H—C—C—C*—C—H
    |  |  |  |
    H  H  H  H
```

   No, there is no carbon atom bonded to four different groups in propan-2-ol, the alcohol which results from the reduction of propanone.

10. A change in the shape of the enzyme so that its substrate no longer fits into the active site.
    High temperatures or extremes of pH could damage the enzyme by disrupting the hydrogen bonds which hold it in a particular shape.

11. Any protein – egg or blood, for example.
    Many people associate high temperature washes with efficient cleaning.

12. They must have a lone pair of electrons with which to form a dative covalent bond.

13. Step1: addition
    Step 2: addition
    Step 3: rearrangement
    Step 4: elimination.

14.

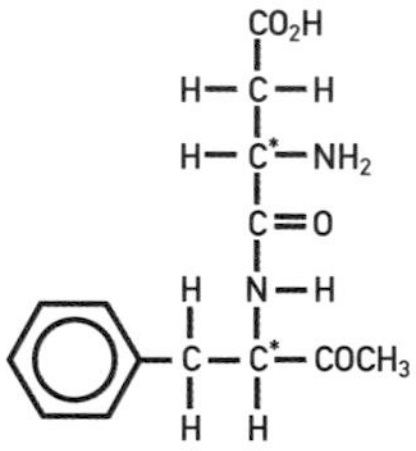

```
                CO2H
                 |
              H—C—H
                 |
              H—C*—NH2
                 |
                 C=O
                 |
           H     N—H
           |     |
 (C6H5)—C—C*—COCH3
           |     |
           H     H
```

Introductory page pictures: Rover Group and Faraday Discussion 105.

Following chemical reactions

What actually happens during a chemical reaction? We normally know the starting materials (or reactants) and the products, but how exactly does one set of chemicals turn into the other? Obviously, bonds break and bonds are made, but we usually have to deduce the details of what goes on in the course of the reaction.

Now, with a new generation of lasers, chemists can watch the minute details of what happens and learn how and why certain molecules interact with one another. This allows us to begin to understand what is going on during important chemical processes such as combustion, atmospheric ozone depletion and **photosynthesis** and even how the first molecules formed in the universe.

Following reactions by flash photolysis

Chemists have been measuring the kinetics of reactions for more than a century by merely mixing reactants together in a container and measuring the concentrations of the reactants and products, perhaps by titration. Although short-lived transition states and intermediates cannot be observed directly, their existence is inferred by monitoring the formation of products and trying to deduce a mechanism using 'common sense', for example using the idea that positively charged ions are attracted to negatively charged parts of the molecule. Only fairly slow reactions can be tackled by this sort of method.

? 2. (a) What is the fastest timescale which can be monitored if a reaction is being followed by doing a titration? Less than a second, a few seconds, a few minutes, half an hour, several hours, several days?

(b) Make a list of as many other techniques as you can which may be used to follow the progress of chemical reactions.

It was not until just after World War II that very fast reactions triggered by visible light could be observed. This was as a result of a revolutionary experimental development. In 1949, Ronald Norrish and George Porter, at the University of Cambridge, developed an exciting technique called flash photolysis. The word photolysis means 'breaking up molecules using light'. This new technique could measure chemical reactions lasting just a few milliseconds or even tens of microseconds (a millisecond is 10^{-3} seconds, and a microsecond 10^{-6} seconds). It earned the two chemists the 1967 Nobel Prize for Chemistry.

In their first experiments, Norrish and Porter used banks of capacitors from war-surplus stores to pass huge electrical currents for very short times through flash lamps containing noble gases (similar to those used in modern flash photography). These short, intense pulses of light were then used to 'pump' large numbers of atoms or molecules in a gas into highly **excited electronic states** and thereby start a reaction. The changes in concentration of these species could be monitored over a period of time as described below.

Figure 4 is a schematic diagram of the apparatus used in a typical flash photolysis experiment. The short light pulse – the photolysis flash – is directed into a glass bulb containing the reactants. A significant number of the reactant molecules absorb this light to form excited versions of themselves, or they break up to create free radicals (highly reactive molecules with unpaired electrons) and other short-lived fragments.

Now, here is the clever part. By firing a second 'probe' flash through the reaction vessel over a series of carefully timed intervals after the photolysis flash, the absorption spectra of the species present at the instant of each probe flash can be obtained. In those early days the visible spectrum was recorded on photographic film; today, an array of electronic photodetectors is used. If one of the species in the reaction vessel absorbs light of a particular wavelength, it is observed as a dark line in the spectrum of the probe pulse and we can identify it and measure its concentration.

Norrish, Porter and their Cambridge colleagues, exploited this flash technique to study a huge variety of problems as wide-ranging as the explosive reaction of hydrogen and oxygen, the combination of separate iodine atoms to form molecular iodine and the photolysis (breakdown using light) of organic molecules (such as ethanal and propanone). One particularly interesting experiment, carried out by Porter in the 1950s has important implications for our present-day understanding of the chemistry of our atmosphere. This was the light-induced breakdown of chlorine dioxide, ClO_2.

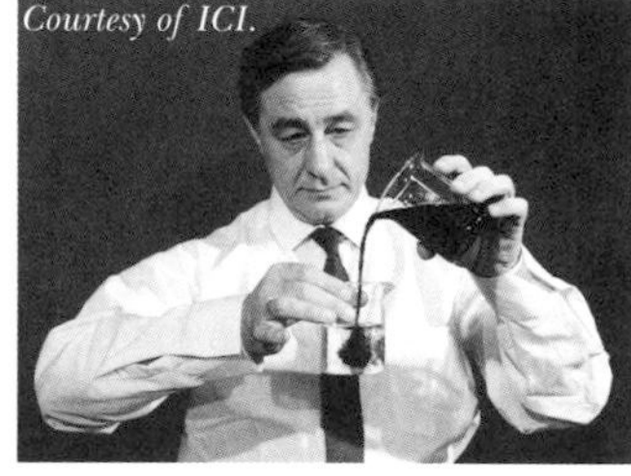

Ronald Norrish and George Porter who developed flash photolysis.

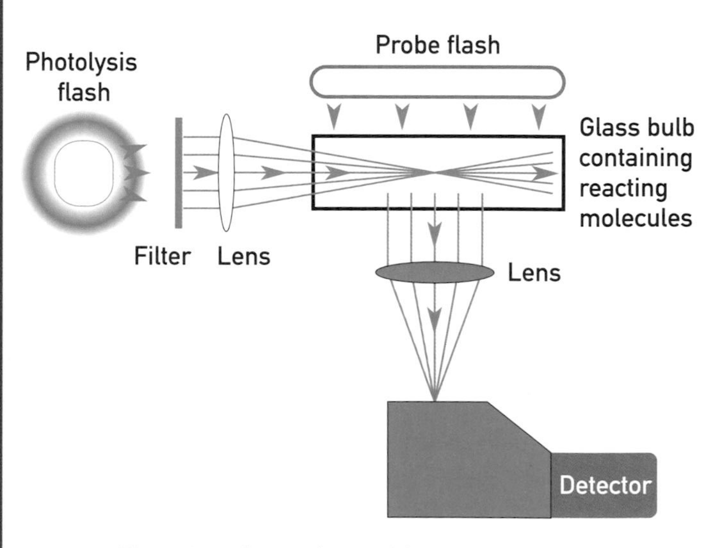

Figure 4. A schematic layout of the apparatus used in a typical flash photolysis experiment. The photolysis and probe flashes are directed into a glass bulb containing the reactant molecules: the photolysis pulse creates the excited species of interest and the probe pulse is used to record absorption spectra of the species present at specific times after photolysis. If a particular molecule absorbs a particular frequency of light then this appears as a dark line in the spectrum of the probe pulse.

Glossary

Excited state: A molecule existing in a higher energy level than the lowest possible (the ground state).

The spectrum of ClO_2. Note how it disappears immediately after the flash because it dissociates.

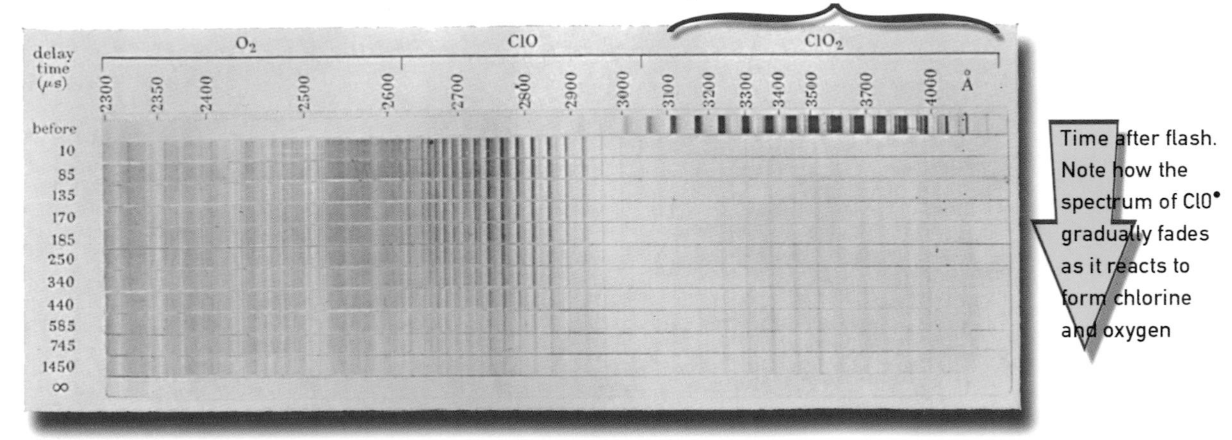

Figure 5. The spectra obtained after flash photolysis of chlorine dioxide. The spectrum of ClO_2 appears before the flash but not after it, showing that it is all decomposed by the flash. The spectrum of $ClO^{\bullet}$ is intense immediately after the flash but then gradually fades as it reacts.
Reproduced with permission from R Norrish, Proc R Soc, Plate 4,1-37, A301, 1967, The Royal Society.

Porter and his colleagues recorded a series of absorption spectra in separate experiments with increasing delays (on a microsecond timescale) between the photolysis flash and the probe flash. Each chlorine dioxide molecule present before the photolysis flash dissociates to form the diatomic radical chlorine monoxide ($ClO^{\bullet}$), and atomic oxygen ($O^{\bullet}$) (the dots indicate unpaired electrons). So the ClO_2 spectrum disappears immediately after the flash.

$$ClO_2(g) \xrightarrow{\text{light}} ClO^{\bullet}(g) + O^{\bullet}(g)$$

The absorption spectum of the $ClO^{\bullet}$ fragments was visible after the photolysis flash and its intensity was seen to decay with time, Figure 5. Its concentration decreases as it reacts to form molecular chlorine and oxygen.

$$ClO^{\bullet}(g) + ClO^{\bullet}(g) \rightarrow Cl_2(g) + O_2(g)$$

It was not until 30 years after these pioneering experiments that people found they were of any practical significance. It turns out that the chlorine monoxide radical plays a role in the breakdown of ozone in the upper atmosphere. Chlorofluorocarbons from aerosols and refrigerator coolants decompose in the upper atmosphere releasing chlorine atoms which react with ozone (O_3) to generate the chlorine monoxide radical, thus removing the ozone.

3. Suggest the equation for the reaction between chlorine atoms and ozone.

In addition to their work on radicals in atmospheric chemistry, Norrish and his students also put a lot of effort into understanding the kinetics of reactions involved in combustion processes – in particular, the role of antiknock agents in petrol.

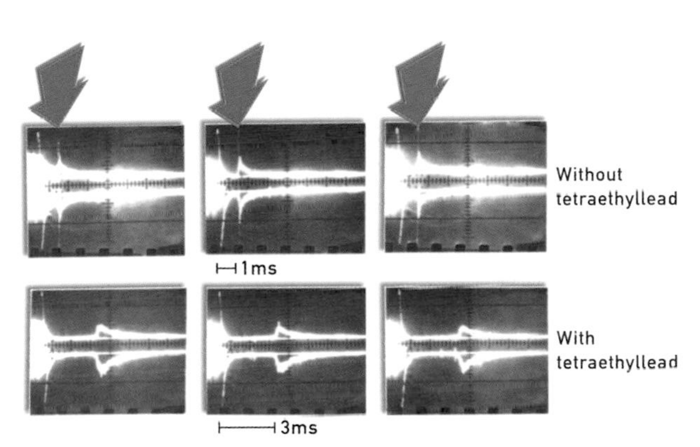

Figure 6. The effect of adding tetraethyllead on a hexane-oxygen explosion reaction. The sharp peaks, marked, on the top oscilloscope traces indicate that the mixture has detonated. These are dramatically depressed in the lower traces.
From 'Fast reactions and primary processes in chemical kinetics', Proceedings of the 5th Nobel symposium.

Controlling the combustion of a mixture of fuel and air inside the cylinder of an internal combustion engine is an important engineering exercise. If the mixture burns too quickly, the piston receives a hard jerk, which is known as knocking, rather than a smooth push. As well as producing an unpleasant sound, knocking decreases the efficiency of the engine. Fuels are usually rated by their octane number – the higher the octane number of the fuel, the better its performance. By adding to fuels certain antiknock agents such as tetraethyllead, it is possible to increase the octane rating (though there are now more environmentally-friendly ways of doing this). An example of one of Norrish's experiments to demonstrate the role of antiknock agents in the combustion of alkanes is shown in Figure 6. The sharp peaks are typical of the detonation of a mixture of hexane and oxygen and are dramatically suppressed by the addition of tetraethyllead.

4. Why has the use of tetraethyllead as a petrol additive been discontinued?

Even faster reactions

Manfred Eigen
©The Nobel Foundation.

Chemists in the early 1950s were limited by the timescale of the flash lamps available (that is the length of time taken for the flash of light to build up and die down). However, the race to study faster and faster reactions was just beginning. The following exchange took place at a Chemical Society meeting in 1954 on fast reactions. (The Chemical Society was one of the forerunners of The Royal Society of Chemistry.) The German chemist Manfred Eigen, who later shared the Nobel Prize with Norrish and Porter, asked Oxford chemist Ronnie Bell how the English language would describe reactions that were 'faster than fast'. Ronnie Bell replied: 'Damn fast reactions, Manfred, and if they get faster than that, the English language will not fail you, you can call them damn fast reactions indeed!'

Enter the laser

Fortunately, while chemists were becoming more excited about the prospect of looking at faster reactions, physicists were busy developing a new light source which was to revolutionise the field of chemical reaction dynamics. This new light source was the laser. The word laser is an acronym for light amplification by stimulated emission of radiation. It produces a highly focused beam of radiation which is not only of a single wavelength but is also coherent (the vibrations of the light waves all coincide, or are 'in phase' with one another).

Uses for lasers

Today, a huge variety of lasers is available, using various gases, solutions of organic materials, inorganic crystals and semiconductors as lasing materials that emit electromagnetic radiation over a wide range of wavelengths from ultraviolet to microwaves. As a result they are extremely useful tools not only in the laboratory but also in everyday applications such as CD players, supermarket checkout scanners and telecommunications. Lasers are also used routinely in eye surgery for removing cataracts and correcting poor eyesight and in plastic surgery for removing birthmarks. During the US 'Star Wars' programme in the 1980s, millions of dollars were spent developing defensive lasers such as the dual-purpose Stingray laser, which generates a wide beam to scan for reflections from enemy sensors and an intense narrow beam to destroy night-vision equipment and gun sights.

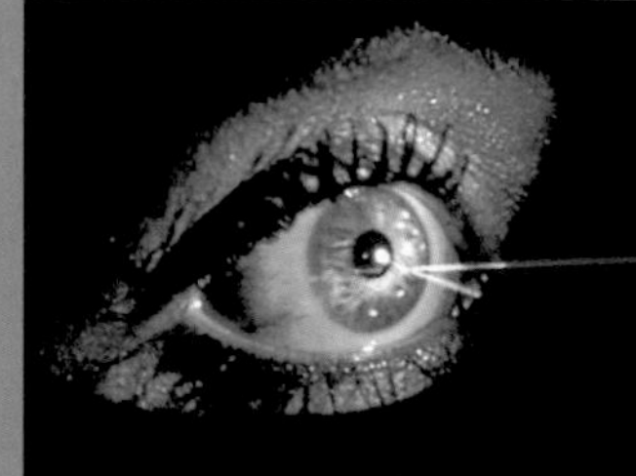

Laser eye surgery.
Alexander Tsiaras/Science Photo Library.

CDs and bar codes are both read by lasers.

The Physics Department, Imperial College/Science Photo Library.

How lasers work

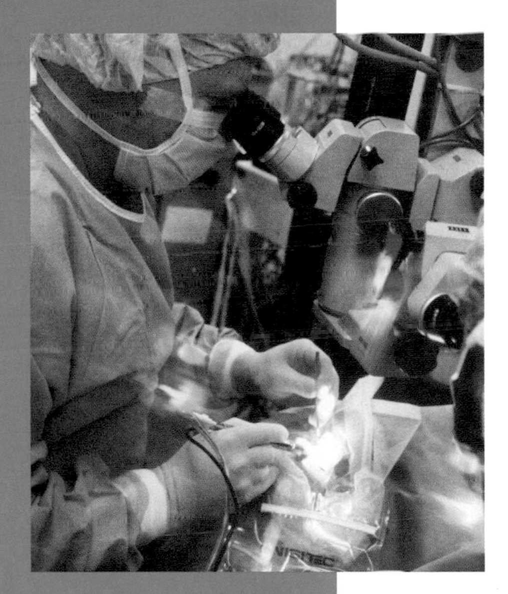

The first step in generating laser radiation is to create a 'population inversion' in a material, that is to pump the constituent atoms or molecules up to excited states, usually by using a light source. Next, a few of these excited species may release energy spontaneously by emitting photons. Each of these photons then interacts with other excited species and stimulates them to emit a photon with the same wavelength. This travels in the same direction and, most importantly, is in phase with, the first. ('In phase' means that all the vibrations of the waves are in step.) This process continues throughout the laser cavity, generating an avalanche of photons. Mirrors placed at either end of the laser reflect the beam backwards and forwards and amplify the intensity still further. By arranging that one of the mirrors is only partially reflecting, laser light is emitted in a highly-focused beam from this end of the cavity.

1. Initally the molecules of the lasing medium are mostly in their ground state.

Excited state

Ground state

Pump

2. The atoms or molecules are then pumped up to their excited states to create a 'population inversion'.

Excited state

Ground state

3. A few of these excited species may emit photons spontaneously. These photons then go on to stimulate emission of additional photons with the same wavelength, direction of travel and phase.

Excited state

Ground state

The principle of the laser.

Tuned and pulsed lasers

Since the energy levels of laser materials are fixed, the output of most lasers occurs at fixed frequencies. However, for some lasers such as those using dyes as lasing materials, it is possible to use different energy levels and hence 'tune' the output of the laser to produce laser light of a specified frequency. Tunable dye lasers are very popular in research laboratories and have had a dramatic impact on chemical spectroscopy and photochemistry studies.

The first commercial lasers produced continuous beams of light. However, it was not long before it was possible to produce short bursts of radiation in pulses lasting only a few nanoseconds (billionths of a second), about one-thousandth shorter than the pulses of light generated using flash lamps. This enabled even faster reactions to be studied – those which were over in less than the time taken for the flash of a conventional lamp to build up and die away.

Nanosecond flash photolysis

Michael Topp and George Porter were among the first chemists to exploit these new nanosecond lasers and perform faster flash photolysis experiments. They worked at the Royal Institution, London in the mid-1960s. The major difficulty they had to overcome was the need to generate two pulses, the photolysis pulse and the probe pulse, separated by a few nanoseconds. Their solution to this problem avoided electronics completely by using an optical delay unit, Figure 7. A pulse of laser light is sent through a beam-splitter which, as its name suggests, splits the light beam into two portions. Some of the light is reflected directly into the reaction vessel forming the first photolysis pulse, while the rest of the light passes straight through the beam-splitter towards a mirror which is mounted on a moveable stage. At the mirror, the light is reflected directly back onto the beam-splitter where it is then directed into a vessel containing a liquid called the scintillation solution. The large molecules in this solution behave in much the same way as the organic dyes used in dye lasers, and provide the probe pulse for monitoring the species in the reaction vessel. This part of the experiment is much the same as the earlier experiments based on flash lamp excitation. Varying the distance between the beam-splitter and mirror changes the distance that the probe light pulse must travel and determines the time-delay between the photolysis and probe pulses. It is quite straightforward
to change the total light path to generate delays in the nanosecond regime since a path of 3 m corresponds to a delay of 10 nanoseconds.

 **5. The speed of light is 3×10^8 m s^{-1}. Confirm that a light path of 3 m produces a delay of 10 nanoseconds (1×10^{-8} s).**

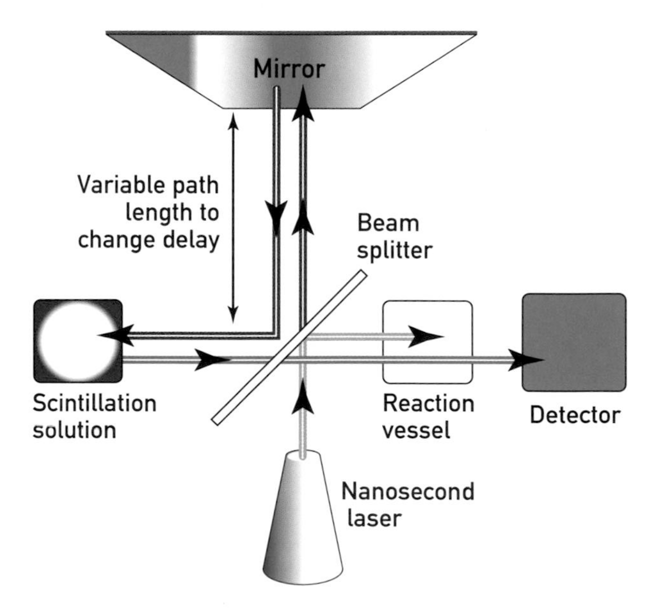

Figure 7. A nanosecond flash photolysis experiment. The lime green line shows the path of the photolysis pulse and the red line shows the path of the delayed probe pulse from the beam splitter to the scintillation solution.

Following reactions with resonance Raman spectroscopy and infrared spectroscopy

Although much of the work on reactions is done with visible light, which tells us about electronic energy levels, chemists also use infrared spectroscopy which tells us about the vibrations of parts of molecules. However, infrared spectra are often complex. Raman spectroscopy is a variation on infrared spectroscopy which enables us to look at a specific part of a large molecule.

In resonance Raman spectroscopy a visible laser acting as a probe is used to shine at the sample light of the same frequency as an absorption of the sample. A portion of this radiation is scattered – most of it at the same frequency as the incoming light. However, some of the scattered light has a slightly higher or lower frequency than the incoming light. What has happened is that the vibrating molecule has absorbed some energy from the incoming beam or passed some of its vibrational energy onto the beam. The difference corresponds to the frequency of vibration of one of the bonds in the molecule – thus providing information about its vibrational properties.

Short-lived intermediates in fast reactions can be investigated using this technique. First the short-lived intermediate is generated by a powerful pulsed laser (flash photolysis), and this is followed by a second, probe, laser. The method is particularly suited to biological molecules.

Glossary

Homogeneous catalysis: Catalysis in which the catalyst and the reactants are in the same phase – all in solution, for example.

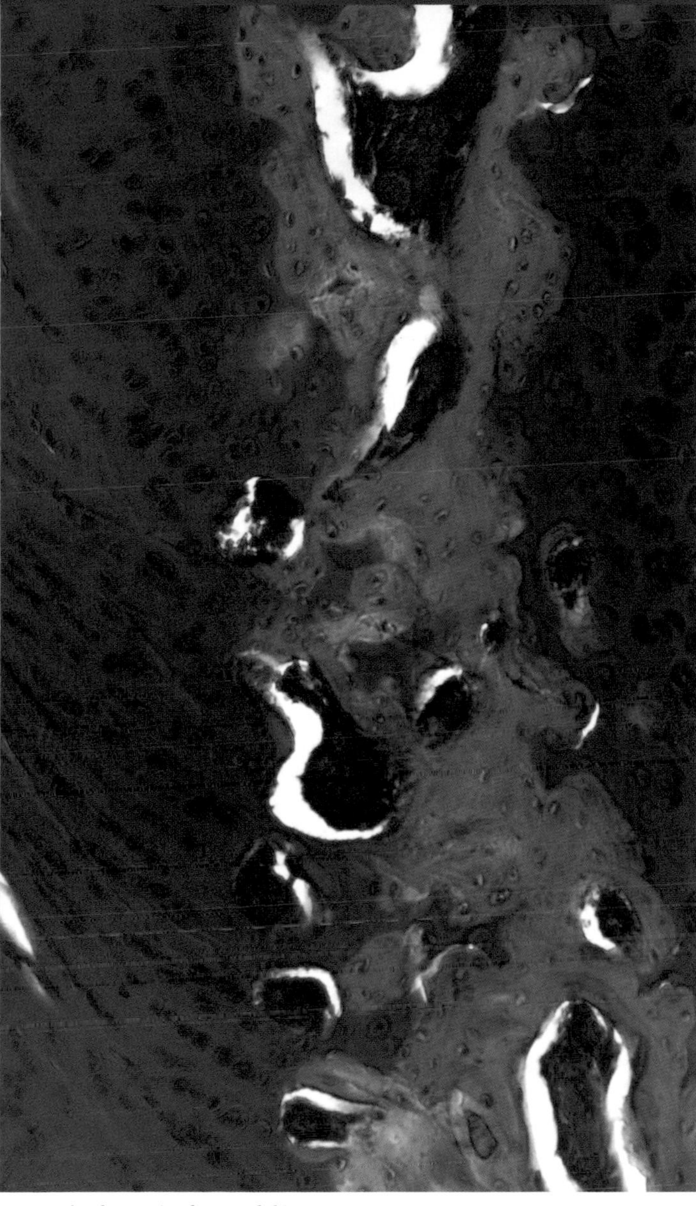

Blood contains haemoglobin.

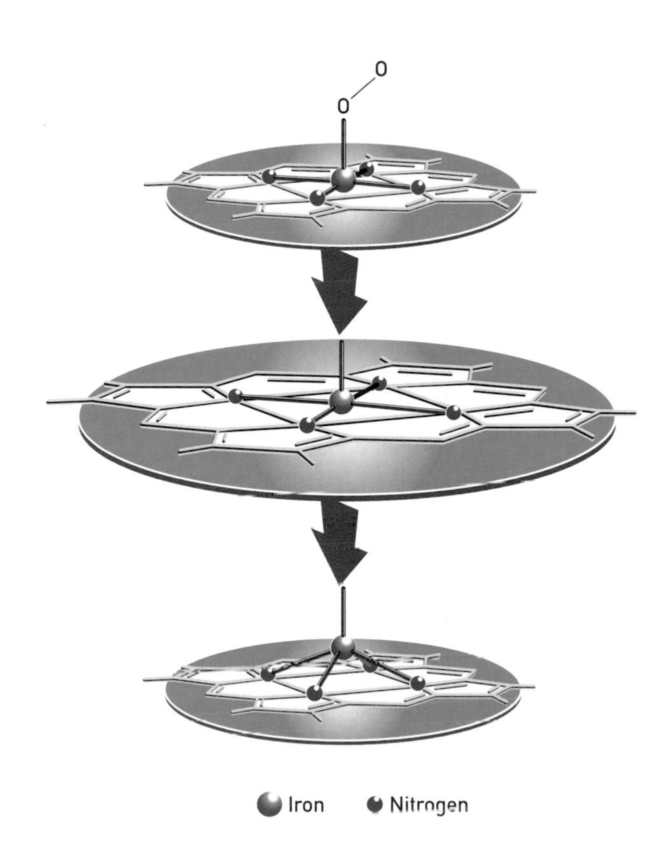

Figure 8. A representation of the change in structure on loss of oxygen from oxyhaemoglobin. In oxyhaemoglobin the iron atom is nearly in the plane of the porphyrin ring and is bound to the oxygen as shown. When the oxygen leaves it increases in size and stretches the ring. The iron atom then moves out of the plane and the ring goes back to its original size.

Even though these molecules may be very large, we are often interested in what is happening in one particular location in the molecule. For instance, the large protein haemoglobin, which carries oxygen in the blood, contains a flat structure based on four rings, called a porphyrin, with a central iron atom. We know that the iron atom lies nearly in the plane of the porphyrin when oxygen is attached (as in the oxygenated form oxyhaemoglobin) but out of the plane when oxygen is not attached, Figure 8. This suggests that the iron atom is a bit bigger when it is not bonded to the oxygen and is thus squeezed out of the porphyrin ring. This is confirmed by resonance Raman spectroscopy which homes in on the iron by concentrating on its vibrations, and ignores those of the rest of the protein molecule.

In some circumstances, ordinary infrared spectroscopy can be used to examine short-lived species too. This is particularly true for transition-metal compounds which contain carbonyl (CO) or cyanide (CN) groups bonded to the metal. If a simple metal carbonyl such as the chromium compound $Cr(CO)_6$ is subjected to a burst of ultraviolet laser light, one carbonyl is removed. Very fast infrared spectroscopy can show that the remaining short-lived species, $Cr(CO)_5$ has a shape like a square pyramid, and it can follow this intermediate's reaction with other molecules. The shape of the molecule can be deduced from the way in which the whole molecule vibrates. Molecules containing carbonyl groups are often important in **homogeneous catalysis** and the technique presents one way of understanding their reactions.

6. What is the general name for groups such as carbonyl (CO) or cyanide (CN) which form bonds to transition metal ions? What feature of their electronic structures is necessary for them to be able to do this?

7. Sketch the shape of the $Cr(CO)_5$ molecule. What shape would you predict for $Cr(CO)_6$?

12. The energy of a Cl-Cl bond is 242 kJ mol^{-1}. Use the relationship $E = h\nu$ to calculate the frequency of radiation needed to just break this bond ($h = 6.6 \times 10^{-34}$ J s). Remember that the bond energy refers to a mole of molecules, so you will need the Avogadro constant (the number of particles in a mole) which is 6×10^{23}.

Is the value you have calculated within the ultraviolet range (approximately 5×10^{14} to 5×10^{16} s^{-1})?

13. The overall radical chlorination of methane proceeds via two other types of step as well as propagation. Name them. Which stage is represented by

$$Cl_2(g) \xrightarrow{\text{UV light}} 2Cl^{\bullet}(g)\ ?$$

The team passed the beam of methane molecules and chlorine **radicals** through polarised infrared radiation the frequency of which is the same as that of the C-H bonds in methane. This makes some of the C-H bonds vibrate – they are said to be 'vibrationally excited'. A hydrogen chloride molecule is formed in the reaction by a collision between a chlorine radical and a methane molecule. The researchers have been able to measure the direction in which this molecule is scattered and relate it to the orientation of the collision between the chlorine radical and the methane molecule.

They have even been able to determine how the hydrogen chloride molecule spins as it moves away after the chlorine-methane collision. It turns out that when the hydrogen chloride molecule is forward-scattered it comes away rotating around an axis which is parallel to its direction of travel, rather like the propeller of an aeroplane. On the other hand, when the molecule is backward-scattered it comes away rotating around an axis which is perpendicular to its line of travel, rather like a frisbee, Figure 12.

These are just a couple of illustrations of how much fine detail about simple chemical reactions can be found by use of a variety of modern techniques.

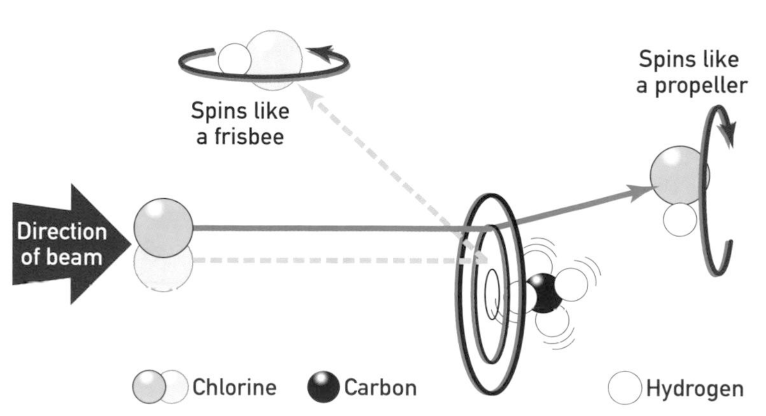

Figure 12. A model for the reaction between atomic chlorine reacting with vibrationally excited methane. When the hydrogen chloride molecule is forward-scattered it comes away rotating like an aeroplane propeller, whereas a backward-scattered molecule comes away rotating like a frisbee. Adapted with permission from W.R. Simpson, T.P. Rakitzis, S.A. Kandel and A.J. Orr-Ewing. Jounal of Chemical Physics, 103 (17), 7313, (1995). ©2000 American Institute of Physics.

The Great Nebula in Orion. ©1982 PPARC, Royal Observatory, Edinburgh and Anglo-Australian Telescope Board.

Glossary

Radical: A species with an unpaired electron.

Chemistry between the stars

Using a combination of molecular beam methods and lasers, it is also possible to study reactions between ions and molecules. Ion-molecule reactions are extremely important in the gas clouds that lie in the vast cold regions of space between the stars. In fact, the most common ion-molecule reaction in the Universe is that between a neutral hydrogen molecule and the charged hydrogen molecular ion (H_2^+) to produce a neutral hydrogen atom and the triatomic hydrogen ion.

$$H_2(g) + H_2^+(g) \rightarrow H(g) + H_3^+(g)$$

A research group at Oxford has recently developed a new experimental method to study reactions such as this one. By using a combination of pulsed ultraviolet laser light and electric fields they are able to generate molecular ions with particular amounts of vibrational and rotational energy. These molecular ions are then allowed to flow through a supersonic jet of neutral molecules where the reaction occurs. It appears that end-over-end rotation of the hydrogen molecular ions hinders their reaction with the neutral species. As the molecules cool, their rotational energy decreases and consequently they become more reactive. This effect is particularly significant in the low temperatures of interstellar gas clouds.

14. What is unusual about the temperature dependence of the rate of the reaction described above?

The future of molecular dynamics

Since the early days of flash photolysis, chemists have improved their techniques so that they can now study reactions one thousand million times faster. This has revolutionised the field of chemical dynamics. Now, chemists are using state-of-the-art femtosecond (1 femtosecond is 10^{-15} s) lasers to watch isolated molecules fall apart or rearrange themselves. Those working at the boundary between chemistry and biology are trying to understand the primary processes in photosynthesis. Those on the boundary between chemistry and physics are using picosecond techniques to watch electrons ionise from atoms and molecules. A combination of pulsed laser and molecular beam methods is allowing scientists to understand the fine details of reactions between two molecules or a molecule and an ion.

One of the ultimate goals of the chemist is to control chemistry at the molecular level. Scientists have recently shown that lasers can be used to control the pathway of a chemical reaction. However, this work still has a long way to go before it is of any practical use. There are still, nevertheless, a great number of questions to be answered about basic molecular dynamics, and when the simple systems are understood it will be possible to move towards a greater understanding of more complex processes, particularly those that occur in biology.

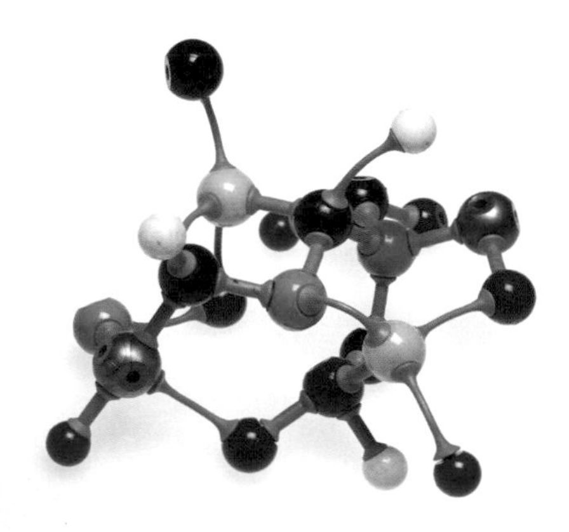

Answers

1. The bond energy is 6.7×10^{-19} J.
   The frequency is 1×10^{15} s^{-1}.
   The wavelength is 3×10^{-3} m.
   The ultraviolet region.

2. (a) A few minutes.
   (b) Many are possible depending on the reaction. Suggestions might include: collecting gas evolved; colorimetry; conductivity measurement; measurement of optical rotation; and dilatometry.

3. $Cl^{\bullet}(g) + O_3(g) \rightarrow ClO^{\bullet}(g) + O_2(g)$

4. The lead compounds in motor vehicle exhausts are causing environmental problems such as brain damage in children exposed to high levels of them.

5. speed = distance / time, so time = distance / speed
   time = $3 / 3 \times 10^{8} = 1 \times 10^{-8}$ s

6. Ligands. They require a lone pair of electrons (with which to form a dative covalent bond).

7. 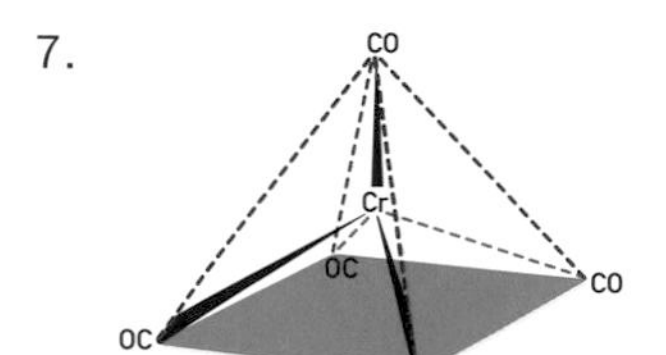

   $Cr(CO)_6$ could be octahedral.

8. 0.3 mm.

9. 1,2-Diphenylethene (*cis*- and *trans*- forms).

10. The π-part of the double bond has broken, leaving the σ-part about which rotation can take place.

11. ${}^{2}_{1}H$

12. 6.1×10^{14} s^{-1}
    Yes, it is within the ultraviolet range.

13. Initiation and termination; initiation.

14. The rate decreases with temperature. Normally the rate increases with temperature.

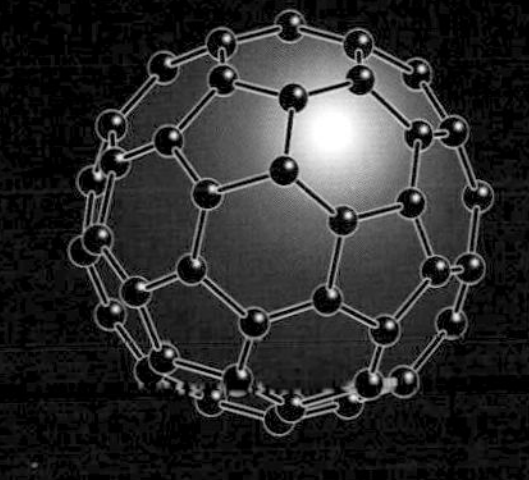

In the past 20 years, researchers have discovered some new chemical compounds with unusual physical properties that are broadening our understanding of the physical world at a fundamental level.

These also look set to provide the next generation of electronic, magnetic and optical devices for the 21st century. These devices could include:

- smaller processor chips – so that computers can become even smaller;
- improved memory storage;
- increased speed of data transfer; and
- faster processing.

21st century materials

In this chapter we look at three examples of these materials:

- **a group of mixed metal oxides that exhibit highly unusual electrical conduction and magnetic behaviour;**
- **tiny clusters of metal atoms about 1 nanometre in size; and**
- **new types of carbon molecules based on 'football-shaped' spheres.**

Today, chemists and physicists work very closely together to explore and develop these exciting areas for future commercial exploitation. As well as having obvious practical importance, research in these areas is forcing theorists to re-think their fundamental ideas on the solid state.

Amazing oxides

Computers and other modern technologies create demands for entirely new kinds of electrical conductors and magnets. Many of these materials are turning out to be solid oxides of common metals such as iron, manganese and copper, with simple compositions. Yet they have complex chemistries and remarkable properties. The atoms in these oxides are arranged in ordered giant structures.

1. Explain what is meant by a giant structure.

A sample of magnetite
Courtesy of Professor P P Edwards

Glossary

Cation: A positive ion.
Mixed metal oxide: An oxide containing two (or more) different metal ions.
Redox reaction: A reaction in which electrons are transferred.
Transition metal: A metal from the central block of the Periodic Table which forms at least one compound with a partly-filled d-shell of electrons.

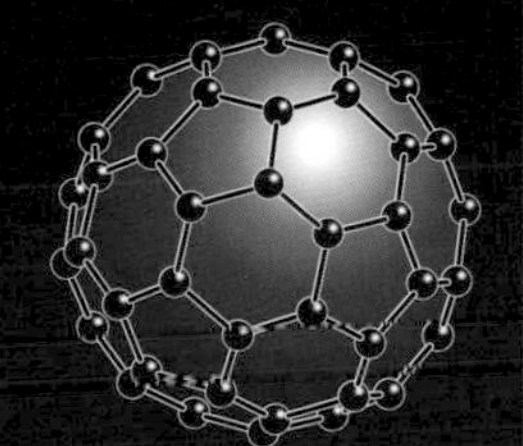

The first magnet

The mineral magnetite has been known since ancient times and was used in the first compasses. In fact, the term 'magnetic' derives from the name of this mineral, which was found in the Magnesia district of Greece. Magnetite has the formula Fe_3O_4. The iron in this compound shows two of the key properties of the **transition metals**: unpaired electrons and variable oxidation states.

2. State three other typical properties of transition metals and their compounds.

3. The electron arrangement of an iron atom may be written $[Ar]3d^6 4s^2$.

(a) Explain what is meant by the shorthand [Ar].

(b) What would be the electron arrangement of (i) the Fe^{2+} ion; and (ii) the Fe^{3+} ion.

(c) Draw diagrams for Fe^{2+} and Fe^{3+} to show the d-electrons in boxes which represent the five d-orbitals. How many unpaired electrons does each of these ions have?

Every electron in an atom or molecule or ion behaves like a tiny magnet, but in most substances electrons are paired up so that their magnetic fields cancel out. This is called diamagnetic (non-magnetic) behaviour. Transition metal **cations** are unusual in having some electrons that remain unpaired, and in magnetite the individual electron magnets are all lined up in the same direction so that the entire solid is magnetic or, to be more accurate, ferromagnetic.

The chemical formula of magnetite can be formally written as $Fe^{2+}(Fe^{3+})_2(O^{2-})_4$ to show the charges on the ions. An Fe^{2+} ion has one more electron than Fe^{3+}, and when the two ions are close together, the extra electron can hop from Fe^{2+} (which now becomes an Fe^{3+} ion) to the Fe^{3+} (which now becomes Fe^{2+}). This is a **redox** (reduction-oxidation) **reaction** in which the reactants are identical to the products. In solid magnetite there are very large numbers of Fe^{2+} and Fe^{3+} ions close to each other so this self-redox process takes place countless times. The transferred electrons hop through the material making it electrically conducting – although less so than typical metals. In fact this conduction is a disadvantage which prevents magnets made of magnetite being used in electrical transformers and motors, because eddy currents would flow in the magnetite and lead to energy losses.

To preserve the useful magnetic properties of magnetite while suppressing the conductivity due to redox transfer of electrons, the Fe^{2+} ions are removed. This can be done by carefully oxidising all the Fe^{2+} to Fe^{3+}, without disturbing the arrangement of the iron ions, to produce γ-iron(III) oxide whose formula is Fe_2O_3. This is widely used in magnetic recording tapes. Alternatively, Fe^{2+} can be replaced by other ions such as Mn^{2+}, Co^{2+}, Ni^{2+} or Zn^{2+} giving a family of materials known as spinel ferrites that are widely used in high-frequency transformers and inductors.

4. Write a balanced equation for the reaction of magnetite (Fe_3O_4) with oxygen to form iron(III) oxide (Fe_2O_3).

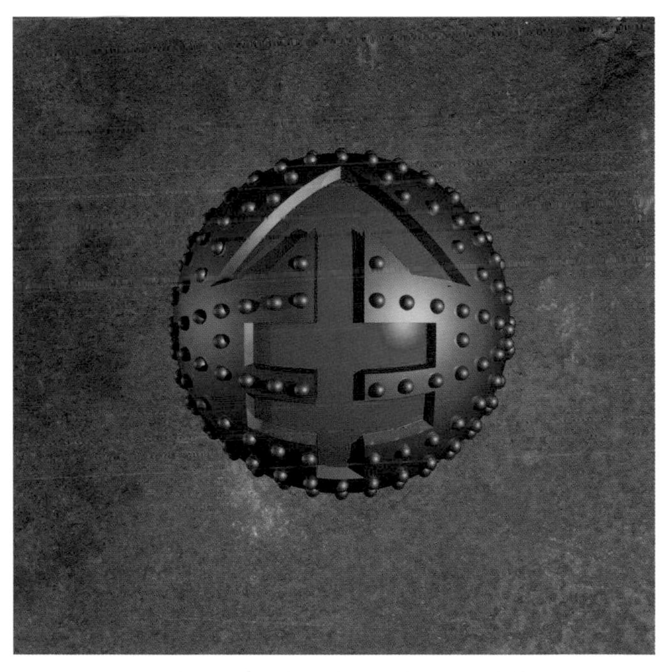

A representation of iron taken from the Royal Society of Chemistry's Visual Elements Periodic Table.

Magnetoresistance

When an electrical conductor such as copper is placed in a magnetic field, its resistance changes slightly. This magnetoresistance effect is a very useful property for sensing magnetic fields and it enables magnetically stored information to be read. Information is written onto computer hard disks by changing the direction in which small particles of a magnetic material such as γ-iron(III) oxide are magnetised. When the disk is moved past a magnetoresistive sensor, the variations in magnetic field due to the magnetised particles on the disk cause changes in the electrical resistance of the sensor.

The resistance changes of most metals when they are placed in a magnetic field are between 1 and 2%, which are too small to be of practical use. However, the resistance changes in a particular family of manganese oxides are spectacular.

The ions in lanthanum manganese oxide, $La^{3+}Mn^{3+}(O^{2-})_3$, form a giant structure known as the perovskite structure, Figure 1. (Perovskite is the mineral form of calcium titanium oxide ($CaTiO_3$) which has given its name to this crystal structure.) If some of the lanthanum is replaced by calcium, the product has the formula $(La_{1-x}Ca_x)MnO_3$.

Lanthanum and calcium are not transition metals so their ions La^{3+} and Ca^{2+} have fixed ionic charges as does the oxide ion, O^{2-}. Substituting some of the lanthanum ions by calcium ions is compensated for by altering the charge on some of the manganese ions (manganese is a transition element so differently charged ions are possible), giving the formula $(La^{3+}{}_{1-x}Ca^{2+}{}_x)(Mn^{3+}{}_{1-x}Mn^{4+}{}_x)(O^{2-})_3$. So for x values other than 0 and 1, a mixture of two differently charged manganese cations is present. This situation is similar to that in magnetite. Electrons hopping from Mn^{3+} to Mn^{4+} make lanthanum calcium manganese oxide a good conductor of electricity. Its magnetism arises from a parallel alignment of all the unpaired electrons.

5. What would be the charge on all the manganese ions in (a) $LaMnO_3$ and (b) $CaMnO_3$? (Remember, lanthanum always forms La^{3+}.) Would you expect these compounds to conduct? Explain your answer.

The $(La_{1-x}Ca_x)MnO_3$ perovskites are better electrical conductors than magnetite but their magnetism is less strong. However, when they are placed in a magnetic field their electrical resistances decrease markedly. Some manganese oxide perovskites give resistance changes of over 99%. The superlative 'colossal' is used to describe them. However, these colossal magnetoresistances require extremely strong magnetic fields of around 5 tesla, whereas the fields generated by aligned magnetic particles on a typical computer hard disk are only about 0.001 tesla. (The magnetic field produced close to a fridge magnet or a laboratory bar magnet is about 0.01 tesla.)

Chemists around the world are currently trying to improve the properties of manganese oxide perovskites so that these large resistance changes can be induced by small magnetic fields. If this is achieved then ultra-small computer disks, holding much more information than conventional media, could follow.

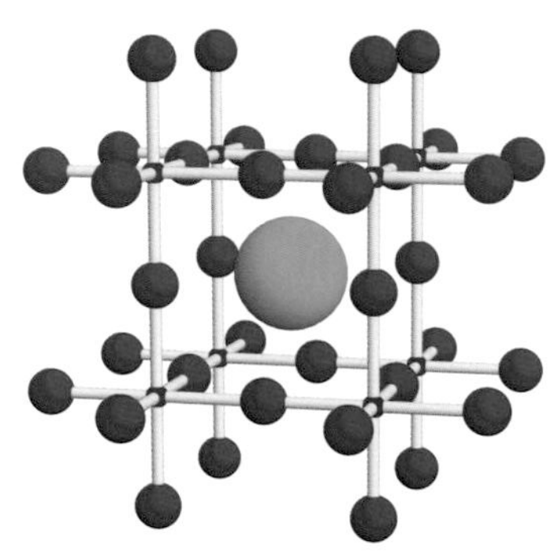

Figure 1. The colossal magnetoresistances of lanthanum calcium manganese oxides $(La_{1-x}Ca_x)MnO_3$ arise from the 'perovskite type' arrangement of their atoms. Electrons move between manganese ions (blue) via the interconnecting oxygens (red) through the network of bonds, while the lanthanum or calcium cations (green) fill the spaces in between.

High temperature superconductors

The most spectacular property of transition metal oxides discovered so far is undoubtedly that of high temperature superconductivity. In 1911, the Dutch physicist Heike Kamerlingh Onnes found that when he cooled mercury down to 4 K using liquid helium, there was a remarkable change; at 4.2 K the electrical resistance of mercury dropped sharply to zero. Onnes named this striking low temperature behaviour superconductivity – the ability to conduct electricity without any resistance, so that an electric current, once started, continues to flow forever.

Materials become superconducting only below a characteristic temperature known as the critical temperature, T_c, (4.2 K for mercury) above which they behave as ordinary metals. The possibility of superconducting wires that would carry electricity without any energy losses was immediately apparent. But the cost of cooling superconductors with liquid helium outweighed the energy saving, except in small devices such as coils for producing high magnetic fields (as in superconducting magnets for particle accelerators and magnetic resonance imaging). Higher T_c values would be required to make superconductors practical. However, in the next 75 years the highest known critical temperature rose only to 23 K for an alloy of the metals niobium and germanium; this still required liquid helium cooling.

For many years no one really understood the bizarre phenomenon of superconductivity. Then in 1957 three American physicists John Bardeen, Leon Cooper and Robert Schrieffer came up with a revolutionary theory that in superconductors the electric current is carried by pairs of electrons, now known as Cooper pairs, whereas single

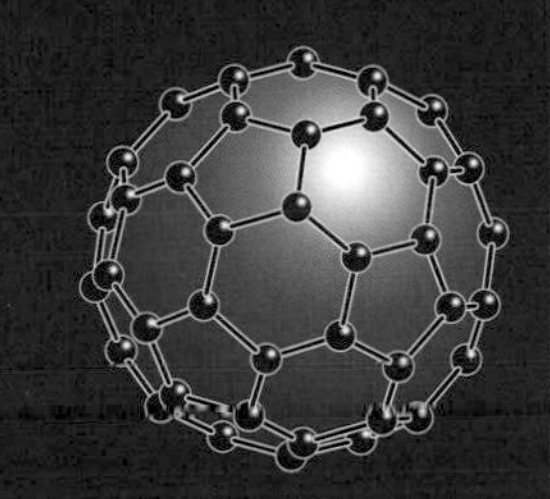

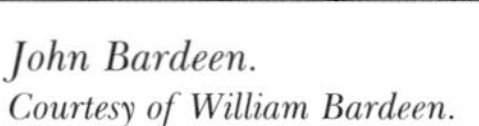

Heike Kamerlingh Onnes. *©The Nobel Foundation.*

John Bardeen. *Courtesy of William Bardeen.*

Leon Cooper. *©The Nobel Foundation.*

Robert Schrieffer. *Florida State University.*

electrons are the current carriers in metals conducting normally. This theory enabled the T_c values for different superconductors to be calculated, and predicted that the highest possible T_c would be around 35 K.

Then in the 1980s things suddenly changed. George Bednorz and Alex Müller, who were working for IBM in Switzerland decided to search for new superconductors among transition metal oxides rather than metals themselves. Their hunch was that Cooper pairs might be formed and thus the compounds might superconduct at high temperatures. In 1986 they reported superconductivity in lanthanum barium copper oxide with a T_c of 30 K.

This hit the headlines early in 1987, starting a rush in laboratories around the world to find other copper oxide-based superconductors. Within a year a new yttrium barium copper oxide had been discovered with a T_c of 93 K, showing that the Bardeen, Cooper, Schrieffer theory was incomplete. This superconductor could be cooled by liquid nitrogen, which boils at 77 K and is much cheaper than liquid helium. (Volume for volume, liquid nitrogen costs less than milk, while liquid helium costs around the same price as table wine.) In 1987, Bednorz and Müller received a Nobel Prize. This followed Nobel awards given to Onnes (1913) and to Bardeen, Cooper and Schrieffer (1972).

The chemistry of the superconducting copper oxides is related to that of the conducting iron and manganese oxides. The lanthanum barium copper oxide discovered by Bednorz and Müller has the chemical formula $(La_{2-x}Ba_x)CuO_4$. The replacement of each La^{3+} ion by a Ba^{2+} is compensated for by oxidising a Cu^{2+} ion to Cu^{3+} according to the formula $(La^{3+}{}_{2-x}Ba^{2+}{}_x)(Cu^{2+}{}_{2-x}Cu^{3+}{}_x)(O^{2-})_4$.

The presence of two differently charged copper ions makes the material conduct. However, the barium-doped compound is not magnetic but instead becomes a superconductor at low temperatures.

Why are these copper oxides not magnetic like magnetite and the manganese oxide perovskites? An important difference is that the Cu^{3+} ions have no unpaired electrons and so are non-magnetic. Why superconductivity occurs instead and with such high T_cs is not yet understood although many physicists believe that there is a deep theoretical connection between magnetism and superconductivity.

MRI scan in progress.

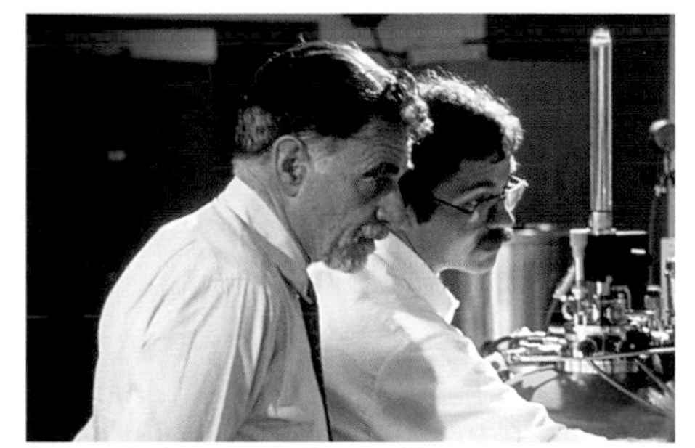

George Bednorz and Alex Müller. *IBM, Switzerland.*

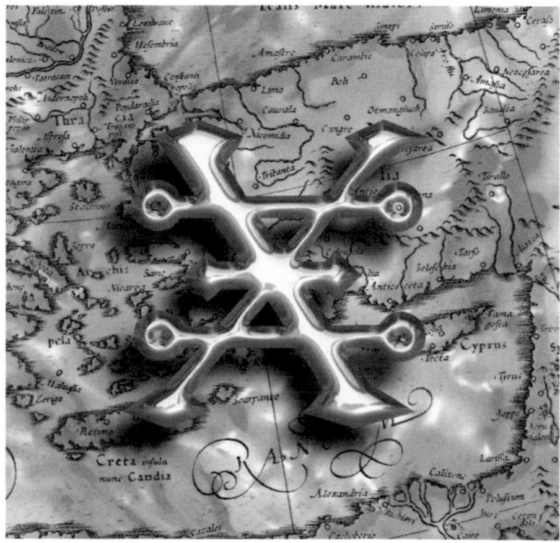

A representation of copper from the Royal Society of Chemistry's Visual Elements Periodic Table. *Images ©Murray Robertson 1998-1999.*

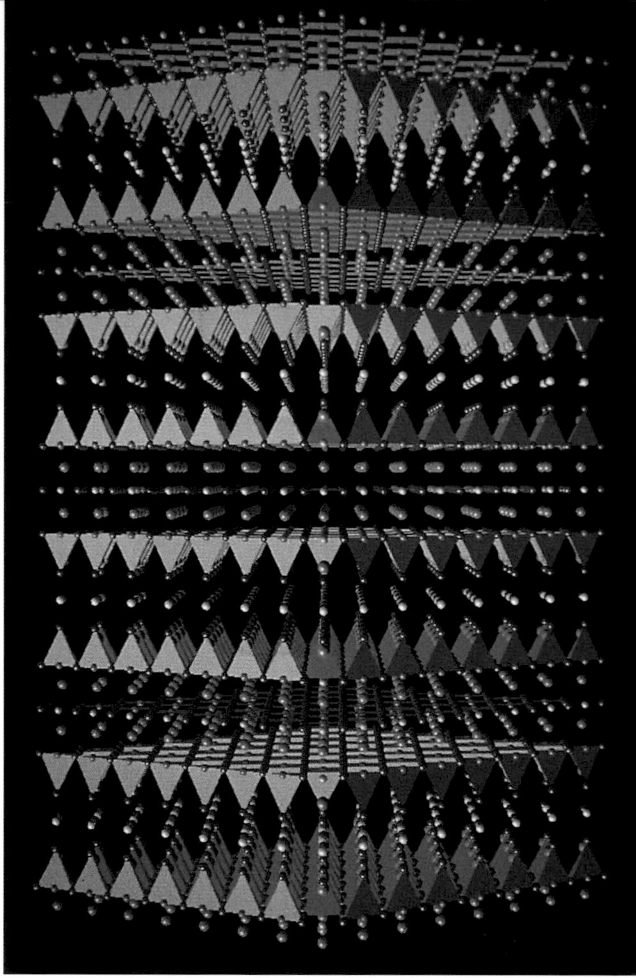

Figure 2. The current highest-T_c material superconducts at 135 K and contains layers of mercury (purple), barium (green), calcium (blue), copper (yellow) and oxide (red) ions stacked in a regular sequence. Courtesy of P P Edwards and G B Peacock.

6. Cu^{3+} is an unusual oxidation state for copper. What are the two more usual oxidation states formed by this element?

The highest-T_c superconductor now known contains mercury, barium and calcium cations in addition to copper and oxygen. The atomic arrangement is shown in Figure 2. It has a T_c of 135 K (-138 °C). This is still a very low temperature in everyday terms (the lowest recorded temperature on Earth is 183 K, or -90 °C, in 1983 in Antarctica), but until recently it would have been considered an impossibly high T_c for any superconductor.

The technological development of high-T_c superconductors has proved difficult because they are brittle **ceramic** materials and hard to make into wires. However, after a decade of effort, kilometre-long sections of superconducting wire are now being produced and tested. The discovery of the high-T_c materials has led to an intense search by chemists for other new superconductors. Several new types have been discovered, although none of these have T_cs above 40 K. Most notable among these new types of conductors are the fullerides, which are described later on in this chapter.

7. Using simple models of the giant ionic and metallic structures below, explain why giant ionic structures tend to be brittle while metallic ones are malleable and ductile (*ie* they retain their new shape when they are deformed). Hint – think about what happens when a row of atoms or ions (as appropriate) is moved one atomic position from its starting point.

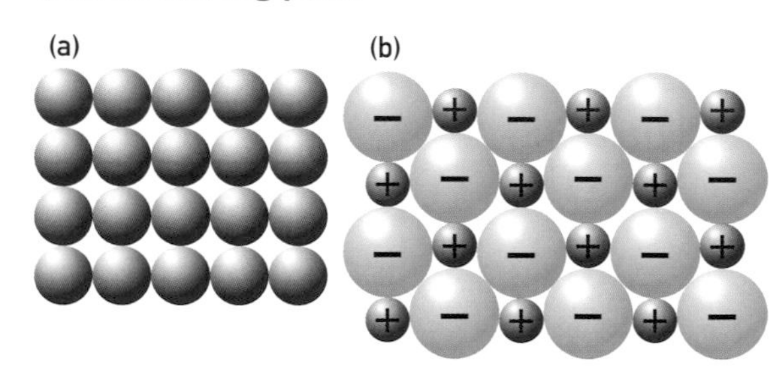

Parts of (a) a giant metallic structure, (b) a giant ionic structure.

Nanoclusters – small but perfectly formed

The second example of materials with novel properties lies in the study of particles and structures with dimensions of around one nanometre, 1 nm, (one thousand-millionth of a metre). The properties of these clusters of atoms have led to the new field of nanoscience.

8. The diameter of a typical atom is about 0.2 nm. Imagine a structure in the shape of a cube 2 nm x 2 nm x 2 nm. How many atoms would it contain?

Glossary

Ceramics: A group of high melting point, heat resistant inorganic materials.

Early predictions of nanoscience

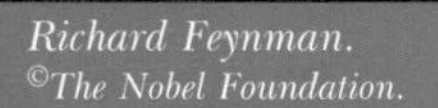

Richard Feynman.
©The Nobel Foundation.

Robert Boyle.
Reproduced courtesy of the Library and Information Centre, Royal Society of Chemistry.

Eric Drexler.

Perhaps surprisingly, several scientists had made predictions about nanoscience well before it became possible.

In 1960, the famous American physicist and lateral thinker, Richard Feynman*, wrote a visionary article called *There's plenty of room at the bottom*, in which he challenged scientists to develop a new field of study where devices and machines could be constructed from components consisting of small numbers (tens or hundreds) of atoms.

The subject can also be traced back to the ideas of Eric Drexler at the Massachusetts Institute of Technology who a decade ago wrote the book *Engines of Creation* which speculated on the future use of nanomachines.

The earliest reference to clusters may even have been made by the Irish scientist Robert Boyle as long ago as 1661. In his book *The Sceptical Chymist*, he speaks of: '.... minute masses or clusters ...., as were not easily dissipable into such particles as compos'd them.'

* As well as being a Nobel prize-winning physicist, Richard Feynman was a noted writer on science and a man of many other talents including safe cracking and playing the bongo drums! You can get a flavour of his character by reading his autobiographical books *Surely you're joking Mr Feynman* and *What do you care what other people think?*

The ideas of Feynman and Drexler (see Box – *Early predictions of nanoscience*) inspired a generation of physicists and chemists to try and make their vision a reality, and today there are hundreds of nanoscience laboratories throughout the world. From nanoscience has sprung the applied field of nanotechnology – in which the aim is to build minute devices for electronic, optical, mechanical and even medical applications.

One of the areas of chemistry which is currently contributing to this field exploits particles consisting of between tens and millions of atoms as the basic building blocks. These nanoclusters are formed by most of the elements in the Periodic Table – even the noble gases.

We will focus on nanoclusters made up of atoms of metallic elements. As well as their possible applications in computer processors and memories, such nanoclusters are also of great theoretical interest because they represent a state of matter that lies in between individual atoms and molecules, and crystalline solids (where the atoms or molecules are fixed in practically infinite regular arrays).

Many of the unusual properties of clusters depend on the fact that a large proportion of the atoms lie on the surface. Clusters that are roughly spherical in shape and contain as many as 10 000 atoms still have nearly 20% of their atoms on the surface. The percentage of atoms at the surface only drops below 1% for clusters of more than 64 million atoms! This high surface-to-volume ratio has made metal nanoclusters important in heterogeneous catalysis.

9. What is a heterogeneous catalyst? Why does the proportion of atoms on the surface matter so much in this context?

The size-dependent behaviour of cluster properties suggests the exciting prospect of using devices made up of arrays of nanoclusters with properties that can be fine-tuned by carefully controlling the size of the component nanoclusters.

10. Try the following exercise to find out what happens to the fraction of atoms on the surface of a cluster as the cluster gets bigger.

(a) Imagine a cube 1 cm x 1 cm x 1 cm. What is (i) its volume in cm^3? (ii) its total surface area in cm^2 (remember it has six faces)? (iii) the ratio of surface area ÷ volume?

(b) Now imagine a cube 2 cm x 2 cm x 2 cm. What is (i) its volume in cm^3? (ii) its total surface area in cm^2? (iii) the ratio of surface area ÷ volume?

(c) Now imagine a cube 3 cm x 3 cm x 3 cm. What is (i) its volume in cm^3? (ii) its total surface area in cm^2? (iii) the ratio of surface area ÷ volume?

(d) Use the same method to find a general expression for the ratio of suface area ÷ volume for a cube of side x cm.

(e) Suggest what happens to the fraction of atoms on the surface of a cluster as the cluster gets bigger.

Nanoclusters have huge potential to impact on computer technology.

Glossary

Inert gas configuration: The electron arrangement of an inert gas – having a full outer shell of electrons.

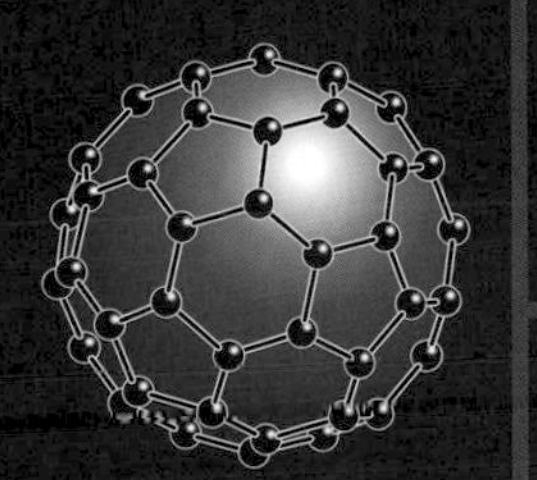

Theories of clusters – magic numbers

In a series of experiments in the early 1980s, Walter Knight and colleagues at the University of California prepared beams of clusters of sodium atoms by heating the metals to give a vapour, which was mixed with a cold inert gas and condensed into clusters of various sizes, Figure 3. They then determined the masses of the clusters using a mass spectrometer.

11. Describe briefly how a mass spectrometer would have been used to measure the masses.

They noticed a periodic pattern of relatively intense peaks in the mass spectra, which corresponded to cluster sizes 2, 8, 20, 40, 58 atoms and so on. The researchers realised that the numbers were the same as the so-called magic numbers originally seen in nuclear structure studies. Atomic nuclei with a combination of protons and neutrons adding up to one of these magic numbers are particularly stable. These magic numbers arise due to the complete filling of shells in the nucleus in a similar way to the filling of electronic shells in atoms to give an **inert gas configuration**. Scientists have become very excited by the idea of a single theoretical model that can be applied over such a wide range of length scales – from the inside of a nucleus *via* the electronic structure of an atom to that of a cluster of hundreds or even thousands of atoms.

Later, mass spectrometry experiments were carried out for even larger clusters of up to 25 000 atoms. Here the scientists saw peaks corresponding to clusters of atoms in various regular geometric arrangements called regular polyhedra (three-dimensional shapes with many sides). These make the clusters stable by reducing their surface areas so that there are no isolated atoms sticking out.

Magnesium, calcium and strontium clusters are icosahedral, having 20 faces, while aluminium clusters are octahedral, with face-centred cubic packing, as in solid aluminium. Some typical geometries are shown in Figure 4.

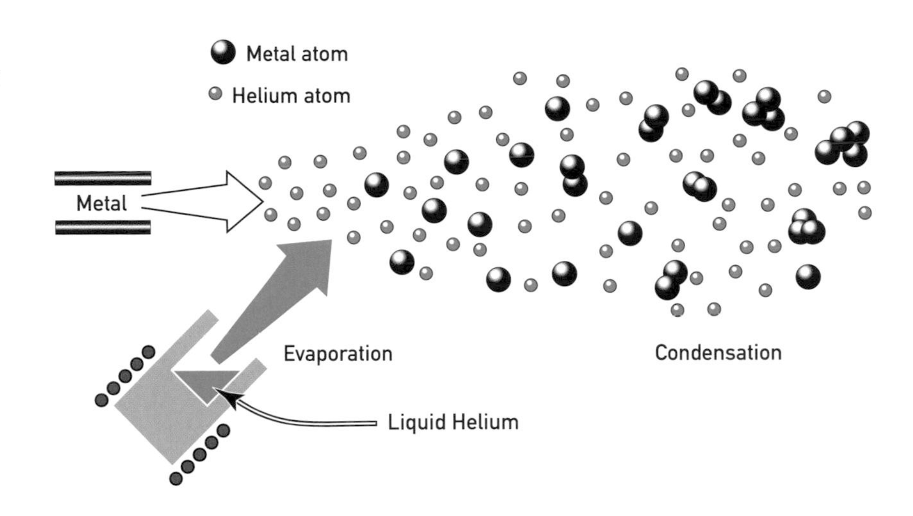

Figure 3. A schematic representation of a nanocluster generation experiment. Metal atoms are evaporated and form clusters in a cold flowing carrier gas. Clusters are further cooled by supersonic expansion to form a cluster molecular beam. Courtesy of R E Palmer.

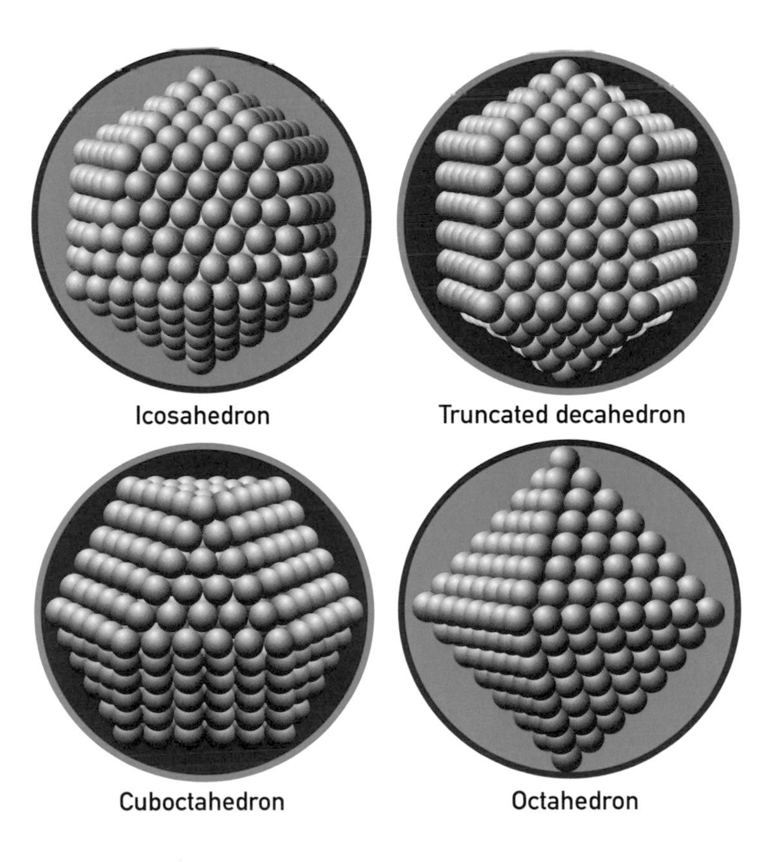

Figure 4. Examples of cluster shell geometries found for rare gas and metal nanoclusters.

A representation of cobalt from the Royal Society of Chemistry's Visual Elements Periodic Table. Images ©Murray Robertson 1998-1999.

Nanomagnets

As described earlier, an electron behaves like a tiny bar magnet. This causes magnetism in atoms and molecules which have single unpaired electrons while those in which all the electrons are paired are essentially non-magnetic. Individual atoms or molecules with unpaired electrons show paramagnetism (the unpaired electron lines up with an external magnetic field) while in bulk metals the unpaired electrons show much stronger magnetic behaviour, lining up together to produce the phenomenon of ferromagnetism as displayed by pieces of iron. Clusters show behaviour that is something in between – they are superparamagnetic. This means that they resemble giant **paramagnetic** atoms. The average magnetic moment per atom is actually higher than in the **ferromagnetic** solid and decreases as the cluster gets larger.

When the cluster size reaches about 500 atoms the magnetic moment reaches the normal value for the bulk solid. Interestingly, rhodium, which is not magnetic in the solid state, has been shown to form magnetic clusters! The magnetism of clusters of iron, cobalt and nickel has been studied by measuring the deflection of beams of clusters in magnetic fields.

Recently, people have become very interested in using such clusters for new types of magnetic devices. The idea is to embed the clusters in a solid. When embedded in metals, or evcn insulators, magnetic clusters (for example, of chromium, iron, cobalt, nickel or mixtures of these metals) are known to show giant magnetoresistance, as described earlier. For cobalt clusters embedded in silver, the change in resistance can be as high as 20%. Such magnetoresistive materials are already being used for magnetic recording and storage of data and show considerable promise for other applications such as magnetic sensors.

Nanowires and nanoswitches

For any useful applications we will need a large number of clusters and some method of storing and transporting them. Unprotected metal clusters tend to merge spontaneously to form bulk metal, so they have to be stabilised. This can be done either by coating them with protective **ligands** so that they cannot merge, or through immobilisation – by tethering them to, or implanting them in, a support. This may be either an inert surface or the inside of a porous material such as a zeolite.

Zeolites are a group of minerals which are riddled with a regular array of interconnecting channels or pores. A group of chemists at the University of Birmingham has managed to synthesise and trap a number of highly reactive charged **alkali metal** clusters inside zeolite cavities. As more and more metal atoms are added, the white zeolite powder turns blue and finally black, and the zeolites become semiconducting as electrons can hop easily from cluster to cluster. The ultimate goal of this work is to form chains of metal atoms in the zeolites and thus assemble dense bundles of conductors or nanowires, Figure 5. Such nanowires could be used in the assembly of electronic devices on the nanoscale.

Nanowires, of course, need nanoswitches. One possible approach to this involves making thin films of gold and silver nanoclusters with surfaces protected by organic molecules such as thiols. (Thiols are alcohols in which a sulfur atom replaces the oxygen atom in the -OH group.) Recent experiments at the University of California, Los Angeles, have shown that a 4 nm thick film of thiol-coated silver clusters becomes metallic (and therefore conducts electricity) when the film is compressed so that the inter-particle spacing is less than 0.5 nm. This works because electrons can tunnel from one silver cluster to another through the organic ligands when the clusters get this close. This process may allow researchers to turn nanoscale devices on and off.

Figure 5. Metal 'nanowires' inside a zeolite framework could be used in electronic devices. Courtesy of P Anderson.

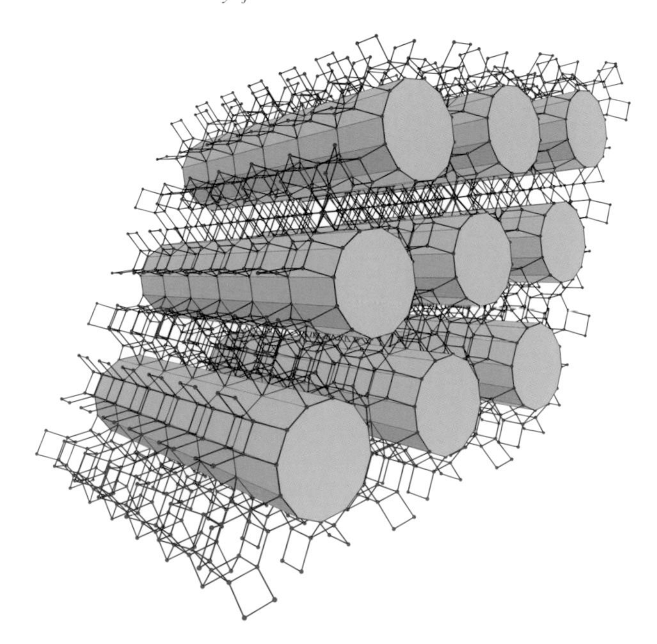

Glossary

Alkali metal: A metal from Group I or Group II of the Periodic Table.
Ferromagnetic: Strongly attracted to a magnetic field.
Ligand: A molecule or ion which forms a dative covalent bond to a transition metal atom or ion via a lone pair of electrons.
Paramagnetic: Weakly attracted to a magnetic field.

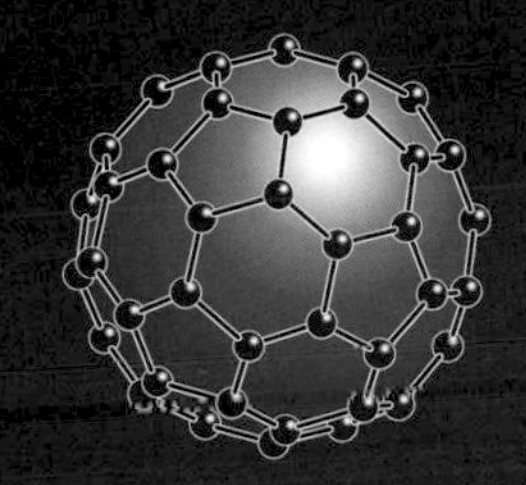

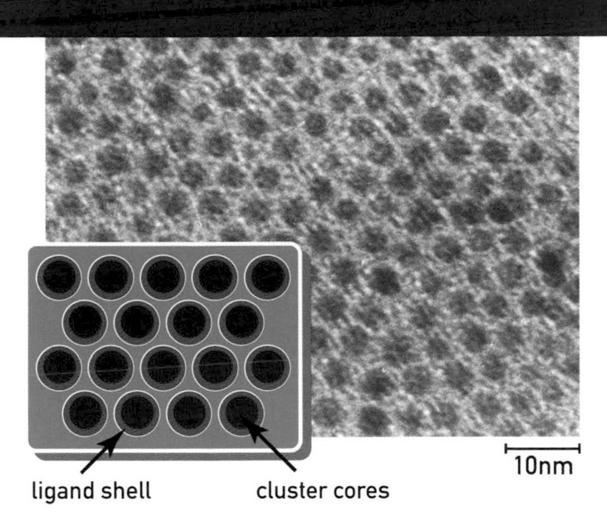

Figure 6. An electron micrograph of a hexagonal film of gold nanoclusters coated with organic thiol ligands. (Inset – schematic representation of the film). Courtesy of C N R Rao.

12. It is often posible to make sulfur analogues of oxygen-containing compounds. These are compounds in which a sulfur atom replaces the oxygen atom (or atoms). Explain why It is possible to make such compounds. Draw the displayed formula of thioethanol.

C_{60} – the most symmetrical cluster

Perhaps the most famous nanocluster of all is the football-shaped carbon molecule, C_{60}, Figure 7. This was identified for the first time by a team of UK and US scientists in 1985, and it represents a third allotrope of carbon; the two best known are graphite and diamond, Figure 8. Since its discovery, C_{60} has caused tremendous excitement among scientists. It has opened the door into a new world of related cage- and tube-shaped carbon molecules, called fullerenes, with new chemistry (and physics) to explore. Like other clusters, C_{60} and its relations have not only provided insights into our understanding of chemical bonding and structure but also offer the promise of new applications in electronics, materials science and even medicine.

Although C_{60} became a subject of serious study only in the mid-1980s, people had speculated earlier on what would happen if you rolled up the hexagonal 'chicken-wire' structure of graphite into a ball. In the 1970s, theorists had suggested that C_{60} should be a stable molecule, and had predicted its properties – although few people seem to have noticed these studies.

13. Until the recent discovery of the fullerenes, all textbooks of chemistry said that the element carbon had two well-known allotropes, diamond and graphite. They are now out of date. Explain what is meant by the term 'allotrope'.

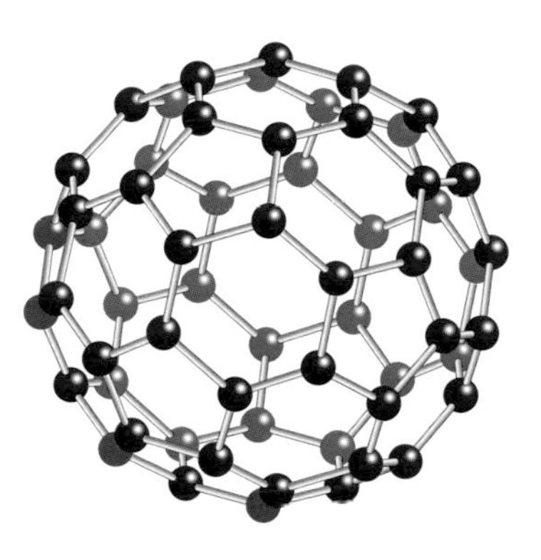

Figure 7. Buckminsterfullerene.

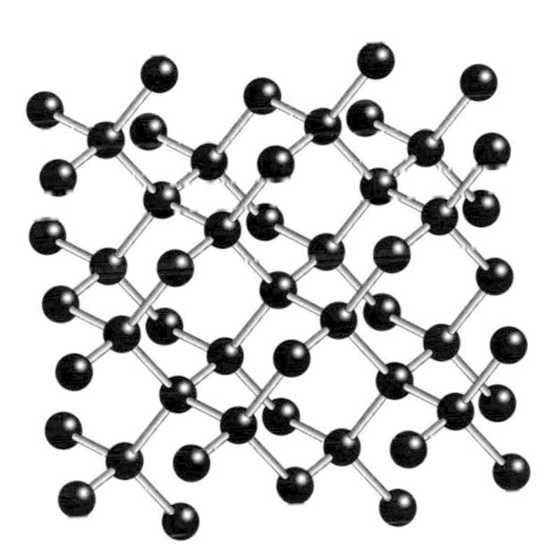

Figure 8(a). Diamond.

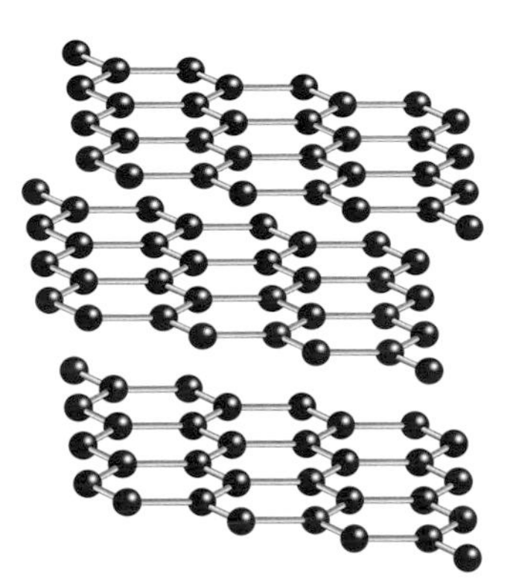

Figure 8(b). Graphite.

Figures 7 and 8.
Courtesy of Crystallographica, ©Oxford Cryosystems.

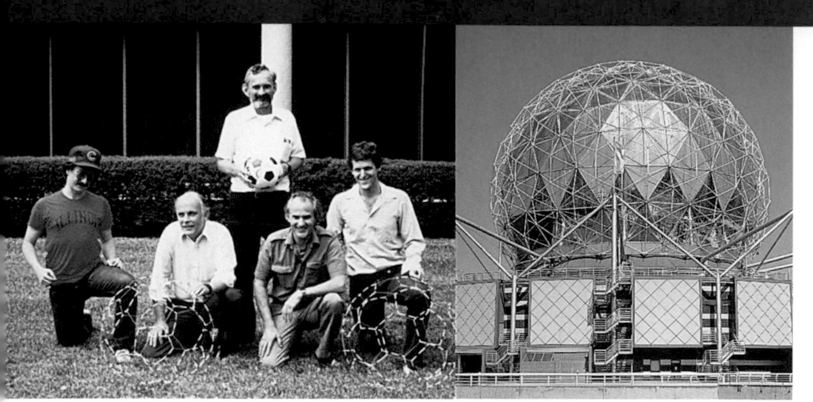

The C_{60} team.
Courtesy of Sir Harry Kroto, University of Sussex.

Geodesic dome.
Courtesy of John Mead/ Science Photo Library.

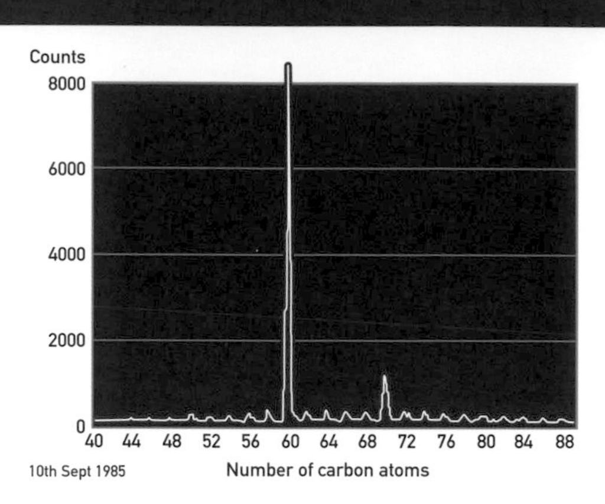

Figure 9. The mass spectrum obtained by vaporising carbon with a laser and allowing the atoms to condense. The peaks for C_{60} and C_{70} can be clearly seen.

The story of a discovery

The full story of the discovery of C_{60} is long and exciting and there is room for only the highlights here. It is well told in the book *Perfect Symmetry* by Jim Baggott (OUP).

The story starts with Harry Kroto (now Sir Harry), a spectroscopist at the University of Sussex. He and his colleagues were interested in very long-chain carbon molecules called cyanopolyynes that might form in the atmospheres of carbon-rich stars. The molecules can be detected, even across the wide reaches of space, by microwave spectroscopy. This technique analyses the electromagnetic radiation absorbed and emitted by these molecules as they rotate and allows information such as bond lengths to be determined.

14. What information does the name 'cyanopolyyne' give about the structure of the molecule?

In 1984 another spectroscopist, Robert Curl at Rice University in Houston, Texas, suggested that Kroto visit the laboratory of his colleague, Richard Smalley. Smalley was using a laser to vaporise atoms from a solid target disc. The beam of atoms was then allowed to cool and condense into clusters in the vapour phase which could be studied by mass spectrometry (rather like the sodium clusters mentioned earlier). When Kroto saw Smalley's apparatus, he realised that it might shed light on his ideas on the creation of carbon chains in the atmospheres of certain stars.

In September 1985, Kroto, Smalley and Curl, together with postgraduates Jim Heath, Sean O'Brien and Yuan Liu set to work using a graphite disc to create carbon clusters. They already knew that clusters with certain 'magic numbers' of carbon atoms were favoured but were surprised to find that one number, 60, dominated the spectrum. Another peak that stood out was that for C_{70}, Figure 9.

The researchers had not then heard of the earlier speculations on C_{60}, but clearly there was something unusual about the structure of C_{60} which made it so stable. They came to the conclusion that C_{60} was a closed spherical cage of carbon atoms similar in shape to the geodesic domes (see photo) designed by the American architect Richard Buckminster Fuller. C_{60} consisted of a hollow spherically-shaped **polyhedron** composed of 20 hexagons and 12 pentagons with carbon atoms sitting at the 60 vertices. When Smalley asked the head of Rice's mathematics department what this shape was called, the answer was: 'I could explain this to you in a number of ways, but what you've got there is a soccer ball!' C_{70}, which is more elongated (in the shape of a rugby ball), has 25 hexagons.

C_{60} was dubbed buckminsterfullerene – later shortened to 'buckyballs' in the popular press – after the architect, while the new class of molecules was given the family name of fullerenes.

The evidence was still circumstantial, although well supported by further experiments and theoretical calculations. The problem was that the researchers could produce only a few tens of thousands of molecules, which was not enough to determine their stuctures.

A breakthrough at last

Kroto's team at Sussex spent five frustrating years looking for a way to make larger amounts of C_{60} so that they could analyse it. Here is where another strand of the story links in, also starting in space. Two physicists, Donald Huffman at the University of Arizona and Wolfgang Krätschmer at the Max Planck Institute for Nuclear Physics in Heidelberg, were interested in interstellar dust which they thought was probably mainly soot-like particles of carbon.

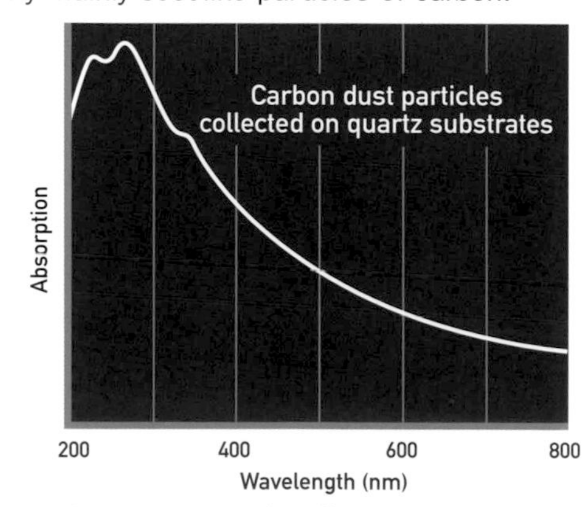

Figure 10. The UV spectrum of Huffman and Krätschmer's samples of 'soot' showed two camel-like humps which are due to C_{60}.

Glossary

Polyhedron: A three-dimensional figure with many faces.

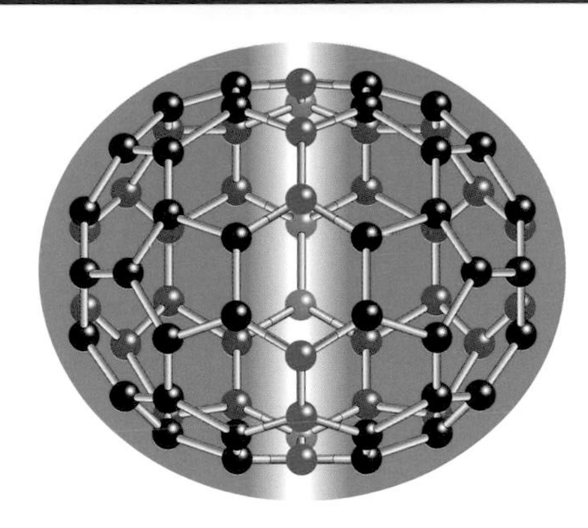

Carbon forms a range of closed cage-like structures – the fullerenes. C_{70} is shown here.
Courtesy of R Smalley, Rice University.

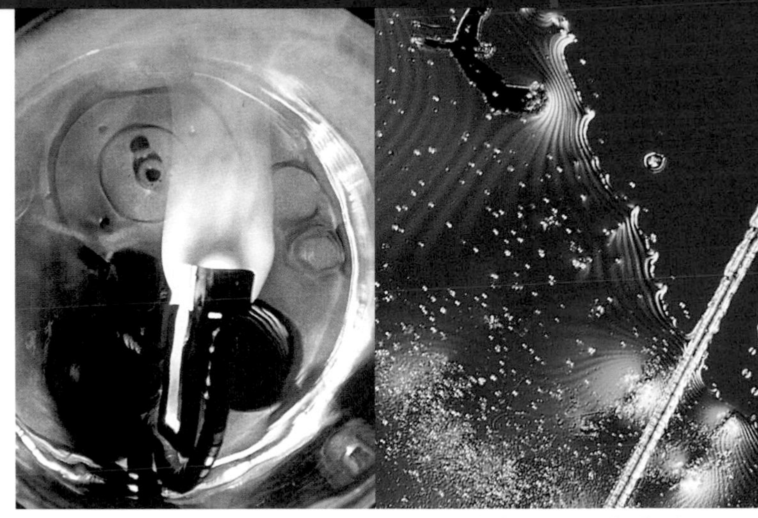

Carbon dust particles floating upwards in the current of helium gas, looking like a cloud of cigarette smoke.
Courtesy of Wolfgang Krätschmer, Max Planck Institute.

Polarised light micrograph of a thin film containing crystals of C_{60}.
Courtesy of Michael W Davidson/Science Photo Library.

In the early 1980s, they started experiments to model the formation of soot in space by evaporating a graphite rod in a bell-jar of low-pressure helium. They collected the soot and measured its ultraviolet absorption spectrum. They noted that there were two strange camel-like humps in the spectrum which they thought must be due to contamination. However, when Kroto and his colleagues announced their supposed C_{60} molecule, Huffman and Krätschmer decided to have another look at the artificially-produced soot, speculating that the 'camel humps' in its spectrum might be due to C_{60}, Figure 10. The Krätschmer-Huffman research teams measured the mass, ultraviolet and infrared spectra. They found that the results fitted the theoretical predictions for C_{60}.

By 1990 the physicists could make quantities of a few milligrams of what they now believed was C_{60}. The soot sample dissolved in benzene to form a red solution, which when evaporated left behind red crystals – a mixture of 75% C_{60} and 23% C_{70} and a few per cent of higher fullerenes. Krätschmer and Huffman called the mixture 'fullerite'. They then went on to measure the crystal structure of fullerite using X-ray and electron diffraction, confirming that the structure of C_{60} was indeed that of a spherical cage.

Richard Smalley.
Courtesy of Leah Bernard-Boggs, Rice University.

Robert Curl.
©The Nobel Foundation.

Sir Harry Kroto.
©The Nobel Foundation.

Donald Huffman and Wolfgang Krätschmer.

15. (a) Why would buckminsterfullerene be more likely to be soluble than graphite or diamond?

(b) Why would buckminsterfullerene be expected to dissolve in a solvent such as benzene rather than, say, water?

When the physicists announced their results in 1990, the researchers at Rice and Sussex had mixed feelings. They were delighted that their early ideas on fullerenes had been vindicated but were disappointed to be pipped at the post. Kroto's team had carried out similar experiments and were only just beaten to publication of results. Nevertheless, the Sussex team extracted C_{60} independently and were able to separate C_{60} and C_{70} by chromatography and take their nuclear magnetic resonance (NMR) spectra. The carbon-13 NMR spectrum of C_{60} revealed the expected single peak (in this highly symmetrical molecule all the carbon atoms are in equivalent positions). The structure of C_{60} was confirmed beyond doubt. In 1996 the discovery of a new form of carbon was recognised through the award of a Nobel prize to the senior members of the original team – Kroto, Smalley and Curl.

16. Why would a proton NMR spectrum of buckminsterfullerene reveal nothing?

17. Make a model of buckminsterfullerene with a ball and stick molecular modelling kit using Figure 7 as a guide (you will need 60 carbon atoms). Use it to satisfy yourself that all the carbon atoms are in fact in identical environments.

Some chemistry of fullerenes

Fullerenes as 'cages'

Early on, during the search to find the structure of buckminsterfullerene, some informative experiments were carried out. The Rice-Sussex team managed to enclose a metal atom in the C_{60} cage by soaking graphite sheets in solutions of metal salts and repeating the laser vaporisation-condensation experiments. Mass spectrometry results revealed that the C_{60} molecules had metal ions attached which could not be removed by irradiating them with an intense laser beam. This result reinforced the idea that the metal atoms were trapped inside the C_{60} structure, Figure 11.

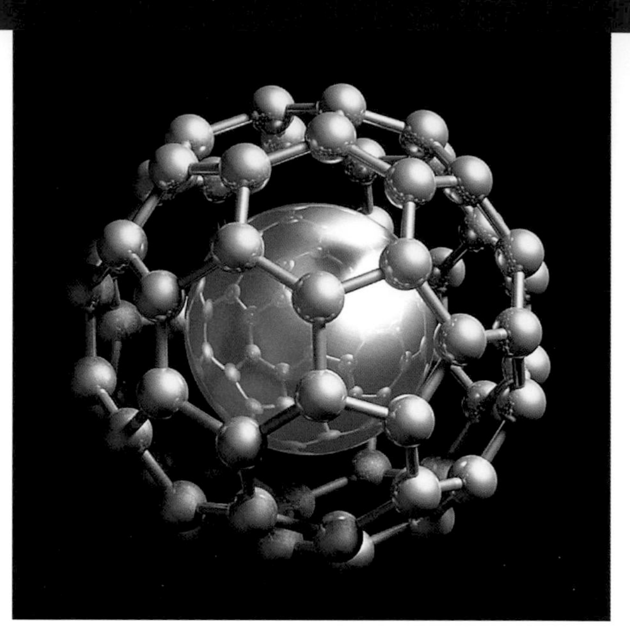

Figure 11. Metal ions can be trapped inside a C_{60} structure.
Ken Edward/Science Photo Library.

18. In a mass spectrometry experiment, what mass would you expect for the molecular ion (parent ion) formed from C_{60}? What molecular ion would you expect for a C_{60} molecule with a potassium atom trapped inside?

What is more, further laser-blasting of the C_{60}-metal compounds ruptured the carbon cage and released two carbon atoms before closing up again. The team found that they could gradually reduce the size of the carbon cage so as to 'shrink-wrap' the metal ions, the final cage dimensions depending on the ionic radii of the enclosed species.

By the early 1990s it was possible to make C_{60} in reasonable amounts and research groups around the world set about exploring its chemistry. Because of its unusual structure, C_{60} was expected to have some fascinating electronic properties. The **electron density** across its surface is uneven, being somewhat higher in the six-membered carbon rings and lower in the five-membered rings. In the solid state, C_{60} molecules pack together in a face-centred-cubic lattice, Figure 12, but are free to rotate at random about their centres, even at low temperatures.

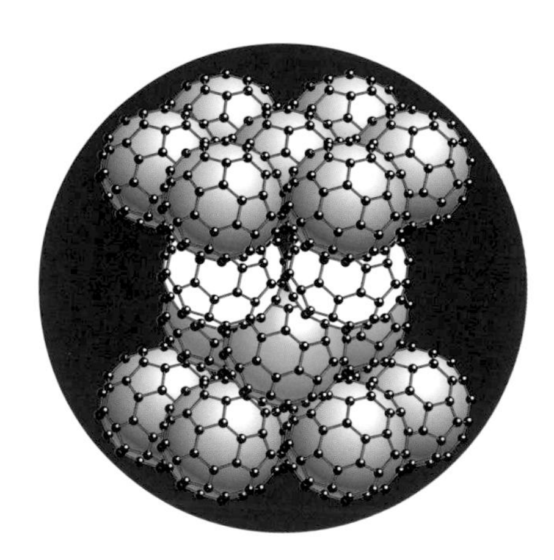

Figure 12. Part of a face-centred-cubic structure.

19. What is the coordination number of C_{60} in the face-centred-cubic lattice?

The electrical conduction of fullerenes

In 1991, researchers at AT&T Bell Laboratories in New Jersey showed that when crystalline C_{60} is 'doped' with metals such as potassium, which can donate electrons, C_{60} forms negative ions and the compound $[3K^+ C_{60}^{3-}]$ is electrically conducting. The highest conductivity is achieved with three potassium ions for every C_{60} molecule. However, the addition of further metal ions turns C_{60} into an insulator. The Bell Labs team further discovered that K_3C_{60} becomes a superconductor below about 18 K. This was improved on by researchers at the NEC Fundamental Research Laboratories in Japan, who substituted potassium with rubidium and caesium to obtain a material with a T_c of 33 K – the highest T_c for a molecular material.

20. Write an ionic equation for the formation of potassium fulleride, K_3C_{60}, from potassium and buckminsterfullerene.

Glossary

Alkene: A compound with a carbon-carbon double bond.

Aromatic molecule: An organic molecule which has a ring or rings with a delocalised system of electrons.

Electron density: The concentration of electron charge in a particular region of space.

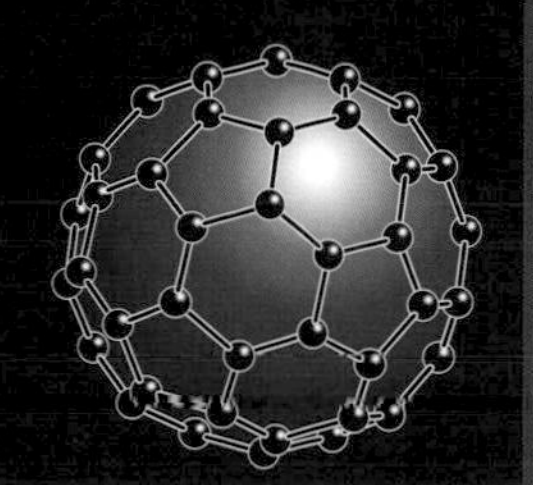

Addition reactions

At first sight, C_{60} seems to contain benzene-like rings. However, chemists have found that it seems to behave more like a 'super-**alkene**' than an **aromatic molecule** such as benzene (C_6H_6). C_{60} can add on osmium tetroxide, amines, hydrogen and various alkyl groups. It also undergoes addition reactions with halogens (fluorine, chlorine and bromine) to form a variety of C_{60}-halogen compounds. UK chemists have prepared C_{60} derivatives in which each carbon atom has an attached fluorine atom sticking outwards. This represents a flourishing new area of organic chemistry.

21. (a) Give the type of reaction (addition, elimination or substitution) and reagent (electrophile, nucleophile or radical) which is most typical of (i) an alkene and (ii) an aromatic molecule.

(b) Explain why buckminsterfullerene *cannot* undergo substitution reactions.

Of particular interest is the formation of C_{60} polymers. Visible or ultraviolet laser light causes C_{60} molecules to form a linear chain with four-membered carbon rings linking the C_{60} molecules, Figure 13.

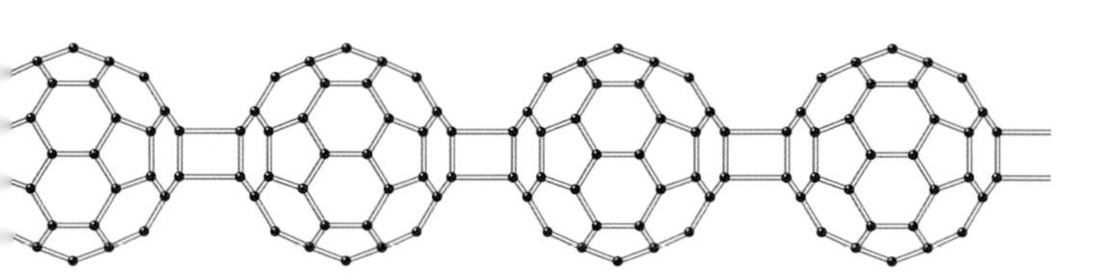

Figure 13. A C_{60} polymer.

Variations on a theme

A whole series of variants on the fullerene theme is believed to exist including giant molecules such as C_{240}, C_{540} and C_{960}, and nested molecules where the cages are arranged like a set of Russian dolls. It is most probable that soot particles formed in ordinary flames contain these kinds of species.

The simplest variant of the C_{60} structure results from substituting one of the carbon atoms with nitrogen to give azafullerene ($C_{59}N$). The result is the very reactive $C_{59}N$ radical (a radical is a molecule with an unpaired electron). This radical rapidly joins with another $C_{59}N$ radical to form a $(C_{59}N)_2$ dimer, Figure 14. This readily reacts with alkali metals to produce an azafulleride ion $(C_{59}N)^{6-}$. The potassium salt of this ion turns out to be electrically conducting. This kind of chemistry opens the way to yet another exciting set of materials with electronic properties potentially as rich as those of the fullerenes.

22. What are the outer electron configurations of (a) carbon and (b) nitrogen? Use these to explain why $C_{59}N$ is a radical. Explain how the dimer $(C_{59}N)_2$ is formed.

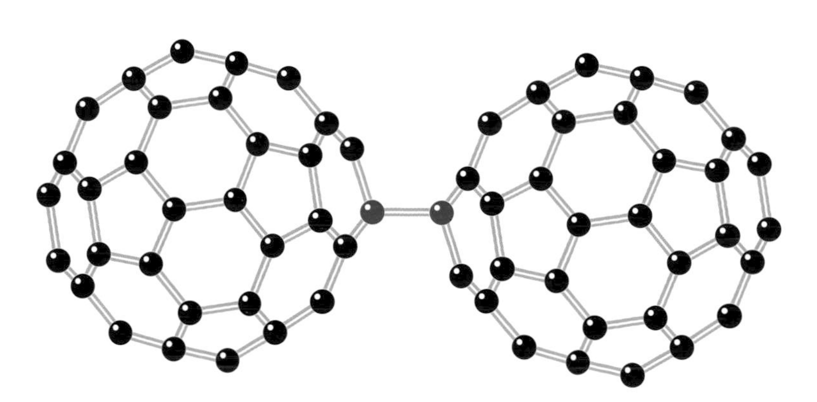

Figure 14. The $(C_{59}N)_2$ dimer.

Nanotubes

Fairly early after the discovery of C_{60}, Japanese scientists prepared tube-shaped carbon molecules – carbon nanotubes which promise a revolution in materials science and electronics, Figure 15. Bundles of nanotubes should have **tensile strengths** between 50 and 100 times that of steel, conduct as well as copper and have electronic properties that could be exploited for quantum devices.

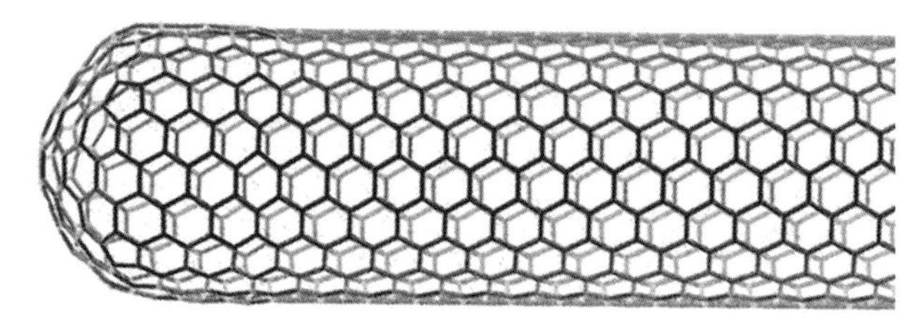

Figure 15. An individual carbon nanotube closed at one end by a fullerene cap containing both carbon pentagons and hexagons. The main cylindrical body consists solely of carbon hexagons.

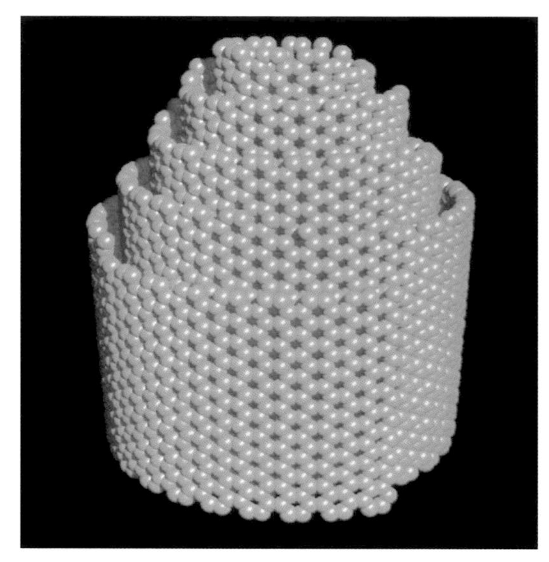

A molecular graphic of a nanotube.
Biosym Technologies, Inc./Science Photo Library.

Applications

There are plenty of ideas for exploiting C_{60} and other fullerenes. C_{60}'s unusual electrical properties mean that it behaves like an 'organic metal'. C_{60} could provide a new kind of conducting or semiconducting material for batteries, transistors and sensors. Scientists have grown highly ordered thin films of C_{60} on other semiconductor substrates such as gallium arsenide, so C_{60} could be used to make a new generation of advanced electronic devices which are based on thin films. Thin films of the potassium fulleride (K_3C_{60}) superconductor can also be made.

C_{60} also has unique optical properties in that it is transparent to low intensity light but nearly opaque above a critical intensity. These 'nonlinear' optical properties mean that C_{60} could be used to protect optical sensors from intense light.

Another set of applications arises from the fact that various chemical **species** can be attached to the surface of fullerenes, or even trapped inside. Finely divided carbon is a common component of industrial **heterogeneous catalysts**, as it has a large surface area. C_{60} balls could be coated with various catalysts, giving a huge surface area for catalytic activity.

An STM image of C_{60} on a surface.
©IBM Zurich Research Laboratory, Switzerland.

Glossary

Heterogeneous catalyst: A catalyst which is in a different phase from the reactants – usually a solid catalyst and liquid or gaseous reactants.

Species: A word which refers to an atom, molecule or ion.

Tensile strength: A measure of the strength of a material when it is being stretched.

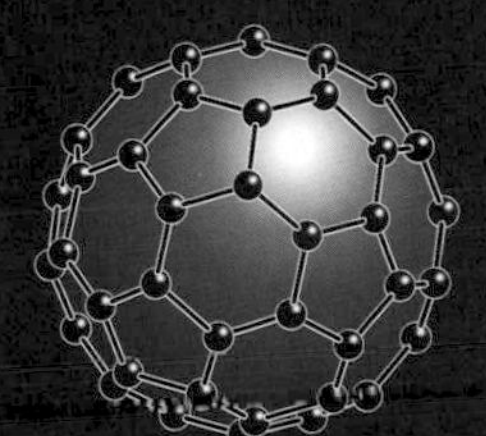

Chemists have managed to squeeze various species such as helium and other inert gases inside the fullerene cages. If the radioactive gas radon was incorporated into C_{60} molecules, they could be used as a 'label' to trace potential pollutants such as crude oil and toxic waste. They could also be used to treat cancer. The radiation from radon preferentially kills cancer cells, so radon-laden fullerenes could be conveyed to tumours by attaching tumour-targeting antibodies to the outside of the fullerene. In medical imaging, the deployment of radioactive tracers trapped in fullerenes would result in lower doses to patients since the fullerene would prevent the tracer from interacting with body tissues other than the one being targeted.

23. How would the radon be used to trace pollutants?

24. Why do the radon atoms need to be trapped inside C_{60} 'cages' to be conveyed to the tumours?

Conclusion

It has been said, in the recent past, that chemistry is a mature subject in which no new fundamental discoveries are likely to be made. The discovery and development of the fullerenes and other nanoclusters, and the high T_c superconductors has shown that chemistry can reveal new types of structures and phenomena in matter which not only extend our understanding of Nature at a very basic level but also will generate new technologies in the future.

Answers

1. A structure made up of repeating units which can be extended without limit in three dimensions.

2. They form coloured compounds. They often show catalytic activity as elements and in compounds. They form complexes.

3. (a) [Ar] = $1s^2\ 2s^2\ 2p^6\ 3s^2\ 3p^6$ – *ie* the electron arrangement of an argon atom.
(b) (i) $[Ar]3d^6$ (ii) $[Ar]3d^5$.
(c)

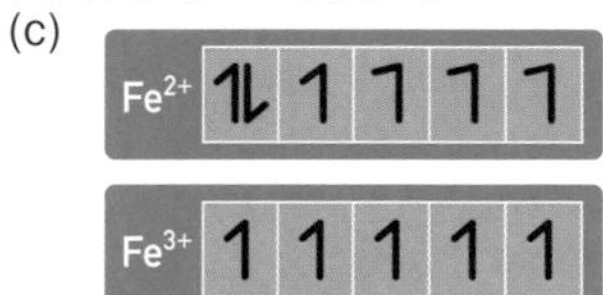

Fe^{2+} has 4 unpaired electrons, Fe^{3+} has 5 unpaired electrons.

4. $4Fe_3O_4(s) + O_2(g) \rightarrow 6Fe_2O_3(s)$

5. (a) 3+. b) 4+. In each case the total charge on the cations must add up to 6+ to balance the charge on the three O^{2-} ions. They would not conduct because symmetrical redox reactions are not possible.

6. Cu^{2+} and Cu^+.

7. If planes of atoms slide over each other in a metallic structure, a displacement of one atom leads to each atom being in an identical environment to its original one, see (a) below.
If planes of atoms slide over each other in an ionic structure, a displacement of one ion leads to an arrangement where ions of like charge are touching. The ions repel one another and so the structure shatters, see (b) below.

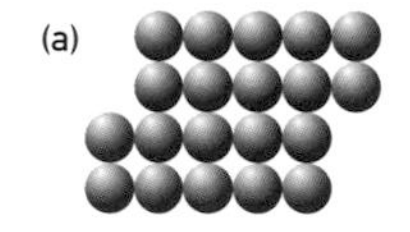

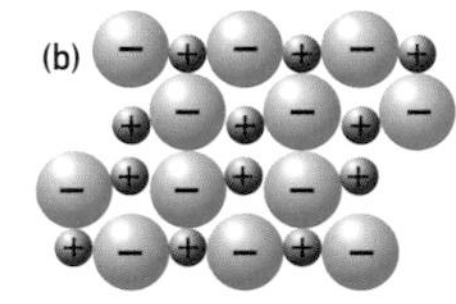

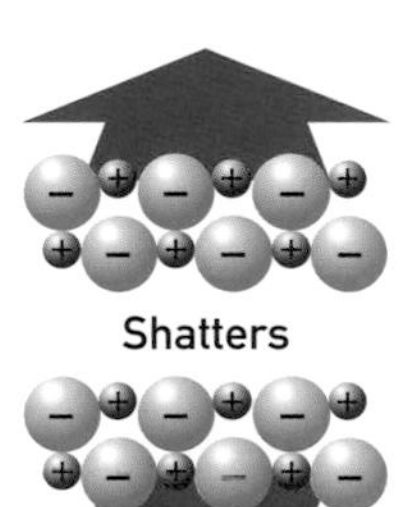

8. The cube would contain 1000 atoms.

9. A catalyst which is in a different phase from the reactants – usually a solid catalyst and liquid or gaseous reactants. Only atoms on the surface come into contact with the reactants.

Introductory page pictures: Biosym Technologies, Inc./Science Photo Library.

10. (a) (i) 1 cm^3 (ii) 6 cm^2 (iii) 6 cm^{-1}.
(b) (i) 8 cm^3 (ii) 24 cm^2 (iii) 3 cm^{-1}.
(c) (i) 27 cm^3 (ii) 54 cm^2 (iii) 2 cm^{-1}.
(d) $6/x$ cm^{-1}.
(e) The bigger the cluster, the smaller the fraction of atoms at the surface.

11. Each cluster would need to be ionised, accelerated by an electric field and deflected in a magnetic field. More massive particles would be deflected less for a given magnetic field strength.

12. Oxygen and sulfur are in the same group of the Periodic Table and therefore have the same outer electron arrangement.

```
    H   H
    |   |
H — C — C — S — H
    |   |
    H   H
```

13. Allotropes are forms of the same pure element which differ in the way in which their atoms are arranged in space.

14. It has several carbon-carbon triple bonds. It has a cyanide (nitrile) group (–C≡N).

15. (a) Buckminsterfullerene has a molecular structure, graphite and diamond have giant structures.
(b) The 'like dissolves like' rule – both benzene and buckminsterfullerene are non-polar.

16. It contains no hydrogen atoms.

17. They are – each carbon is part of a pentagon and a hexagon.

18. 720, 759.

19. 12.

20. $3K(s) + C_{60}(s) \rightarrow (3K^+\ C_{60}^{3-})(s)$

21. (a) (i) Electrophilic addition
(ii) Electrophilic substitution.
(b) It has no hydrogen atoms which can be substituted.

22. (a) C: $2s^2\ 2p^2$,
(b) N: $2s^2\ 2p^3$. Nitrogen has an odd number of electrons and therefore one must be unpaired. The unpaired electrons on the nitrogen of each monomer combine to form a covalent bond joining the two.

23. Its radiation can be detected at a distance by a Geiger Müller tube, for example.

24. Radon is an inert gas and therefore will not form chemical bonds: it must be physically trapped.

The world of liquid crystals

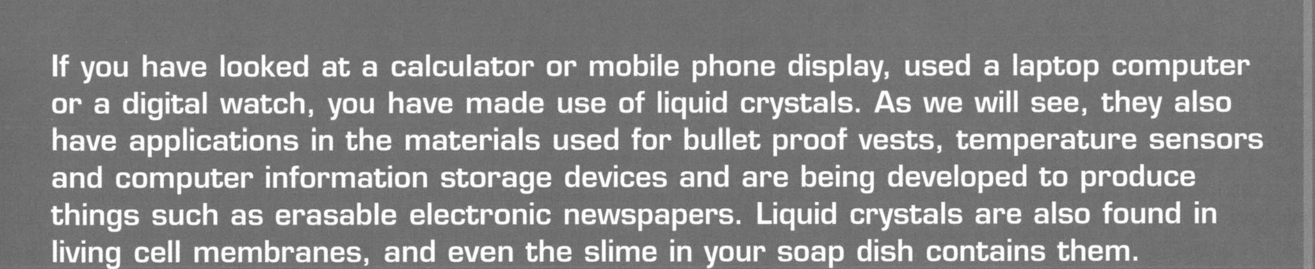

If you have looked at a calculator or mobile phone display, used a laptop computer or a digital watch, you have made use of liquid crystals. As we will see, they also have applications in the materials used for bullet proof vests, temperature sensors and computer information storage devices and are being developed to produce things such as erasable electronic newspapers. Liquid crystals are also found in living cell membranes, and even the slime in your soap dish contains them.

Mankind has known about liquid crystals for just over a hundred years, yet spiders have been spinning webs containing these molecular curiosities since time immemorial.

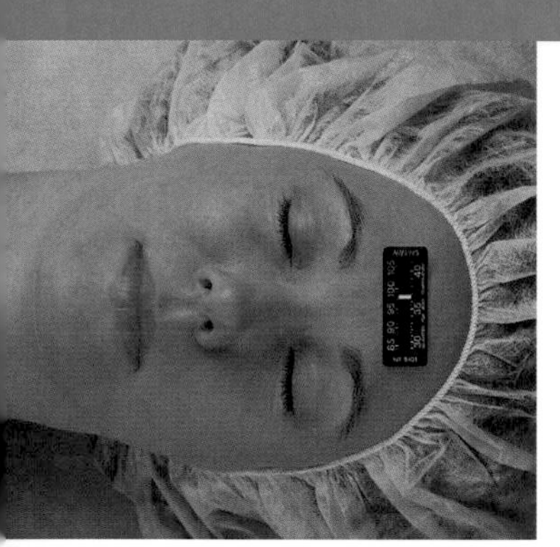

Courtesy of Sharn Inc.

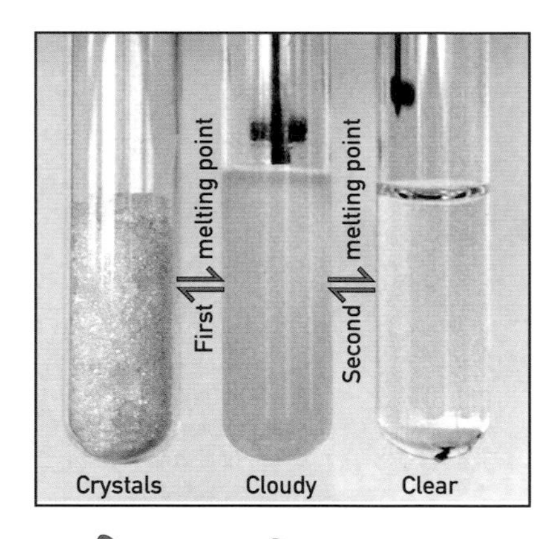

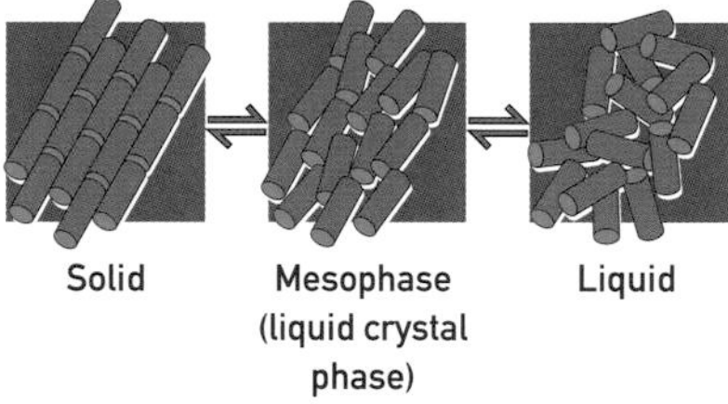

Figure 1. In the liquid crystal state, rod-shaped molecules can move about but still point in the same general direction.

Courtesy of Sharp Electronics (UK) Ltd.

What are liquid crystals?

When we think of a liquid, we picture its molecules as being arranged at random, while the molecules in a crystal are in a highly ordered arrangement. So, it seems odd to talk about a *liquid crystal*. However, it turns out that many substances with rod-shaped molecules have a state, called a mesophase, between liquid and crystal where the molecules have lost their regular arrangement but still tend to point in the same direction, Figure 1. This is the basis of the liquid crystal state – the molecules have lost their positional order (they can move about) but they still have directional order (they all point in essentially the same direction).

The discovery of liquid crystals

The phenomenon of liquid crystallinity was first discovered by an Austrian botanist, Friedrich Reinitzer in 1888. He was trying to measure the melting point of a compound he had just made, cholesteryl benzoate, and found that it seemed to have two melting points. At 145.5 °C (418.5 K) the crystals formed a cloudy liquid and at 178.5 °C (451.5 K) the liquid became clear.

1. Suggest why Reinitzer (a biologist) might have been interested in derivatives of cholesterol. Hint: Where is cholesterol found in living things?

2. Cholesterol has the structure

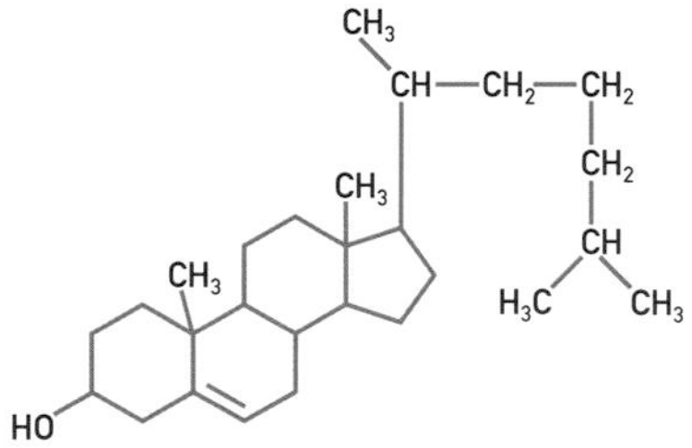

CH$_3$–CH(–CH$_2$–CH$_2$–CH$_2$–CH(CH$_3$)–CH$_3$) side chain on a steroid ring system with two CH$_3$ groups and HO group

Suggest how you might make cholesteryl benzoate from cholesterol. What functional group in the cholesterol is taking part in your suggested reaction?

3. (a) Suggest why chemists often measure the melting points of compounds they have just prepared.

(b) What does the fact that a substance has a sharp melting point imply?

Polarised light and refractive index

Light consists of electric and magnetic fields which vibrate in all directions at right angles to the direction in which the light beam is moving. When light passes through a substance called a polaroid, all the vibrations except those in a certain plane, say the vertical plane, are cut out and the light is 'vertically polarised'. When polarised light passes through a second polaroid filter placed at right angles to the first, all the remaining vibrations are cut out and no light passes through. The lenses of polaroid sunglasses work like this, and such glasses are often sold with a small extra polaroid filter to demonstrate this effect. A pair of polaroid filters at right angles to one another are referred to as 'crossed polaroids'.

The refractive index of a transparent material is a measure of the amount by which light is bent as it travels from air into the material.

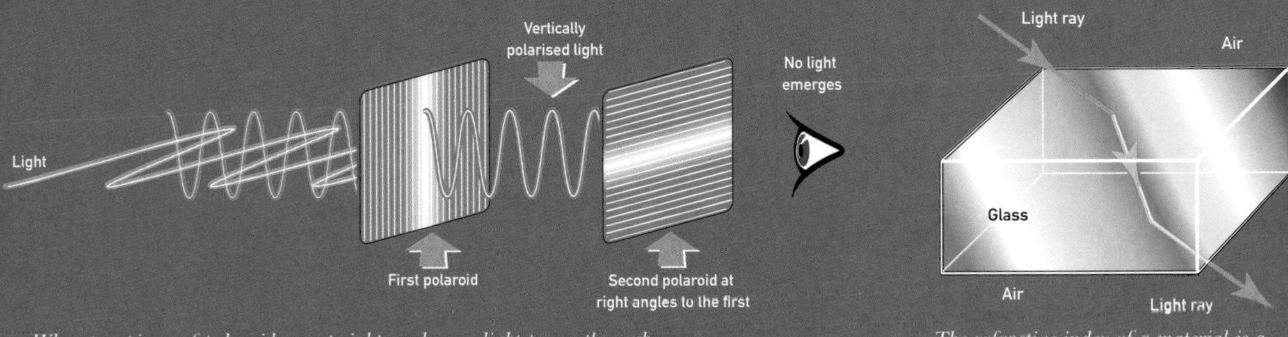

When two pieces of polaroid are at right angles, no light passes through.

The refractive index of a material is a measure of how much a ray is bent as it moves from one medium to another.

Courtesy of Polaroid Eyewear (UK) Ltd.

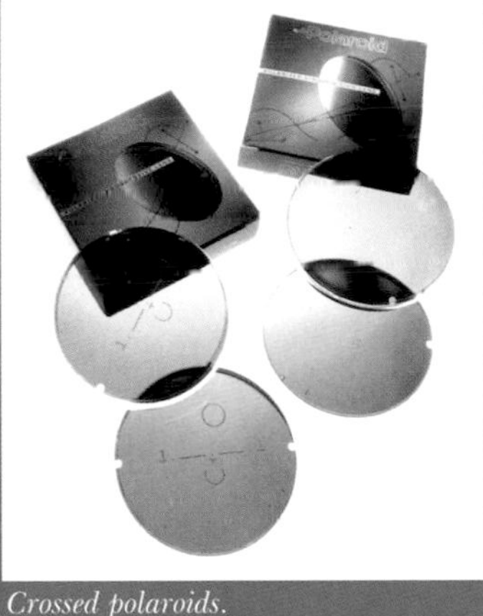

Crossed polaroids.
Courtesy of Polaroid Eyewear (UK) Ltd.

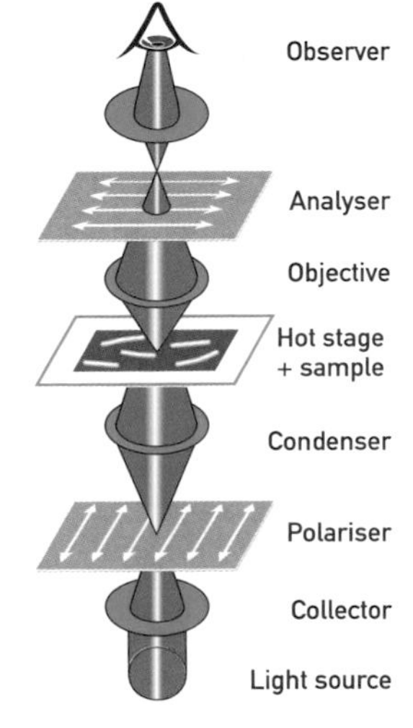

Figure 2. A polarised light microscope.

A colleague of Reinitzer's, Otto Lehmann, examined the liquid 'in-between' phase of cholesteryl benzoate with a polarised light microscope in which the sample being examined is sandwiched between a pair of crossed polaroids, Figure 2. He was surprised to find that it had different refractive indices in different directions – a property called birefringence. This is a property normally associated with ordered crystalline solids, where a direction can be specified in terms of the layers of atoms, molecules or ions in the crystal, and so Lehmann was astounded to find it shown by a liquid. The observation implied that the molecules in the liquid were arranged in an orderly fashion. This was the origin of the term 'liquid crystal'.

The fourth state of matter

We can now explain Reinitzer and Lehmann's observations. Cholesteryl benzoate molecules are rod-shaped overall, rather like pencils, Figure 3. In the solid state, they line up like pencils packed tightly into a pencil box which is just the right size to hold them. When heated up past the first melting point, the molecules are able to move around but they all still point in the same direction. This is the liquid crystal state. It is rather like pencils in a slightly bigger box – the pencils can move from place to place, but the box is too narrow for them to turn round, so they all have to point in the same direction. At the second melting point (the clearing temperature of the cloudy liquid), the 'pencil box' is large enough to enable the molecules to turn around as well as move from place to place, Figure 4.

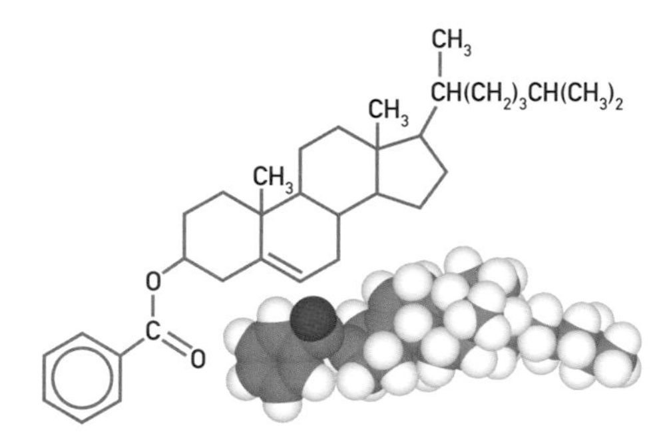

Figure 3. The rod-like molecular structure of cholesteryl benzoate.

Other liquid crystals

The structures of liquid crystal molecules are quite varied, but they have some features in common. The molecules are usually rod-shaped, Figure 5, and they have an uneven distribution of electrons. This leads to **intermolecular forces** which, over a certain range of temperatures, are strong enough to cause the molecules to line up in the same overall direction but not strong enough to hold them firmly in one place.

In the early years of this century, the German chemist Daniel Vorländer established that the rod-shaped molecules which formed the liquid crystalline state had flat, rigid sections consisting of a planar, electron-rich **aromatic** ring or rings joined by short linkages usually containing double bonds. These cores have attached to them one or two carbon chains, Figure 5.

Vorländer also showed that the clearing temperature, when all order is lost, is linked to the shape of the molecule. Rod-like molecules which can pack closely together have high clearing temperatures while molecules with side chains on the rods have lower clearing temperatures.

4. Suggest why molecule (c) in Figure 5 would be expected to have a flat core.

5. What types of intermolecular force are caused by an uneven distribution of electrons?

There are two main types of liquid crystal phase. The one described above for cholesteryl benzoate, where the molecules tend to point in the same direction, is called nematic, and the direction in which they point is called the director. The second main type of liquid crystal is called smectic. Here the molecules are arranged in layers.

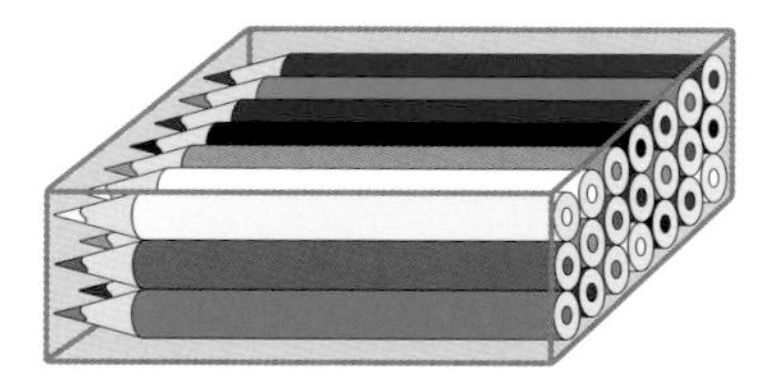

Figure 4. Tightly packed pencils are not free to move ('solid').

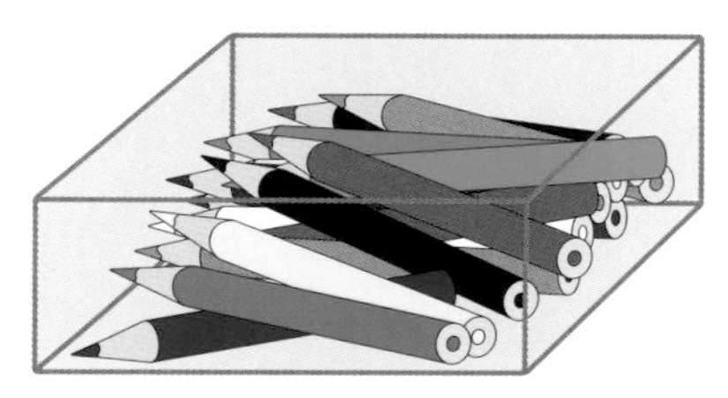

More loosely packed pencils can move around in the box but all point in the same general direction ('liquid crystal').

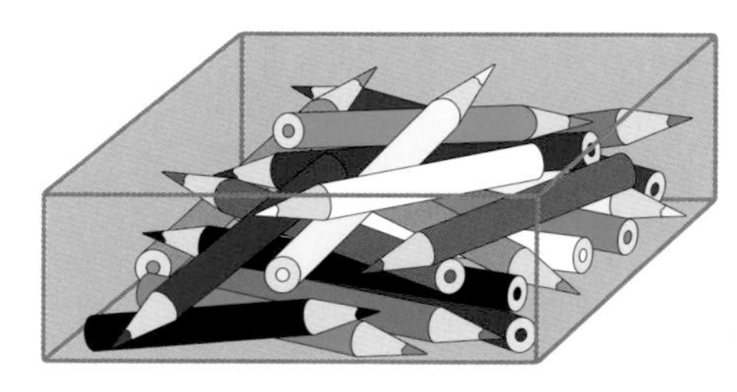

If the box is larger than the pencils, all order is lost ('liquid').

Glossary

Aromatic: Describes molecules based on flat, usually six-membered, rings which have a 'doughnut' of electron density above and below the ring.

Intermolecular forces: A group of forces, significantly weaker than covalent bonds, which act between molecules. They include hydrogen bonding, dipole-dipole bonds and van der Waals forces.

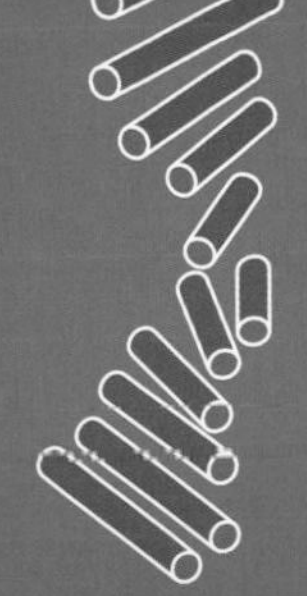

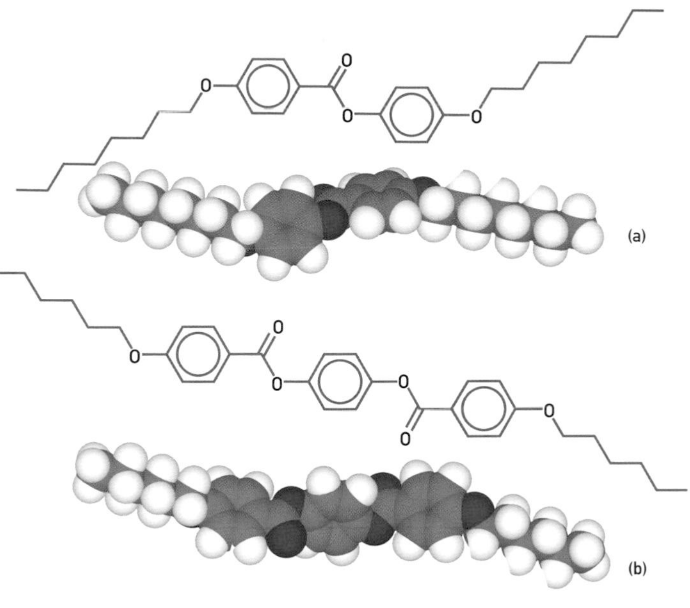

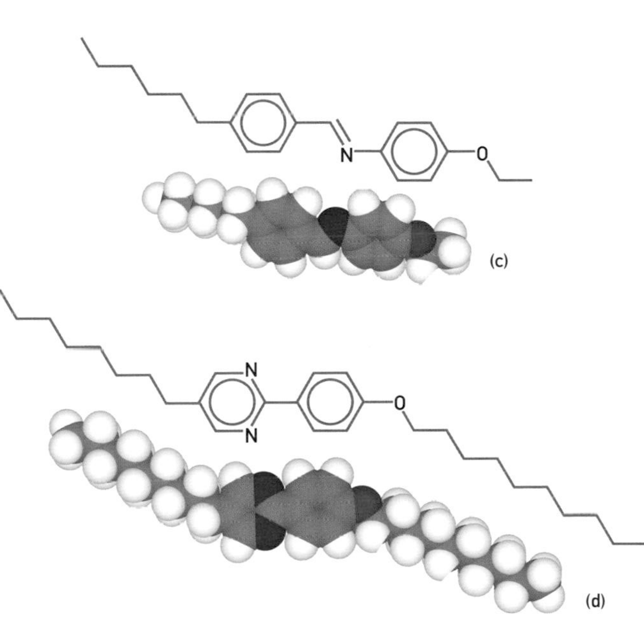

Figure 5. Some examples of rod-shaped liquid crystal molecules showing the cores which have linked aromatic rings ⬡. The rod shapes are seen most clearly from the space filling models.

Courtesy of Sharp Electronics (UK) Ltd.

The birth of a billion dollar industry

In the 1920s and '30s, researchers began to study the effect of magnetic and electric fields on liquid crystals. They found that liquid crystal molecules tend to lie parallel to a surface rather than at right angles to it. This is because the molecules are attracted to the surface by intermolecular forces. However, they found that a magnetic field or an electric field could force the molecules to lie at right angles to the surface – an effect called the Fréderickz transition after its discoverer, the Russian, Vsevolod Fréderickz. This effect, Figure 6, is at the heart of liquid crystal displays but it was not developed for about 40 years until there was a demand from the growing computer, communications and electronics industries for new visual display devices to replace bulky cathode ray tubes.

To give yourself a feel for how liquid crystal displays (LCDs) compare in size with cathode ray tubes (CRTs), remember that a desktop computer monitor contains a CRT while a laptop one is LCD-based.

The first LCD display devices were made in the 1960s but had short lifetimes because the liquid crystal compounds then available tended to decompose within a few weeks.

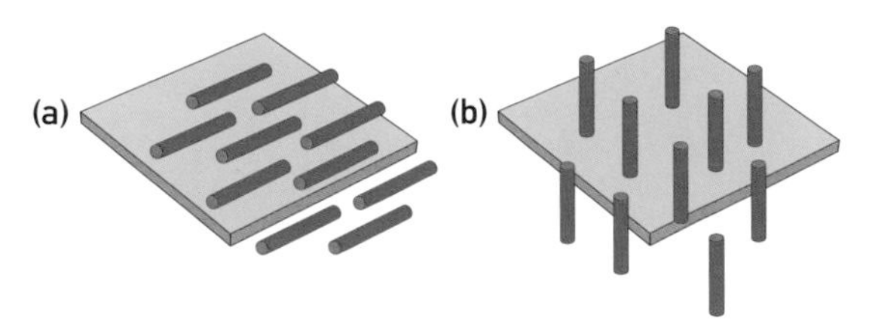

Figure 6. The Fréderickz transition.
Liquid crystal molecules tend to lie parallel to a surface (a) but an electric or magnetic field can force them to lie at right angles to it (b).

The twisted nematic display

The first really practical liquid crystal display device (which is still used today) is called the twisted nematic display, Figure 7. It was developed in 1970 by teams working independently in the US and Switzerland.

In the device a thin layer of liquid crystal is sandwiched between two plates of glass coated with indium tin oxide (a mixture of 90% In_2O_3 and 10% SnO_2) which makes the glass conduct electricity, Figure 7.

Light passes through a polaroid filter, enters the sandwich and then leaves it by passing through a second polaroid filter at right angles to the first.

At first sight we would expect no light to pass through the device because the polaroids are crossed. However, there is one further effect. The upper glass plate has a chemical treatment which makes the rod-shaped liquid crystal molecules close to it line up in a particular direction. The lower plate has been treated so that the liquid crystals near it line up at right angles to this direction. This means that as we move through the liquid crystal the direction in which the molecules line up twists through 90°. This has the effect of rotating the plane of polarisation of the light passing through the device through 90° too, so that light will in fact pass through the device. This is the 'off' state of the device and it looks bright.

If an electric field is now applied across the plates, the liquid crystal molecules line up parallel to the field and no longer rotate the plane of polarisation of the light. In this 'on' state, light cannot pass through the device and it appears dark. On switching off the field, the liquid crystal molecules revert to their previous arrangement and the device goes back to its bright 'off' state.

An actual display device such as a calculator or computer screen consists of an array of many such devices, each switched individually, which make up the numbers and letters of the display.

Courtesy of DuPont.

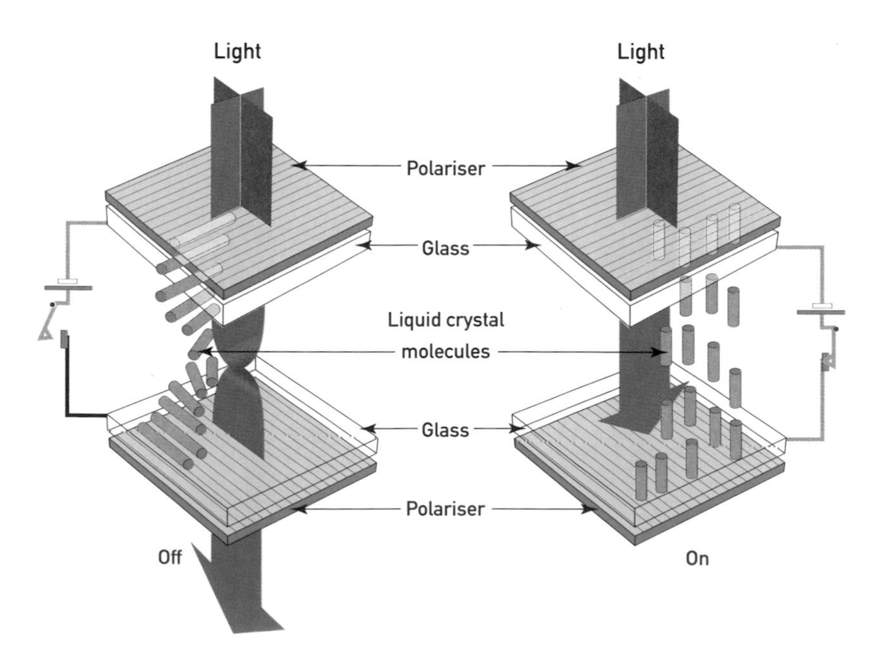

Figure 7. Schematic of the design and operation of the twisted nematic liquid crystal display device.

Glossary

Alkane: An alkane is a saturated hydrocarbon *ie* one with C-C and C-H single bonds only.
Nitrile: The nitrile functional group (sometimes called the cyanide group) is -C≡N.

George Gray.

Courtesy of DuPont.

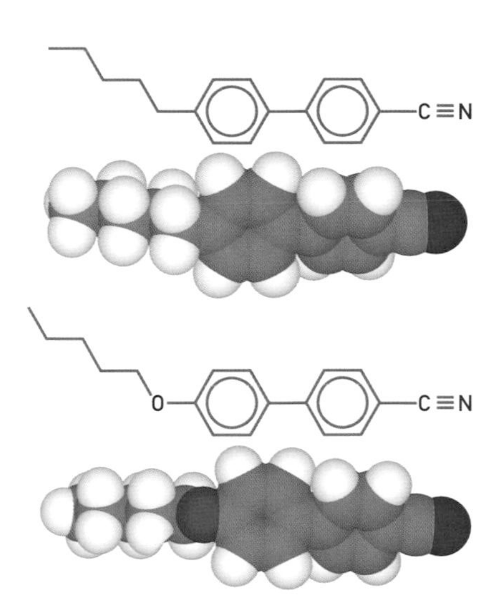

Figure 8. Two of George Gray's molecules which revolutionised the liquid crystal industry. Check that you can identify the biphenyl group, the alkane chain and the nitrile group.

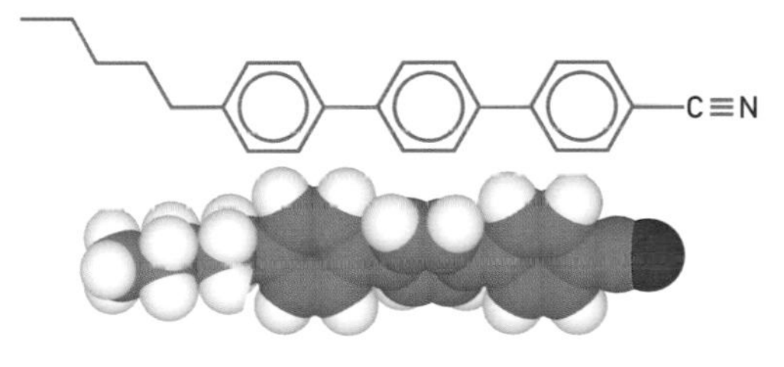

Figure 9. Gray's terphenyl-based molecule.

The chemistry of liquid crystals

To be able to produce liquid crystals suitable for displays, chemists needed to do two things.
1. Understand which factors of the molecular structure give rise to liquid crystal behaviour.
2. Tailor the molecules they made to give them the properties required for useful applications. For example the molecules must be stable and not decompose for, ideally, many years. They must be in the liquid crystal phase over a large temperature range, say -10 °C (263 K) to +60 °C (333 K), to be commercially useful – a calculator which will not work on a very hot or a very cold day will not sell well. They also had to have molecular stuctures which would line up parallel to an electric field to be useful in a twisted nematic device.

Many of these problems were solved in the 1970s by a team led by the British chemist, George Gray, of Hull University. His initial grant for this work was a mere £2117 per year – not much for work which led to today's multi-billion pound display industry!

It was already known that rod-shaped molecules consisting of a core of linked aromatic rings produced liquid crystal behaviour. Side chains attached to the molecule reduced the clearing temperature (above which liquid crystal behaviour is lost) because the molecules cannot pack so well together.

By 1972, Gray's team had produced molecules based on biphenyl (two linked benzene rings) with **alkane** chains and a **nitrile** (cyanide) group, Figure 8. The rigid biphenyl group produced the rod-like shape, the alkane chain was chemically stable, and the polar nitrile group allowed the molecule to interact with the electric field.

6. Explain why an alkane chain is chemically stable.

7. Draw a nitrile group, -C≡N and mark its polarity (δ^+ and δ^-). Suggest how this group would tend to line up in an electric field.

One problem remained however, the temperature range over which these molecules showed liquid crystallinity was still too narrow. So, mixtures of more than one compound were tried. Eventually, Gray devised a mixture of compounds including a terphenyl (with three linked benzene rings), Figure 9, which gave a clearing temperature of 59 °C (332 K) compared with 52 °C (325 K), which was the best achieved with biphenyl compounds alone. This mixture, called E7, allowed the twisted nematic display to become commercially viable, although E7 has since been replaced by better mixtures.

8. Why would you expect a terphenyl-based compound to have a higher clearing temperature than a biphenyl-based one?

The polymer dispersed liquid crystal (PDLC) display

Recently, a new type of nematic display device has been developed. This is the polymer dispersed liquid crystal (PDLC) display. It works slightly differently to the twisted nematic display and has some potentially exciting applications. The device consists of small droplets of a liquid crystal embedded in a matrix of transparent **polymer**. This is created by dissolving the liquid crystal in **monomer** molecules which are then polymerised. As the polymer's relative molecular mass starts to increase, the liquid crystal separates into droplets. The directors of these droplets are randomly aligned, and, as light passes through the PDLC film, the difference in refractive index between the droplets and polymer matrix causes the light to be scattered. Thus the film appears opaque. This is the device's off-state, Figure 10.

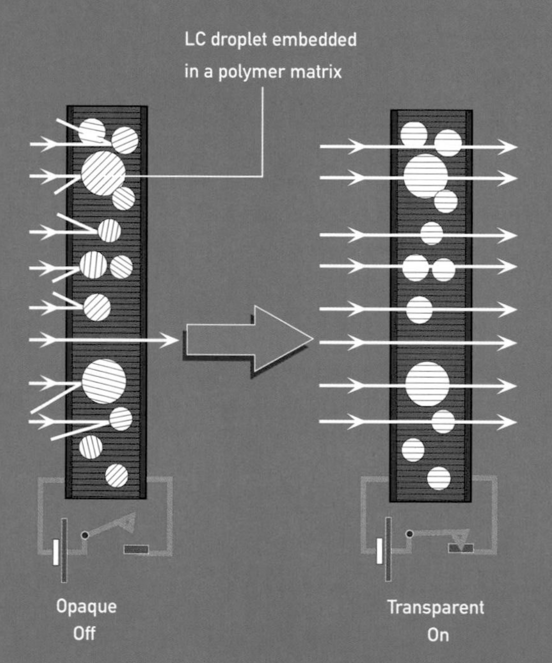

Figure 10. The PDLC display.
Off – the directors of the liquid crystal molecules are randomly aligned. Light is scattered.
On – the directors of the liquid crystal molecules line up with the electric field. Light passes through.

When an electric field is applied across the film, each droplet's director lines up in the same direction. Now, if the refractive index of the liquid crystal measured along the director matches that of the polymer matrix, light is no longer scattered and passes through the film. The electric field therefore switches the film from being opaque to transparent. The proposed applications of PDLCs range from projection and large-area displays to secrecy windows which may be switched electrically from opaque to transparent, Figure 11.

Figure 11. Secrecy window.

Glossary

Monomer: A small molecule, many of which can be linked together to form a polymer.
Polymer: A long chain molecule made by linking together many smaller molecules called monomers.

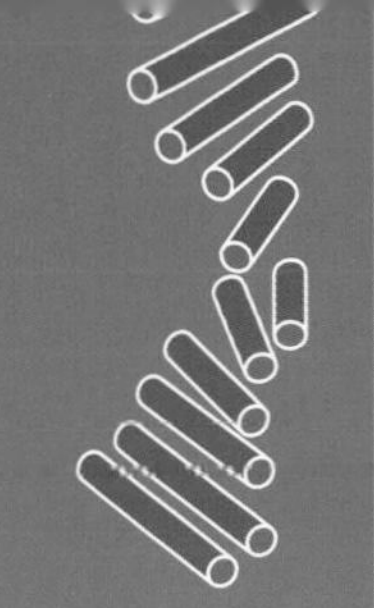

Discotic liquid crystals – molecular wires and electronic noses

Rod-shaped molecules are not the only ones which form a liquid crystal phase, disc-shaped ones can too – hence the term 'discotic'. These molecules usually have flat cores based on aromatic rings with several carbon chains attached. One type is the hexaalkanoyloxybenzenes discovered by Sunil Chandrasekhar and his team in Bangalore, India, Figure 12. In the liquid crystal phase, the molecules behave rather like coins in a box, Figure 13. There is no order in the *position* of the coins but their faces all point in the same *direction*.

Sunil Chandrasekhar (right) and George Gray (left).

In some cases, the molecules arrange themselves in stacks or columns and are so closely packed that the aromatic π-orbitals on adjacent benzene rings can overlap. So electrons can flow easily along the lengths of the columns but not from one column to the next – so that the columns form 'molecular wires'. One possible application is as gas sensors, sometimes called molecular noses. If the liquid crystal absorbs gas molecules, the disc-shaped molecules are forced further apart and the π-orbitals overlap less well. This reduces the electrical conductivity, Figure 14.

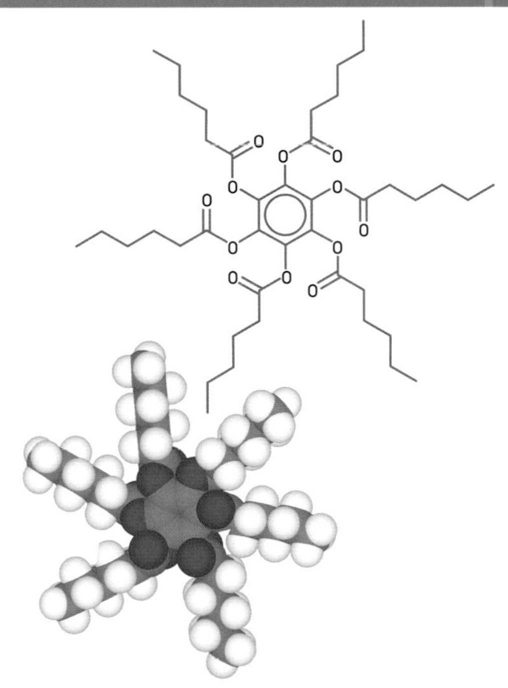

Figure 12. A disc-shaped liquid crystal molecule – a hexaalkanoyloxybenzene.

Figure 13. Disc-shaped liquid crystal molecules behave like coins in a box – their faces all point in the same direction.

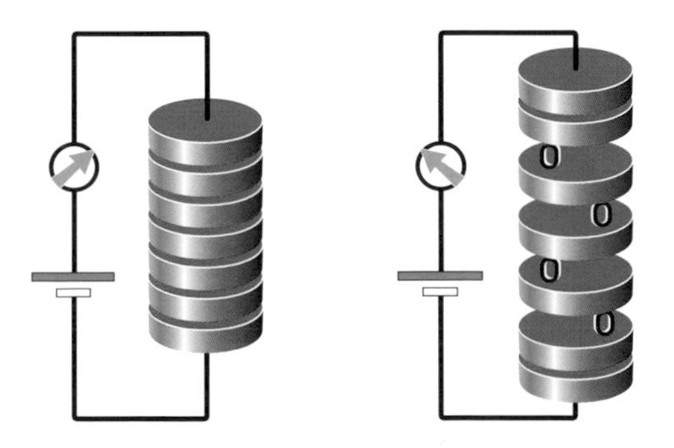

Figure 14. Disc-shaped liquid crystal molecules stack in columns close enough for the π-orbitals to overlap, so they conduct electricity. Absorbed gas molecules, **O**, *force the molecules further apart and the electrical conduction drops.*

Glossary

Bragg's Law: This is a mathematical expression which relates the angle at which the maximum intensity of X-rays is reflected from a crystal to the wavelength of the X-rays and the spacing between layers of atoms in the crystal.

Chiral liquid crystals – temperature sensors

Cholesteryl benzoate, the first known liquid crystal, has a distinctive property – it exists as optical isomers. Pairs of optical isomers have the same molecular formula but have different arrangements of atoms in space such that one is a mirror image of the other but the two are not identical. Such molecules are described as chiral (meaning 'handed', as in left and right handed). Left and right shoes are non-identical mirror images, while a sphere is identical to its mirror image. In organic chemistry, chiral molecules can be identified because they have one or more carbon atoms which have four different groups bonded to them. These carbon atoms are called chiral centres.

9. Use a molecular modelling kit to make a model of bromochlorofluoromethane (in which the carbon atom has four different groups bonded to it). Keep the model and make a second version in which the positions of say the bromine and the chlorine are exchanged. Try lining the two models up atom for atom and you will see that the two models are not the same. You should also be able to see that one is a mirror image of the other. You have made a pair of optical isomers. You may be able to do this exercise on screen if you have access to a suitable computer molecular modelling package.

10. Look again at the structural formula of cholesteryl benzoate, see Figure 3. There are actually eight chiral centres in the molecule. Indicate each of them with a * on a copy of the formula.

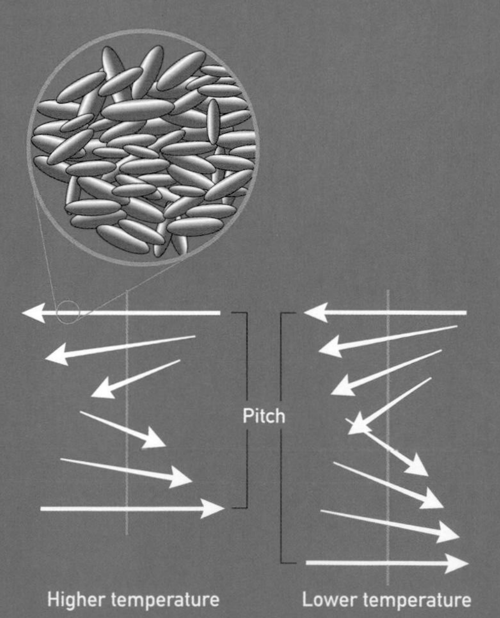

Figure 15. The Chiral Nematic Phase – the directors of the liquid crystal molecules trace out a spiral. This has a longer pitch (is 'looser') at lower temperatures.

The liquid crystal phase of chiral molecules like cholesteryl benzoate is called a chiral nematic phase. It has a particular property in that the chirality leads to the directors of the molecules having a spiral arrangement, Figure 15. This happens because the intermolecular forces between the chiral molecules tend to make each molecule line up at a slight angle to the one next to it and this angle adds up as we move through the liquid crystal structure.

The pitch (repeating distance) of this spiral is the distance one has to move through the liquid crystal before the director is pointing in the original direction again. This pitch depends on temperature, increasing as the temperature drops. This is because at lower temperatures, the molecules have less thermal energy and move less. So the angle between adjacent molecules becomes smaller, more molecules are required before the director returns to its original direction and so the pitch of the spiral increases.

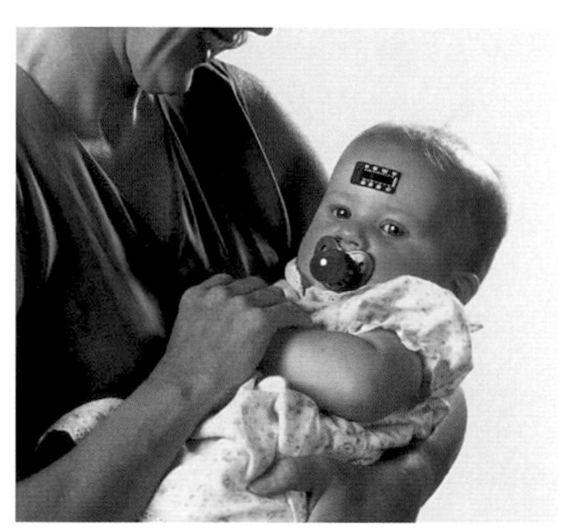

Figure 16. Liquid crystal thermometer in use. Courtesy of Sharn Inc.

This type of liquid crystal reflects only light of wavelength equal to the pitch of the spiral. This selective reflection of light of a particular wavelength by the regular helix is similar to the **'Bragg's Law'** reflection of X-rays from a crystal consisting of regularly spaced atoms. Light of wavelengths other than the pitch of the spiral **interferes destructively** and is cancelled out.

Since the pitch of the spiral depends on temperature, the wavelength, and therefore colour, of light reflected from the liquid crystal depends on temperature. The crystals reflect long wavelength red light at low temperature and shorter wavelength blue light at higher temperature.

These substances are used in thermometers such as those used by pressing them onto the skin, Figure 16. Other applications are in 'thermochromic' fabrics which change colour as the temperature changes, and as coatings on aircraft parts to monitor hotspots. It has also been suggested that cats' eyes in roads could be developed to reflect a particular colour at temperatures below freezing.

Destructive interference: This is the cancelling out of two waves when they meet in such a way that a crest of one wave coincides with the trough of the other.
Dipole: A molecule has a dipole if it has a non-symmetrical distribution of charge so that one end has an excess of negative charge (δ^-) and the other has an excess of positive charge (δ^+).
Pixel: One dot of many which make up a picture or other image.

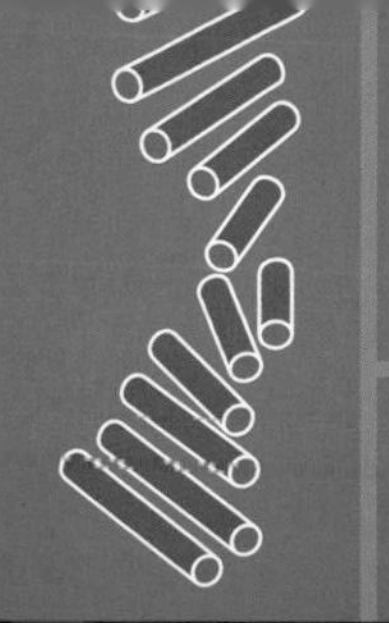

Courtesy of Sharp Electronics (UK) Ltd.

Courtesy of Sharp Electronics (UK) Ltd.

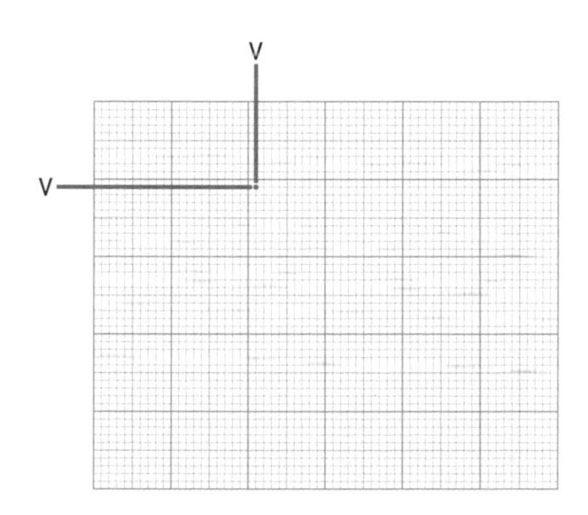

Figure 17. Multiplexing – a particular pixel is identified by applying a voltage to the appropriate row and column.

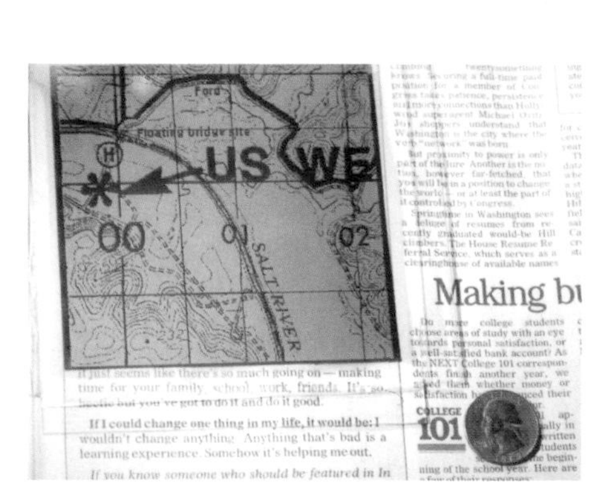

Figure 18. A small-scale cholesteric liquid crystal display image showing excellent contrast.
Courtesy of JW Doane and B Taheri.

Smectic liquid crystals (those in which the molecules are arranged in layers) may also be composed of chiral molecules, and a similar spiral effect is found in the direction in which the molecules point – a so-called chiral smectic C* phase.

Each layer of this phase has a permanent **dipole** because the molecular dipoles line up in the same direction. These dipoles can be made to 'flip' rapidly in direction by using an electric field, and this forms the basis of a display which can be switched up to 1000 times more quickly than a twisted nematic display. Such displays may eventually be used in flat colour TV screens which can be hung on the wall.

To form such a screen many displays are 'multiplexed' together in a rectangular array of rows and columns with rows of electrodes along the top and sides. Each display represents one dot or **'pixel'** of the final image. A particular pixel is selected for switching by applying a voltage to the appropriate row and column. This technique is already used in laptop computer displays, Figure 17.

Recently, Bill Doane and his colleagues at Kent State University, Ohio have invented a new use for cholesteric liquid crystals in displays. The cholesteric phase shows two optical textures (the characteristic patterns seen under a polarising microscope) one of which strongly reflects light while the other does not. One texture is observed when the molecules are arranged at random and the other when they are ordered. The two textures are stable but can be inter-converted using an electric field. The display device based on this effect has a black, light-absorbing back. Therefore, regions that reflect light appear bright while those that do not, appear dark. This display gives excellent contrast, Figure 18, and because both textures are stable the information can be stored for reading at some later date. These cholesteric liquid-crystal displays will play an important role in the development of erasable electronic newspapers.

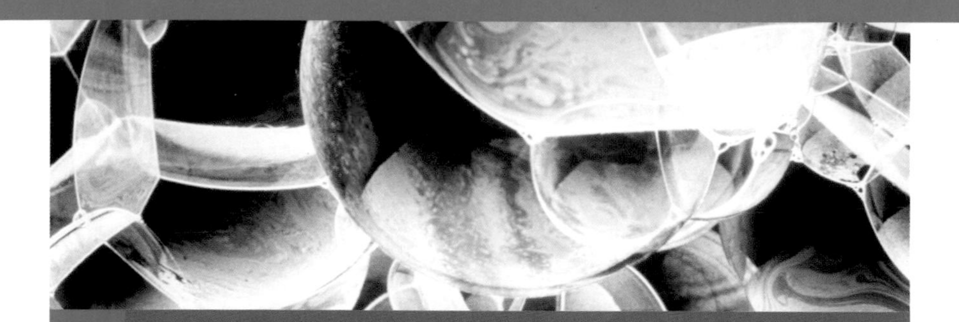

Soaps and detergents

It turns out that among the most common examples of liquid crystals are soaps and detergents in water. In fact, the slime you find in your soap dish is a liquid-crystal phase, and every time you wash your hands or do the dishes, liquid-crystalline phases form in the dirty water. Indeed, it could be argued that the first liquid crystals to find application were the soaps used thousands of years ago by, amongst others, the Phoenicians.

Soaps and detergents are composed of tadpole-shaped molecules containing two distinct regions, one polar and one non-polar. The head of the 'tadpole' is strongly water-loving, or hydrophilic, while the tail, normally a carbon chain, hates being in contact with water and so is hydrophobic. Above a critical concentration, these molecules form liquid crystal phases in which the molecules line up so that only the hydrophilic parts of the molecules are exposed to water. This shields the hydrophobic segments.

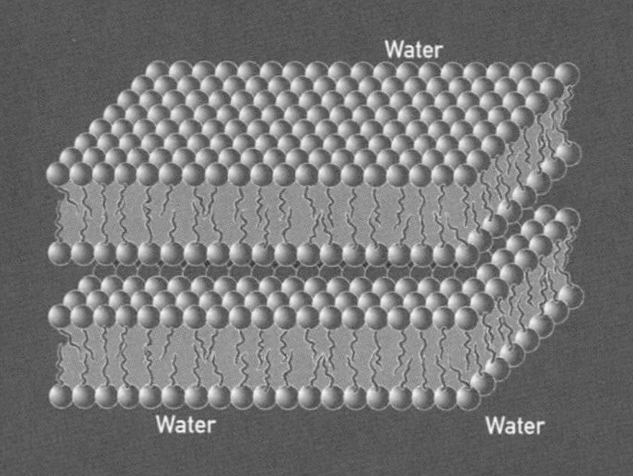

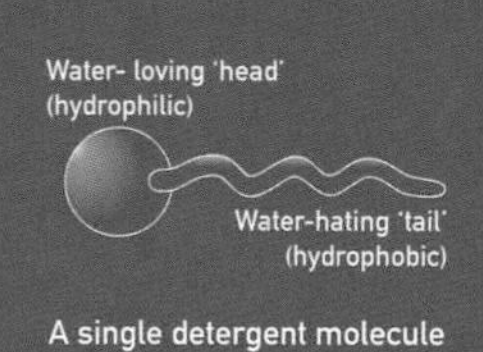

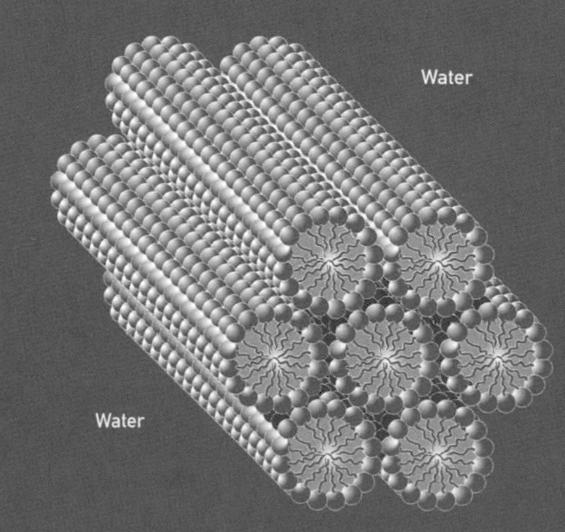

Types of molecular organisation shown by liquid crystal detergent molecules.

Courtesy of DuPont.

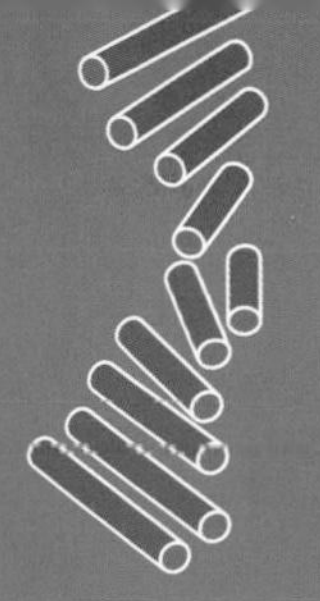

Glossary

Hydrogen bonding: The strongest type of intermolecular force. It takes place between slightly positively charged hydrogen atoms (which are covalently bonded to an oxygen, nitrogen or fluorine atom) and another oxygen, nitrogen or fluorine atom which has a slight negative charge.

Protonate: Donate a proton (a H^+ ion) to.

Liquid crystal polymers – the new wonder materials

You may have heard of the polymer Kevlar® which is used in bullet proof vests, tyre reinforcement, boat sails and ropes. This, too, is a liquid crystal. Kevlar® is a polymer, poly(*p*-phenyleneterephthalamide) which has rigid rod-shaped molecules because of the linked benzene rings, Figure 19. As the polymer comes out of solution, its molecules tend to pack together like logs floating down a river, Figure 20, forming a liquid crystal phase described as lyotropic. Lyotropic means that the existence of a liquid crystal phase is linked to the concentration of the molecules. If the concentration is low the molecules will not line up, nor will logs in a river.

Hydrogen bonding between the Kevlar® molecules, Figure 21, makes them line up parallel to one another in a very ordered arrangement. This means that when pulled along the length of the molecules, the fibres are very strong as it is covalent bonds which are being stretched. (Weight for weight, the fibres are around 10 times stonger than steel.) The strong hydrogen bonding between the molecules means that the polymer is difficult to melt and to dissolve. Solvents such as concentrated sulfuric acid are needed to dissolve Kevlar®. These solvents **protonate** the nitrogen and oxygen atoms and therefore disrupt the hydrogen bonding between the chains.

Vectra® is a similar polymer but based on less regular-shaped monomers. Its molecules therefore pack together less effectively and the polymer is relatively easy to melt and can be processed by conventional methods. These polymers are stable and expand little on heating. They are used for mounting the optical components in CD players and in utensils for use in microwave ovens.

11. Draw a short section of a Kevlar® molecule and show how it can be protonated.

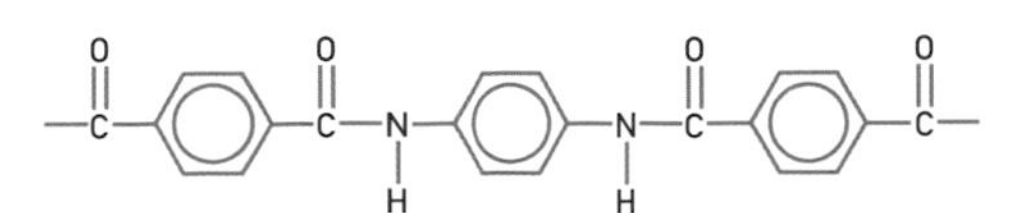

Figure 19. The formula of part of a Kevlar® molecule.

Courtesy of DuPont.

Figure 20. Logs floating down a river tend to line up in the same direction.

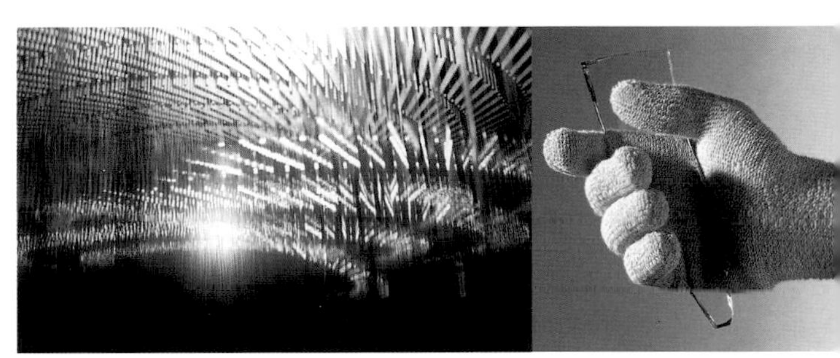

Vectra® is used in drift chamber detectors in particle accelerators such as the Large Electron-Positron collider at the European particle physics laboratory, CERN in Geneva.
Courtesy of CERN.

Kevlar® gloves can help to reduce hand injuries. *Courtesy of DuPont.*

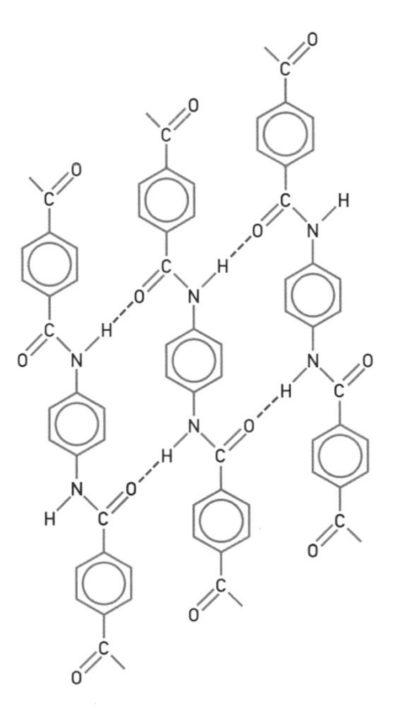

Figure 21. Hydrogen bonding between Kevlar® molecules.

The properties of Kevlar® fibres, and other polymers processed from liquid-crystalline solutions, fall short of those predicted theoretically. This is partly because **defects** become trapped in the structure during processing. To resolve these difficulties, research now focuses on spiders' webs!

Spiders produce silk which they use for the radial threads of webs and also as a safety rope in case of unexpected falls. The mechanical properties of this silk outperform those of any synthetic fibre. For example, Kevlar® is slightly stronger but considerably less resilient. Furthermore, these silks have the highly desirable property of being **biodegradable** and the spider can recycle them.

People have suggested that liquid crystallinity plays a central role in the processing of the silk. The water-soluble silk molecules, fibroin, have a globular structure and are stored by the spider as an aqueous solution in its gland. In the duct leading to the spinneret, these globular molecules form rod-like units which assemble into a liquid-crystal phase. The low viscosity of this phase allows the solution to be spun so fast that the molecules change into a more linear shape that can pack more effectively into a crystalline structure. This structure is insoluble in water. It is the combination of interconnected crystalline and non-crystalline regions in the silk that give the fibres their exceptional properties. Nature has therefore developed a room temperature process for converting water-soluble single molecules into a high-performance insoluble fibre without the need for chemical change!

Spider's web.

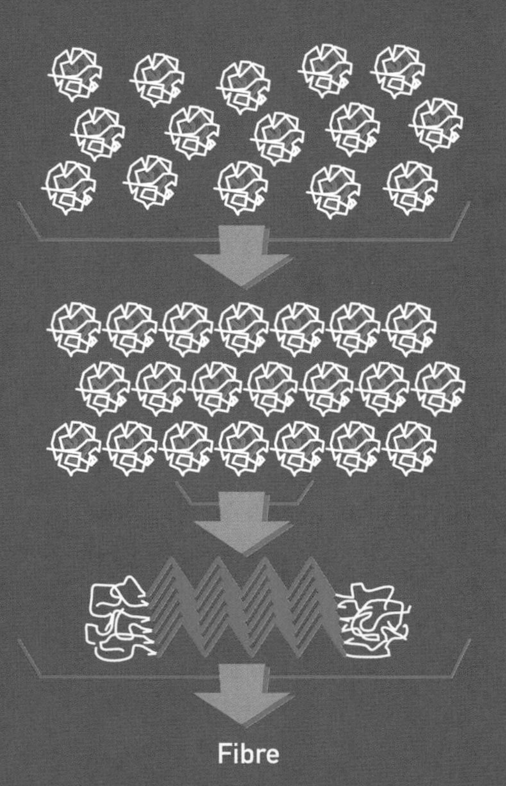

Changes in the molecular structure during the spinning of silk by a spider.

Glossary

Activation energy: The minimum energy that a species must acquire before it can react.
Biodegradable: Decomposed by natural processes in the environment.
***Cis*:** Two groups bonded to the atoms at either side of a double bond are *cis* if they are on the same side of the molecule.
Defect: An irregularity in an ordered arrangement of atoms or molecules.
***Trans*:** Two groups bonded to the atoms at either side of a double bond are *trans* if they are on opposite sides of the molecule.

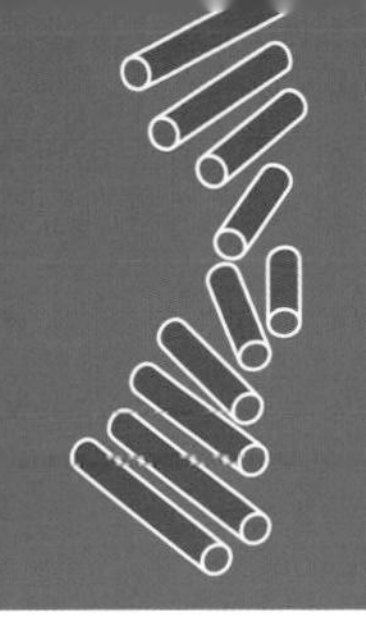

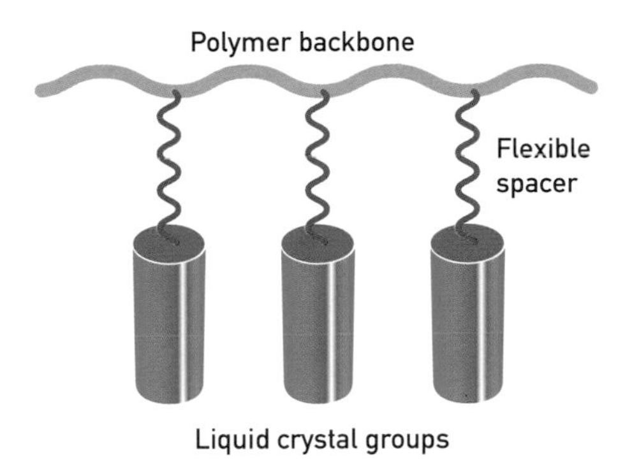

Figure 22. A side-group liquid crystal polymer.

Side-group liquid crystal polymers – for optical memories

Side-group liquid crystal polymers are made by attaching liquid crystal units as side chains onto a polymer backbone, Figure 22. A flexible spacer separates the liquid crystal units from the backbone and this is critical to the properties of the polymer. These properties result from two opposing tendencies – that of the polymer backbone to become randomly coiled, and that of the liquid crystal groups to line up. Thus the resulting polymers can show glass-like properties. They can also be made to respond to light or optical signals.

The ability to respond to light is brought about by incorporating molecules such as azobenzene into the polymer. These molecules exist as ***cis-*** and ***trans-*** isomers, Figure 23, the *cis-* form being angular and the *trans-* form linear. The *trans-* form promotes liquid crystallinity while the *cis-* form does not. Shining laser light on to the molecules provides the **activation energy** to break the π part of the double bond and convert the *trans-* form (which is liquid crystalline and opaque) into the *cis-* form (which is transparent). Once the laser illumination stops, the *cis-* form of the azobenzene changes rapidly back into the *trans-* form, but the side chains of the polymer remain 'frozen' in the non-liquid crystalline form.

So a laser can be used to 'write' on the polymer. The 'writing' can be erased by heating the polymer, upon which the side chains reform the liquid crystalline form. This type of polymer has potential as an erasable information storage system for computers because a laser can be used to write dots on it representing the ones and zeros of a binary numbering system.

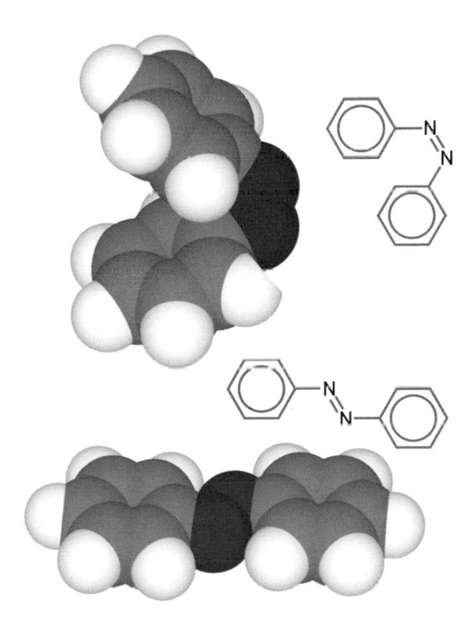

Figure 23. Cis- *(above) and* trans- *(below) azobenzene. In the* cis-*isomer, both the substituents are on the same side of the double bond, in the* trans-*form, they are on opposite sides.*

12. Explain why *trans*-azobenzene is more likely to promote liquid crystallinity than the *cis*-form.

The future of liquid crystals

The use of liquid crystals for various types of displays is well established. The applications of liquid crystals will become ever more diverse, encompassing areas such as information storage, molecular wires and sensors. In addition, as our understanding of the role of liquid crystals in nature increases, this will lead to developments in new materials for application in biological and pharmaceutical areas. Liquid-crystal science will clearly remain an active and exciting area to be involved in during this century.

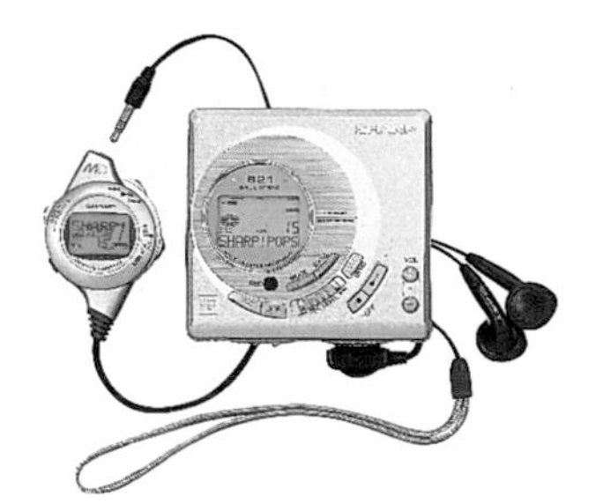

Courtesy of Sharp Electronics (UK) Ltd.

Courtesy of Sharp Electronics (UK) Ltd.

Answers

1. Cholesterol is a biochemical found in blood plasma and cell membranes.

2. React cholesterol with benzoic acid (benzenecarboxylic acid) or benzoyl chloride. The –OH group of the cholesterol reacts.

3. (a) To identify them and confirm their purity.
   (b) It implies that the substance is pure.

4. There would be a delocalised π-orbital spreading across both aromatic rings *via* the double bond.

5. Dipole-dipole bonding if the uneven electron distribution is permanent. Van der Waals (instantaneous dipole-induced dipole) forces if the uneven electron distribution is transient.

6. It has only C-C and C-H bonds which are both relatively stong and non-polar.

7. $^{\delta+}C \equiv N^{\delta-}$. The group would line up with the N atom towards the positive plate and the C atom towards the negative plate.

8. The three linked benzene rings would make the molecule more rigid and rod-shaped than one based on just two rings.

9. Students may need help arranging the two models to show that they are (a) non-identical and (b) mirror images.

10. Note that this leads potentially to 2^8 (256) isomers, not all of which will occur in nature.

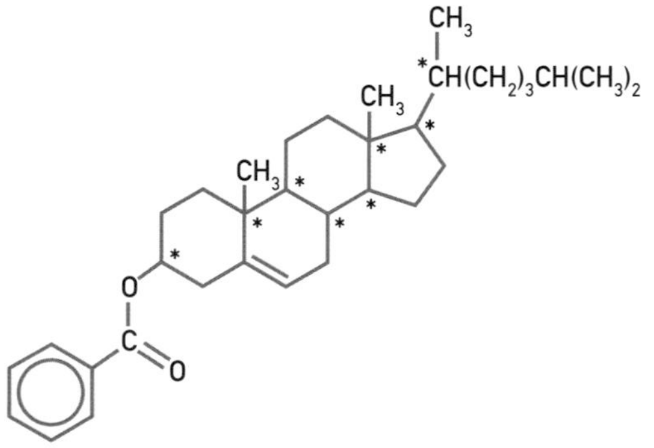

11.

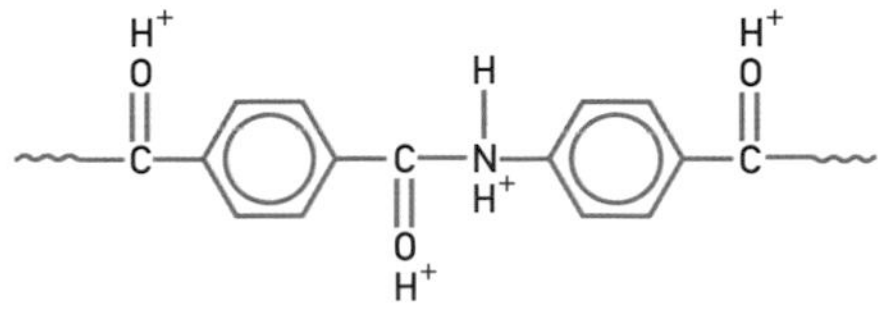

12. *Trans*-azobenzene is a much more linear molecule than the *cis*-form.

Introductory page pictures: Sharp Electronics (UK) Ltd. and CERN.

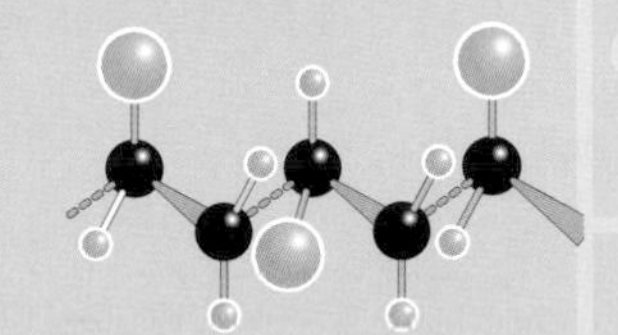

The age of plastics

Synthetic polymers are chain-like molecules constructed from smaller chemical units. They are the basis of many of the materials we use everyday. Indeed, it is hard to imagine living without them. Yet polymers, or plastics as they are more commonly called, are a relatively recent development that has resulted from the efforts of some imaginative and determined chemists.

When you woke up this morning, you were probably lying on a mattress made from several different synthetic polymers including, polyester fabric, polyurethane foam, nylon thread and various adhesives. Your sheets, pillows, quilt or blankets were probably made with synthetic polymers. These materials are everywhere in our homes and workplaces – in paints, in PVC window frames, in soft furnishings, toiletries, packaging, electrical insulation and furniture. We rely on polymers for clothing, for transport and even for surgical implants.

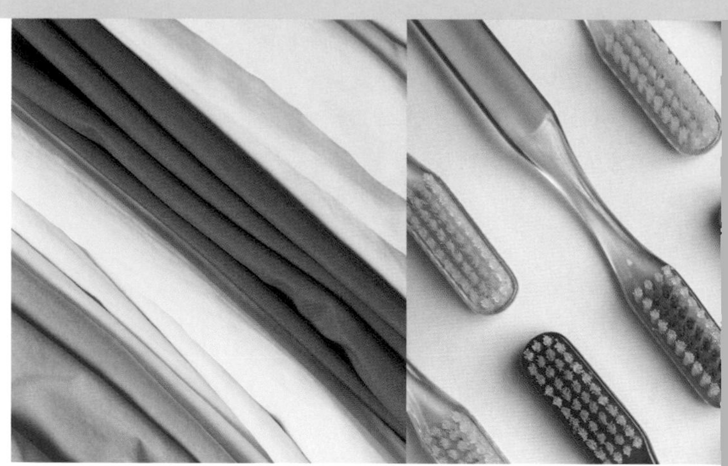

25% of nylon goes into fibre application.

Toothbrush bristles made from nylon.

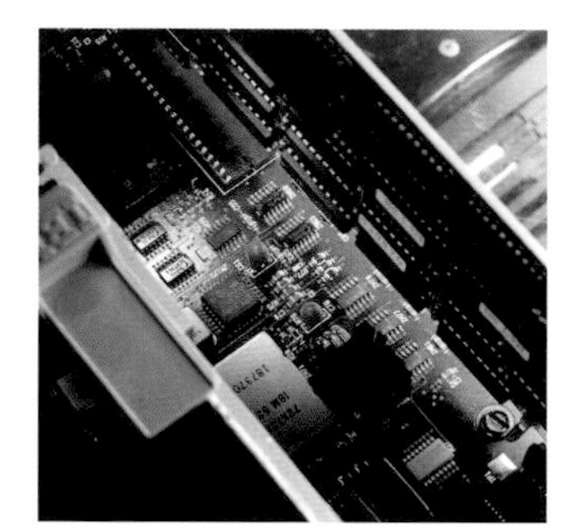

Plastic is widely used in the electrical and electronics industry. Courtesy of DSM.

Courtesy of DuPont.

Some applications of polymers
Household furnishings Carpets, curtains and wallpaper
Electrical fittings Wire insulation, casings for electrical goods, printed circuit boards
Household fittings Drain pipes, kitchen fittings, window frames, mastics
Surgical prostheses Implants, tooth fillings, contact lenses
Transport Bicycles, cars, trains, planes, space-craft
Sports materials Clothing, shoes, athletics tracks
Domestic Utensils, containers, non-stick coatings
Paints and surface coatings
Furniture
Fabrics
Packaging
Toiletries and cleaning materials
Clothing

Glossary

Semiconductor: A material with an electrical conductivity between that of a conductor and an insulator.

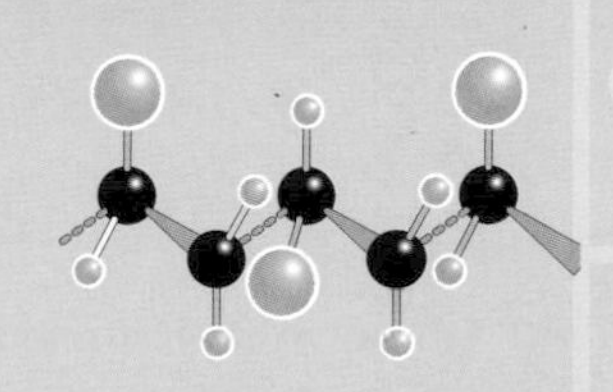

Dentistry and contact lenses are typical applications of polymers.

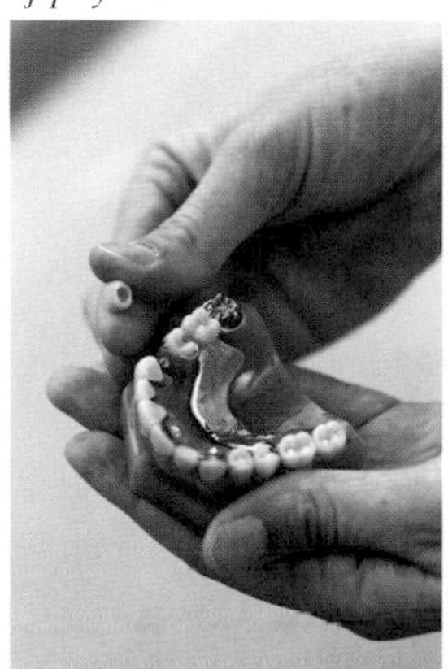

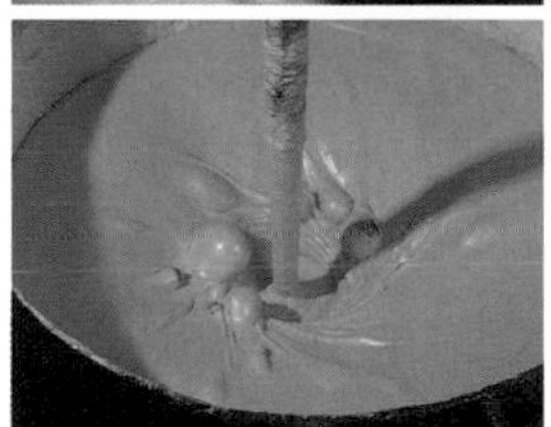

Paints and surface coatings are familiar products of the polymer industry.

The amount of plastics we use is enormous and growing. In many cases, plastics have displaced wood, stone, glass, leather, natural fabrics and metals from their traditional uses. Plastics have the advantage of being light, strong and hard-wearing and are often cheaper than the material they replace. They are mostly easy to process, take up colours well and can be chemically tailored to have specific properties. A huge range of polymers is now available with a wide variety of chemical structures. Today, chemists are still developing ever more sophisticated plastics, including new polymers that possess the electrical properties of metals and of **semiconductors**.

What are polymers?

Polymers are quite different from other compounds in that they are built up from smaller molecules (**monomers**) linked in long chains. Usually, one polymer is built from just one or two different types of monomer. A typical polymer sample is made up from a collection of chains of widely differing lengths. Most common polymers have 'backbones' based on carbon which readily forms very stable carbon-carbon **covalent** bonds. However, other elements also form polymers. These include silicon combined with oxygen (the resulting polymers, called silicones, have many everyday applications), sulfur, nitrogen and some metals, Figure 1.

The simplest carbon polymer, polythene, is also one of the most familiar. It is sometimes called polyethylene or, more correctly, poly(ethene). Its constituent monomers are ethene molecules. Poly(ethene) consists of a string of linked carbon atoms with each carbon joined to two hydrogen atoms, Figure 2. Some of the molecules in poly(ethene) are extremely long, having more than 50 000 carbon atoms; they are truly giant molecules.

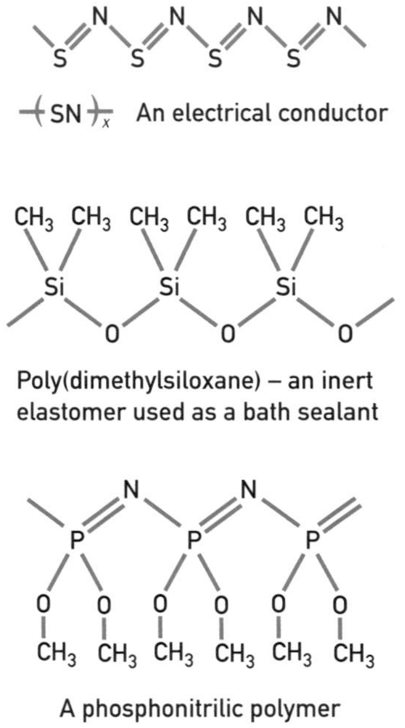

$-(SN)_x-$ An electrical conductor

Poly(dimethylsiloxane) – an inert elastomer used as a bath sealant

A phosphonitrilic polymer

Figure 1. Some inorganic polymers.

1. Use a ball and stick molecular modelling kit to make a section of a poly(ethene) molecule, say 12 carbons long. Rotate the carbon-carbon bonds and notice how many different shapes it can adopt.

Poly(ethene)'s string-like structure gives it its familiar physical properties. In the solid, some parts of the molecule are neatly packed together in a **crystalline** array; how much of the molecule is crystalline depends on how it is made. We can think of this neatly packed arrangement as being rather like uncooked spaghetti in a jar.

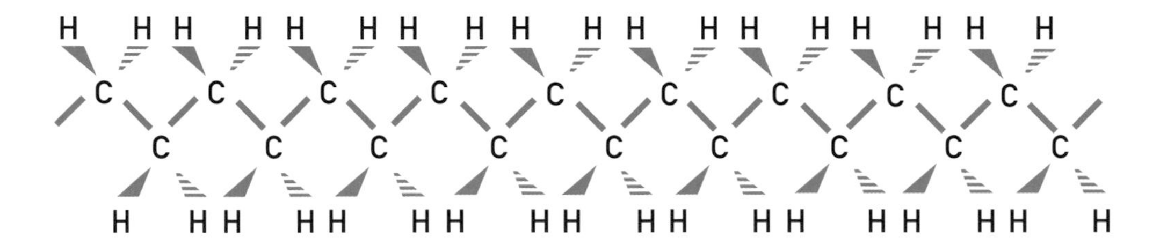

Figure 2. A fragment of a poly(ethene) molecule shown in three dimensions. The C–H bonds shaped like wedges come out of the paper, whereas the C–H bonds shown as dashes point into the paper.

Glossary

Covalent: Describes a chemical bond formed by sharing electrons between a pair of atoms.

Crystalline: Describes a substance which has a regular arrangement of atoms, molecules or ions.

Monomer: A small molecule used as a building brick for making a polymer.

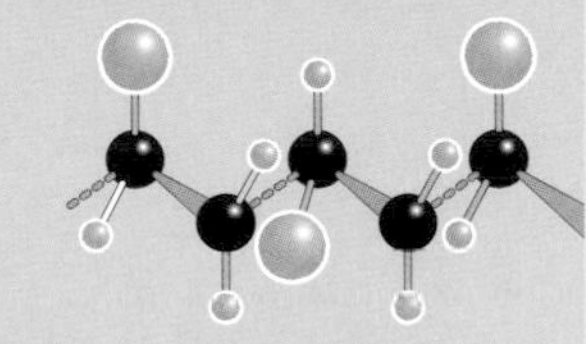

Shopping bags, a common use of plastics.
Courtesy of British Plastics Federation.

When poly(ethene) is heated it first becomes soft, and then melts into a sticky, viscous fluid. (Anybody who has left a plastic washing-up bowl on a hot ceramic hob will have seen the messy results of this process.) As the plastic gets hot, the stiff, extended chains soften and begin to wriggle about like cooked spaghetti. In fact, the chain movements are rather livelier than cooked spaghetti – more like a mass of writhing worms. When the molten poly(ethene) cools, it reverts to the highly-ordered crystalline form again and recovers its original properties. This behaviour allows technologists to shape the material into anything we might want – a plastic bag, a washing-up bowl or a hip-socket replacement. Materials with this kind of property are called thermoplastics, which simply means they can be shaped when hot.

2. (a) What kind of intermolecular forces operate between two molecules of the polymer poly(ethene)?

(b) Individually these forces are weaker than covalent bonds. State roughly how many times weaker (to the nearest factor of 10) one of these bonds is compared with a covalent bond.

(c) In view of this weakness, explain why poly(ethene) is a relatively strong material.

$$H_2C{=}CH_2 + H_2C{=}CH_2 \xrightarrow[\text{and temperature}]{\text{High pressure}} \bullet CH_2{-}CH_2{-}CH_2{-}CH_2\bullet$$

$$\bullet CH_2{-}CH_2{-}CH_2{-}CH_2\bullet + H_2C{=}CH_2 \longrightarrow \bullet CH_2{-}CH_2{-}CH_2{-}CH_2{-}CH_2{-}CH_2$$

$$\bullet CH_2{-}(CH_2CH_2)_n{-}CH_2\bullet$$

Where n is very large

Figure 6. An alternative picture of what happens in ethene polymerisation. At high temperatures and pressures, ethene molecules slam into each other and bond to form new chemical species with unpaired electrons, or radicals at each end. These readily react with more ethene molecules eventually to form poly(ethene).

They readily react with any ethene molecule that comes within bonding distance to build up the molecular chain, Figure 6.

This **free radical** description of the polymerisation allows us to explain why branches form in poly(ethene). The free radical chain ends are so enormously reactive that if an appropriate collision occurs, the chain-end free radical ($-CH_2^\bullet$) will simply grab any hydrogen atom within reach – including one in the middle of an already existing chain. This involves breaking the hydrogen-carbon bond and creating a free radical on another carbon atom, which could be anywhere in the chain. This can now grow another sequence of ethene units as before, forming a branch, Figure 7.

The higher the temperature and pressure of reaction the more frequent the branches in the product poly(ethene) will be. Since the type of branching and its frequency influences the properties of the material, we can begin to see how chemists can tailor the reaction to make products with specific properties required for particular applications. For poly(ethene) made in this way the greater the relative proportions of long straight sequences of $-CH_2-$ units (which can pack together well) the more crystalline and hard the product is. This is because ordered crystalline chain arrangements are harder than disordered ('amorphous') chains, and regular straight-chain segments are more easily arranged in a crystal **lattice** than segments with branches.

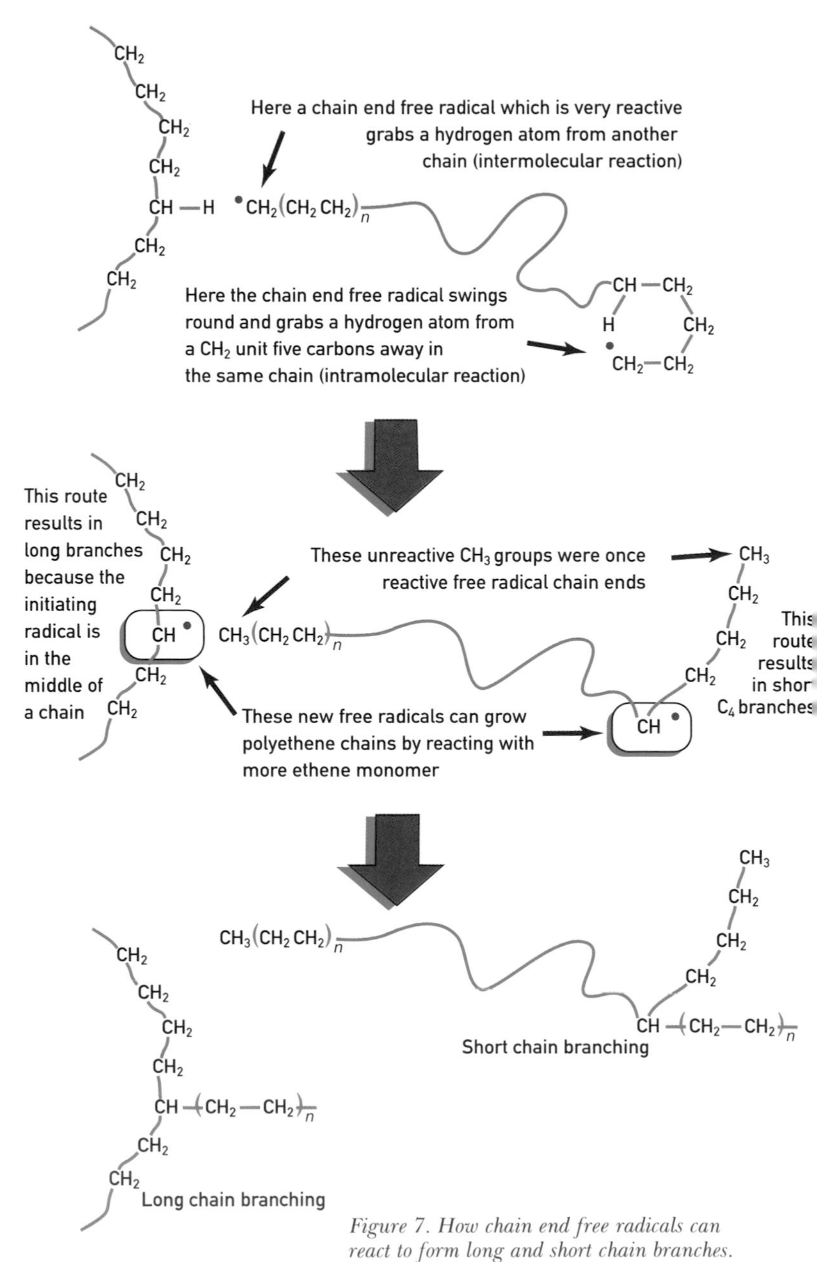

Figure 7. How chain end free radicals can react to form long and short chain branches.

Glossary

Free radical: An atom or group of atoms with a single unpaired electron.
Lattice: A regular, three-dimensional arrangement of atoms, molecules or ions.

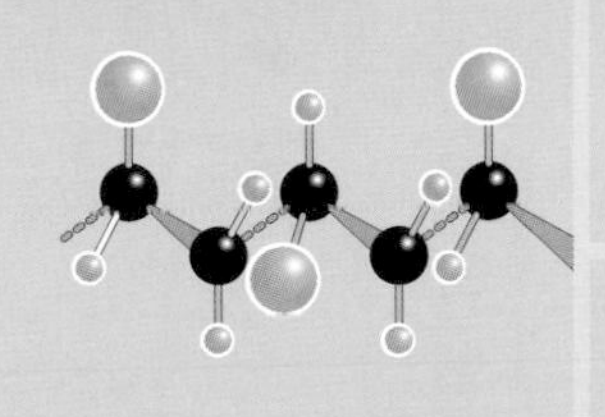

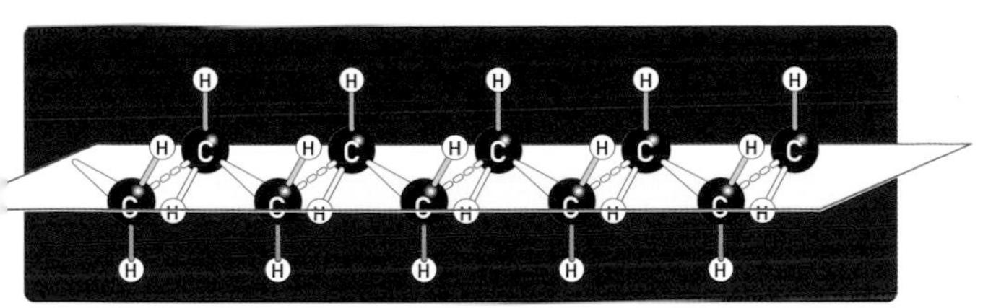

Figure 8. Part of a linear chain of poly(ethene).

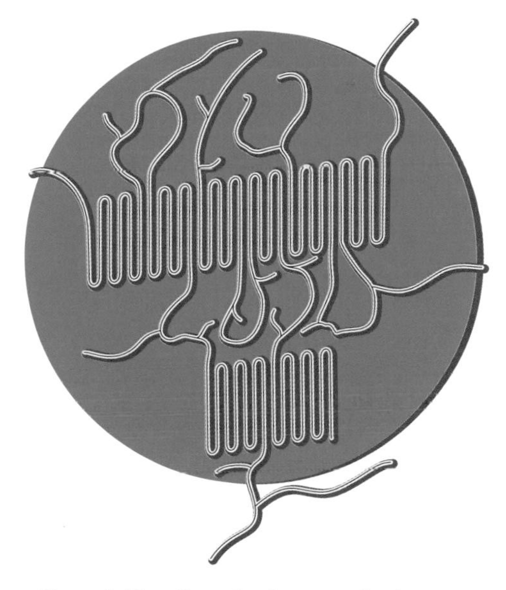

Figure 9. Two-dimensional cartoon showing crystalline and non-crystalline (amorphous) regions, and long and short-chain branching. The lines represent poly(ethene) molecules.

The crystalline parts of poly(ethene) are associated with so-called linear chains. In these, the carbon atoms are actually arranged in a zig-zag fashion with the carbons in a plane, Figure 8. These chains are folded backward and forward on themselves in a regular way, Figure 9.

Figure 9 also shows that the chain branches are located in the disordered ('amorphous') regions. One molecular chain can pass through a chain-folded, crystalline part of the sample into a disordered part and then into another crystalline part (or back into the crystalline part it emerged from). This complexity in the structural organisation is what gives poly(ethene) its variable and valuable properties.

The uses of poly(ethene)

World War II broke out soon after the discovery of poly(ethene) and this influenced its early applications (see Box – *A secret military material*).

A secret military material

Within a relatively short period after its discovery, poly(ethene) was to have an influence on the outcome of World War II. By the end of 1939, ICI was selling its new material to a few carefully selected customers. Within a year the selection of customers was tightly restricted and the need to produce more material became immense. That year, three UK cable companies took 17 500 pounds (7945 kilograms) of ALKETH with a further 1800 pounds (817 kilograms) going to a manufacturer of candles. With the fall of France, however, the requirement for candles in the trenches was deemed to be of low priority and virtually all production was directed into electrical insulation. The high-frequency cables required for radar were the major application, although no mention of this is to be found in any of the available records of that period. Production spiralled from 8 grams in 1935 to about 8 tonnes in 1938, 557 tonnes in 1942 and 1441 tonnes in 1944. Sadly much of this part of the history of the development and exploitation of polythene was lost to view in the cloak-and-dagger atmosphere that surrounded what became, between 1935 and 1950, a strategic military material.

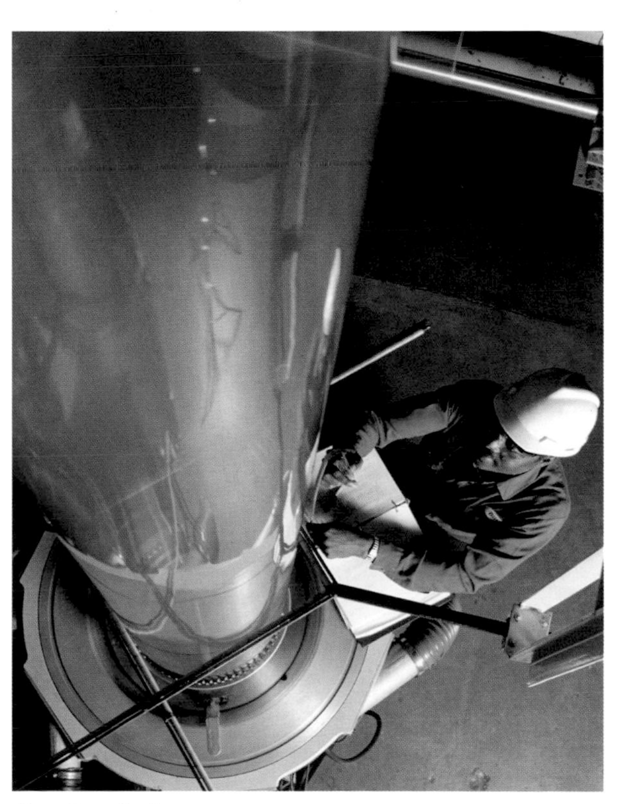

Extruding plastics.

After a dip in demand following the end of hostilities, new markets for this cheap, versatile material began to appear. By the end of 1946 ICI had three production units giving a total output of 4000 tonnes a year. By about 1950 the uses of the material had expanded into packaging (bottles, film, bags) and it was beginning to displace metals from various market niches. This was to be one of the themes of polymer science during the next 50 years.

In the early days of poly(ethene) production, the applications were fairly undemanding objects like washing-up bowls and buckets. It required new chemistry before the material could challenge the tensile strength of steel cables, the bursting pressure required for pipes for gas and water mains, and eventually meet the stringent requirements for surgical implants. How this happened is the next part of our story.

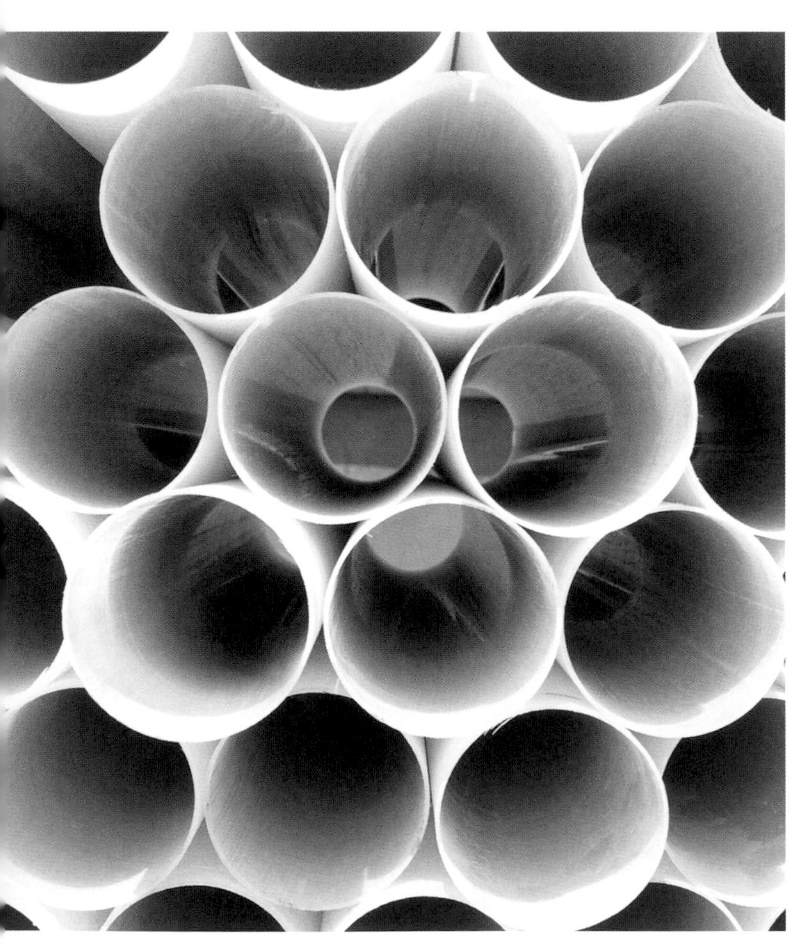

Poly(ethene) is used to make pipes.

A mature technology?

By about 1950, many of the business people involved with the companies producing poly(ethene) assumed that its production and applications had reached the stage of 'mature technology' – that is that there was nothing more to learn technically and that the only further interest would be in driving down production costs. Nothing could have been further from the truth.

In 1950 the company decision-makers had not heard of the work of Karl Ziegler's group in the Max Planck Institute in Mülheim, Germany. Ziegler was fascinated by the formation and reactions of metal-carbon bonds. In the early 1950s his research group was studying aluminium **alkyls**. The key compound they studied was a trialkyl aluminium, R_3Al, in which a single atom of aluminium has three hydrocarbon groups (R) attached to it via aluminium-carbon bonds. Ethene reacts with this compound by inserting itself into the aluminium-carbon bond.

$$R_3Al + CH_2{=}CH_2 \rightarrow R_2Al\text{-}CH_2\text{-}CH_2\text{-}R$$

This process required temperatures of 100 to 120 °C (373 – 393 K) and was carried out in sealed pressure vessels called autoclaves, although at much lower pressures than those used by ICI. Just like ICI's earlier work, the research had its element of excitement, even danger, because aluminium alkyls are pyrophoric – that is, they catch fire spontaneously when in contact with air.

9. Suggest the possible products when an aluminium alkyl, $(CH_3)_3Al$, reacts with air.

Ziegler's group found that from time to time they obtained a white powder which was clearly poly(ethene), but which was more crystalline and had a higher melting point than the material produced by the ICI route. The aluminium trialkyl was catalysing ethene polymerisation. This discovery promised to extend the range of poly(ethene)'s exploitable properties. The researchers also observed that 'traces of colloidal nickel accidentally left in a reactor' and/or 'trace impurities in the steel reactor' made the poly(ethene)-yielding reaction go rather better than was the case in immaculately clean conditions.

These astute observations led them to search for better systems to catalyse the polymerisation of ethene, and within a few years, they showed that they could produce poly(ethene) using a catalyst made by reacting trialkyl aluminiums with **transition-metal** compounds such as zirconium or titanium tetrachloride. What was even more amazing was that this process produced an essentially linear (with no, or very few branches), high molecular mass polymer at room temperature and at about one atmosphere pressure.

This poly(ethene) was different from the form made at high pressure and high temperature by ICI. Not only was it more crystalline but it also had a higher molecular mass, higher density, and better machining and wear properties. It was therefore called high density polyethylene (HDPE).

Glossary

Alkyl: An organic group formed by the removal of a hydrogen atom from a saturated hydrocarbon, *eg* ethyl, C_2H_5.

Composite: Describes a material made from two or more other materials and which combines the properties of both.

Transition-metal: A metal which forms compounds in which it has a partly-filled d-shell of electrons.

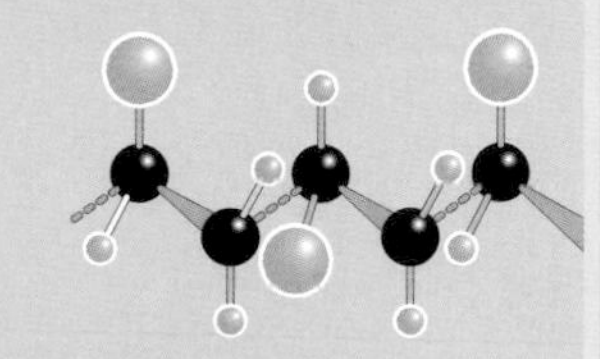

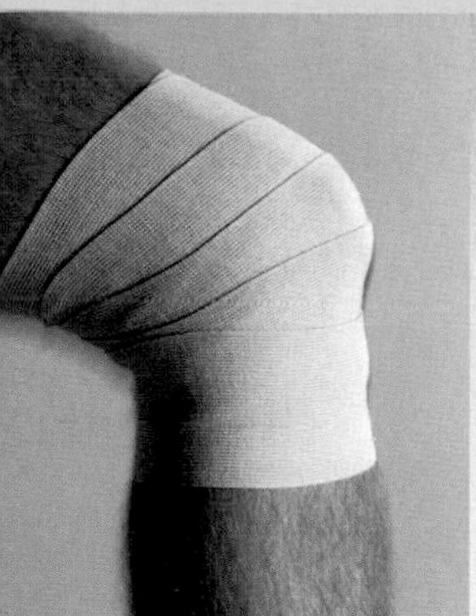
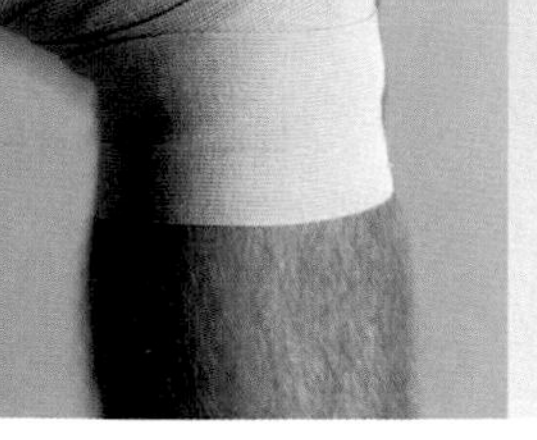

Plastic components are used in some knee surgery.

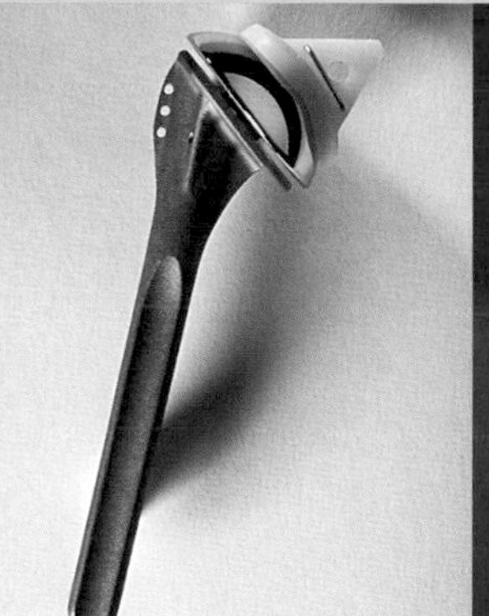
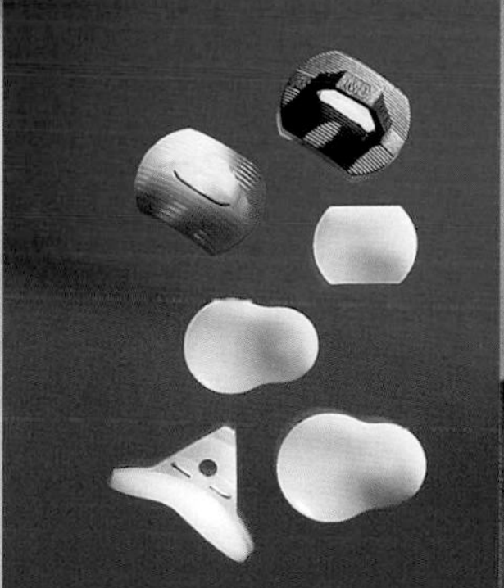

Artificial hip, knee and knuckle replacements.

Karl Ziegler who developed catalysts for polymerisation. Courtesy of GDCH.

10. Explain why you might expect more-linear poly(ethene) chains to produce a material of higher density.

Ziegler's method allowed the makers to avoid the costly requirements of high temperature and high pressure. It also offered the potential of a different material – superior in some of its properties – at lower cost, which could be used in all sorts of exciting new ways.

They found that films of the same or higher strength, could be made much thinner than with the conventional material – a fact with obvious economic and environmental benefits. Very high molecular mass samples are suitable for making parts of artificial joint replacements. When mixed with calcium hydroxyapatite and suitably processed, HDPE makes a satisfactory substitute for bone in restorative surgery. Furthermore, the long, straight, virtually defect-free form of poly(ethene) obtained using the Ziegler catalyst offers the possibility of producing superbly ordered molecular arrangements of polymer chains which allows the production of fibres and therefore ropes that compete with steel hawsers and have the advantage of being light and rust-free.

One new form of the material is a sort of **composite** rather like glass-reinforced plastic ('fibre glass'). It has fibres, in which the poly(ethene) molecules are highly aligned, embedded in a matrix of conventional poly(ethene). It has properties that make it suitable as protective shields – for delicate electronic components as well as the sensitive parts of people engaged in sports such as cricket and American football!

Giulio Natta who developed catalysts for polymerisation.

Poly(propene)

Another major development initiated by Ziegler's discovery took place in the laboratories of the Polytechnic Institute of Milan in a research group directed by Giulio Natta. What Natta's group did was change the monomer from ethene to propene ($CH_3CH{=}CH_2$). Propene polymerises to produce a material, poly(propene), with the same basic structure as poly(ethene) except that one of the hydrogen atoms on alternate carbon atoms in the chain is replaced by a methyl, $-CH_3$, group, Figure 10.

This introduces complications that do not exist for poly(ethene) in that the methyl groups can be arranged in different ways with respect to the carbon chain backbone. Natta named the polymer in which all the methyl groups are on one side of the extended chain the isotactic form; if the groups alternated from side to side he called it syndiotactic, and the form in which the groups were randomly distributed on either side he called atactic. Different catalysts for the polymerisation reaction gave different forms in predictable ways. This was very important from a practical point of view because the isotactic and syndiotactic forms, which are called stereoregular polymers, have much more useful properties than the atactic form. Stereoregular poly(propene) is a semicrystalline thermoplastic similar to poly(ethene). In this polymer the chains in the crystalline parts are not stretched out in a zig-zag way as in poly(ethene) but curl up as regular spirals.

Poly(propene) is the next most common polymer in use after poly(ethene), and is found in all sorts of applications from the fascias of cars to thermal underwear. Fittingly, Natta shared the 1963 Nobel Prize with Ziegler for his work using the Ziegler catalysts to 'stereoregulate' the polymerisation of propene – the Ziegler-Natta synthesis.

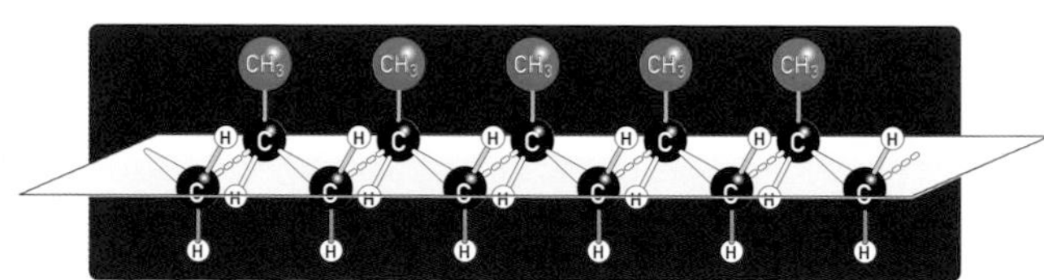

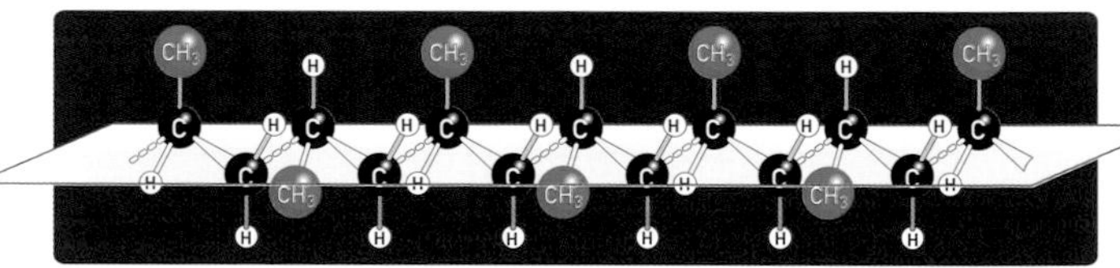

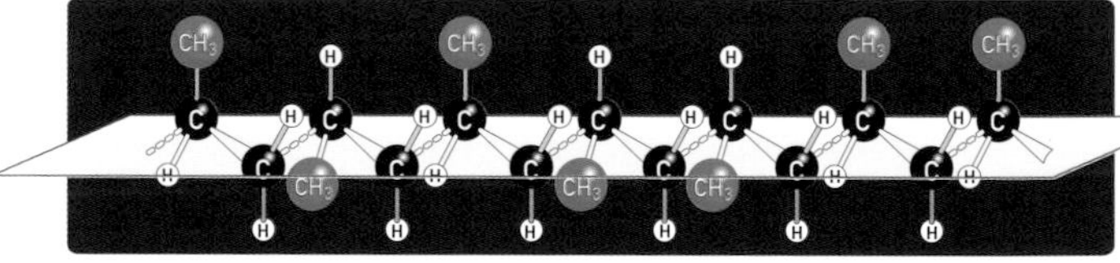

Figure 10. The three types of poly(propene). Isotactic and syndiotactic are described as stereoregular.

Car bumpers are made from polypropene.
Courtesy of DSM.

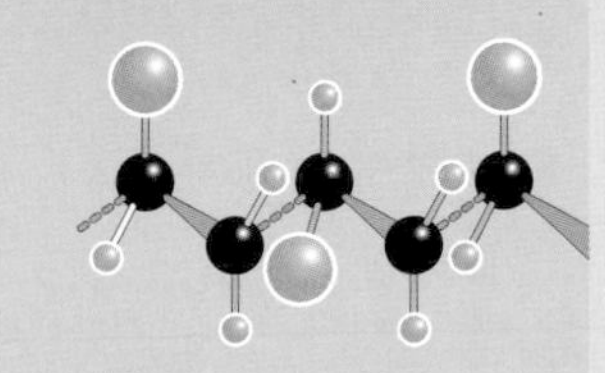

Ziegler's key patent dates from 1953, but the process of innovation did not stop there. In the intervening years there has been great progress in catalyst design, both homogeneous and heterogeneous which give improved control of polymer structure and properties – see Chapter 3 – *Chemical marriage brokers*. It seems reasonable to assume that the science and technology of making and processing this very variable and valuable material still has a long way to progress.

11. Explain the term 'homogeneous catalyst' and contrast it with a heterogeneous one. Give an advantage and a disadvantage of each type of catalyst.

Nylons

Nylons are polymers whose monomers are held together by amide linkages, -CONH-, made by the reaction of an amine group ($-NH_2$) with a carboxylic acid group (-COOH), Figure 11. Nylon is in fact a trade name which has come into general use.

The amide links are also found in nature in the vast and complex range of natural polypeptides and proteins.

$$-\overset{O}{\overset{\|}{C}}-O-H + H-\overset{H}{\overset{|}{N}}- \longrightarrow -\overset{O}{\overset{\|}{C}}-\overset{H}{\overset{|}{N}}- + H_2O$$

Figure 11. Formation of an amide linkage.

The first nylon was invented and developed by the Du Pont company in the US during the late 1930s. The scientist who laid the foundation of this part of polymer chemistry was Wallace Carothers. His nylon was made by the so-called salt dehydration method outlined in Figure 12. In this process, a diamine, $H_2N(CH_2)_6NH_2$, reacts with a dicarboxylic acid, $HOOC(CH_2)_4COOH$, to give a salt by transfer of hydrogen ions, H^+, to the base units, $-NH_2$. When this salt, $H_3N^+(CH_2)_6{}^+NH_3\ {}^-OOC(CH_2)_4COO^-$, is heated, it gives off water and forms amide bonds which hold the polymer together. This polymer is called nylon -6.6 because there are six carbon atoms in both original monomers.

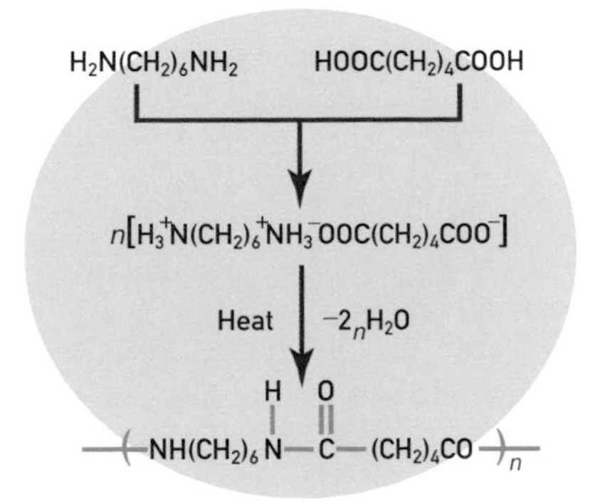

Figure 12. Salt dehydration method for making nylon-6.6.

12. Poly(ethene) and poly(propene) are called addition polymers whereas nylon is a condensation polymer. Explain the meanings of the terms addition polymer and condensation polymer.

Nylon is not a just a single material – there is a whole family of materials called nylons each of which is identified by numbers at the end of its name. Generally, in between the amide links are $-CH_2-$ units. The numbers represent the number of carbon atoms in each monomer (with the number in the diacid written first).

13. Draw the monomers required to make nylon-4.6.

Some nylons are slightly different as they are made from a single monomer with an $-NH_2$ group on one end and a -COOH group on the other so that both the acid group and the amine group are part of the same molecule. An example is nylon-6. This is made from the cyclic monomer called caprolactam in which the $-NH_2$ group has reacted with the -COOH group to form a ring. Polymerisation takes place when the ring first opens and a linear polymer is formed with five carbon atoms between each amide, Figure 13. Nylons such as -6.4 and -6.10 cannot be made by polymerisation from this type of monomer. This is because with a single monomer only, there must always be the same number of $-CH_2-$ groups between each amide group. Nylon made from caprolactam is called nylon-6, not nylon-6.6.

14. Write the stuctural formula of the linear compound from which caprolactam is derived.

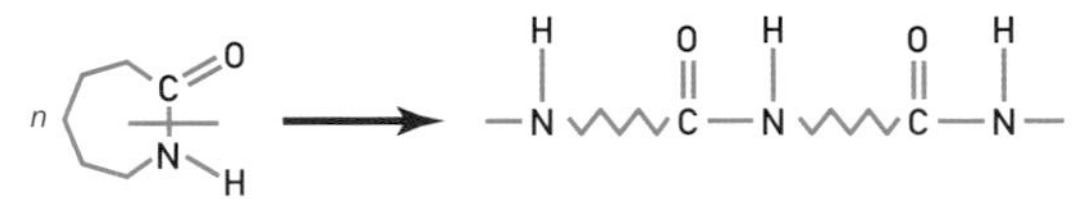

Figure 13. Nylon-6 is made from the cyclic monomer, caprolactam.

15. Draw the structural formulae of nylon-6.6 and nylon-6 showing at least three monomer units in each. How do the arrangements of the amide groups differ?

Different nylons have different properties. The longer the chains of $-CH_2-$ groups, the lower the nylon's melting temperature and the less it absorbs water.

Some uses of polyurethanes.

16. At first sight it may seem odd that *more* -CH_2- groups in the repeat unit lead to a *lower* melting point; usually higher molecular mass means higher melting point. Try to explain this apparent oddity using the hints below.

(a) What types of intermolecular forces operate between (i) -CH_2- groups; (ii) amide (-CONH-) groups?

(b) Which of these intermolecular forces is stronger?

17. Use the same approach as above to explain why more -CH_2- groups lead to lower water absorption.

As well as fibres, nylons are used in light engineering applications such as pram wheels, cams and gears and, when reinforced with fibre glass, for more demanding applications including bike wheels. They display low shrinkage in the mould so can be easily manufactured into objects with precise dimensions.

All the nylons-X, from nylon-2 to nylon-11, have been made, the most important commercially being nylons -3, -6 and -10. Of the nylons-XY, the -4.6, -6.6, -6.9, -6.10 and -6.12 versions have been developed for commercial use. Nylons-6 and -6.6 account for about 80% of all nylon manufacture at present and 25% of that goes into fibre applications.

There are many other polyamides including the structures shown in Figure 14 which contain rings of carbon atoms. The aromatic polyamides, the two structures at the bottom of Figure 14, are known as polyaramids. They give fibres of exceptional strength, crystallinity, and stability with melting points above 400 °C. Such materials find many specialist applications as fire and impact-resistant fabrics. They are extremely strong so are used as aerospace materials and in bullet-proof vests.

—NH—⬡—CH_2—⬡—$NHCO(CH_2)_{10}CO$—

—$NHCH_2$—⬡—$CH_2NHCO(CH_2)_6CO$—

—NH—⌬—NHCO—⌬—CO—

Nomex(DuPont); Conex (Teijin)

—NH—⌬—NHCO—⌬—CO—

Kevlar (DuPont); Twaron (Akzo)

Figure 14. Some more examples of polyamides. The two lower structures are called polyaramids – different manufacturers use different trade names.

Courtesy of DuPont.

Glossary

Crosslink: A covalently bonded link between two polymer chains.

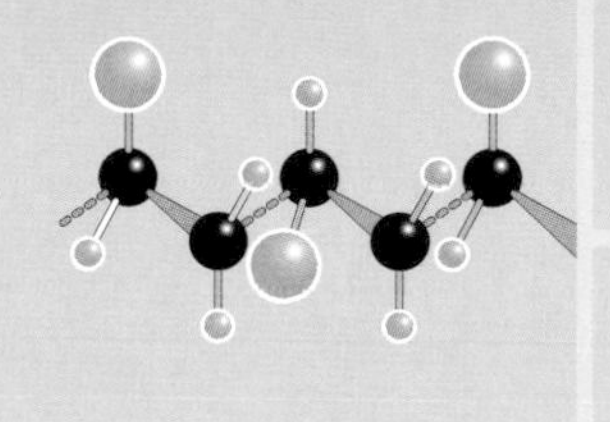

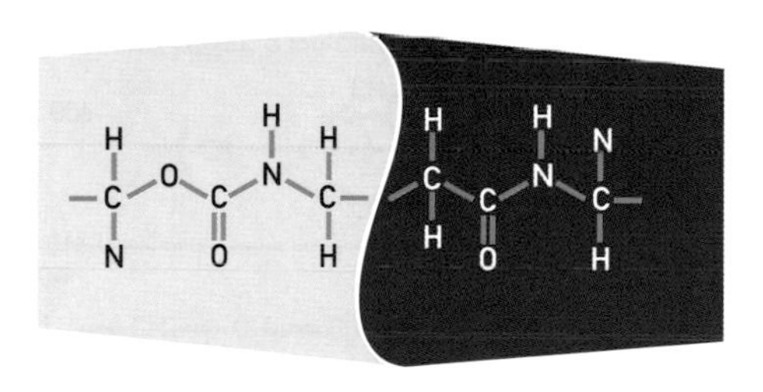

Figure 15. A comparison of the linking units of polyurethane (left) and nylon.

Polyurethanes

Related to the nylons are the polyurethanes which are said to have been invented in an attempt to get round the nylon patents. This seems plausible when you compare the linking units of the two polymers, Figure 15.

One method of the many invented for the synthesis of polyurethanes involves the addition of alcohol units, $-CH_2OH$, to isocyanates, $-CH_2N=C=O$. Isocyanates are very reactive and also very toxic. So, although the chemistry works very well, it has to be carried out under very well-controlled conditions.

$$O{=}C{=}N{-}(CH_2)_x{-}N{=}C{=}O + HO(CH_2)_yOH \longrightarrow \left[\overset{O}{\overset{\|}{C}} - \overset{H}{\overset{|}{N}}(CH_2)_x\overset{H}{\overset{|}{N}} - \overset{O}{\overset{\|}{C}} - O(CH_2)_yO \right]_n$$

The polyurethanes are softer and lower-melting than comparable nylons and find application in encapsulation, coatings and elastomers (stretch fibres such as Lycra, used in swimwear and tights). As well as the formation of linear polymers described above, it is possible to produce **crosslinks** between the chains by adding a little triol or tri-isocyanate. This leads to a three-dimensional polymer network. If a gas (carbon dioxide or an inert gas) is released or injected during polymerisation, this leads to an expanded honeycomb structure that is relatively stable and produces a foam. The type of foam – rigid or soft – is determined by the nature of the polymer network and the size of the holes in the foam. Such foams are used in applications varying from the rigid interiors of doors and surf-boards to cavity wall insulation and elastic foams for furniture.

Answers

1. A vast number of shapes (called conformations) is possible even with such a short section of chain.

2. (a) Van der Waals forces.
   (b) Roughly 100 times weaker.
   (c) The strength of the material is due to the fact that to break it, vast numbers of intermolecular interactions must be broken.

3. M_r of $-CH_2O-$ is 30.
   The number of moles of chains in the sample is half the number of moles of end groups.
   The average M_r of a polymer chain is the mass of the sample in grams divided by the number of moles of chains.
   The average M_r of the sample divided by 30 gives the average chain length (*ie* the number of $-CH_2O-$) units in a chain.

4.

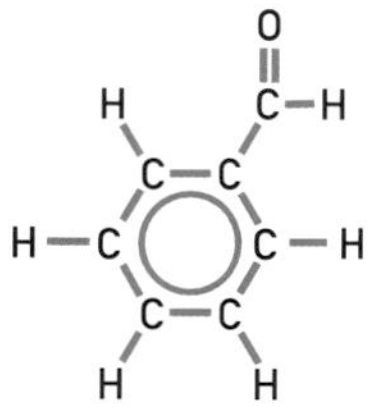

5. A batch process is one in which the product is made in a series of fixed quantities rather than in a continuous flow.

6. $C_2H_5OH \rightarrow H_2C{=}CH_2 + H_2O$.

7. Crude oil is first distilled, then some of the fractions are cracked. Ethene is one of the products of cracking.

8. (a) The electron density in the π-orbitals is further away from the space between the nuclei than it is in a σ-orbital, therefore it is less effective at holding the nuclei together.
   (b) The energy of the π-bond is 265 kJ mol^{-1}.
   (c) 'Homolytically' means that the bond breaks so that one electron of the bonding pair goes to each of the atoms involved in the bond.

9. Carbon dioxide or carbon monoxide (depending on the amount of oxygen), water and aluminium oxide.

10. The more-linear chains are able to pack together more efficiently.

11. Homogeneous catalysts are in the same phase as the reactants (*eg* both liquids or both gases), heterogeneous catalysts are in a different phase to the reactants (usually solid catalyst and liquid or gaseous reactants). Homogeneous catalysts are more effective in that all the catalyst is exposed to the reactants but they can be difficult to remove from the products for re-use. Heterogeneous catalysts have only the atoms or molecules on the surface exposed to the reactants but can usually be easily removed from the products by filtration.

12. In addition polymerisation, the monomers just add together, whereas in condensation polymerisation, a small molecule such as water is eliminated during the polymerisation reaction. No such molecule is eliminated in addition polymerisation. So addition polymers have exactly the same empirical formula as the monomer from which they are made.

13. $HOOC(CH_2)_2COOH$, and $NH_2(CH_2)_6NH_2$.

14. $NH_2(CH_2)_5COOH$.

15.

Nylon — 6

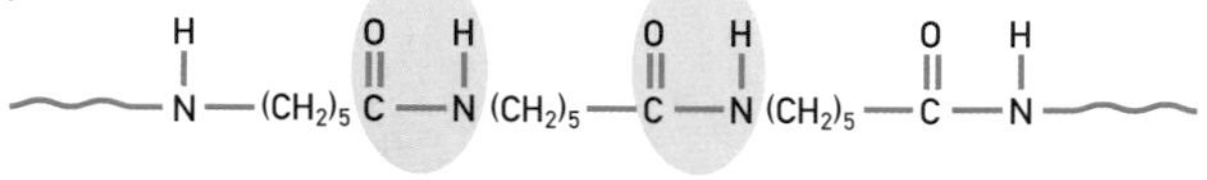

Nylon — 6.6

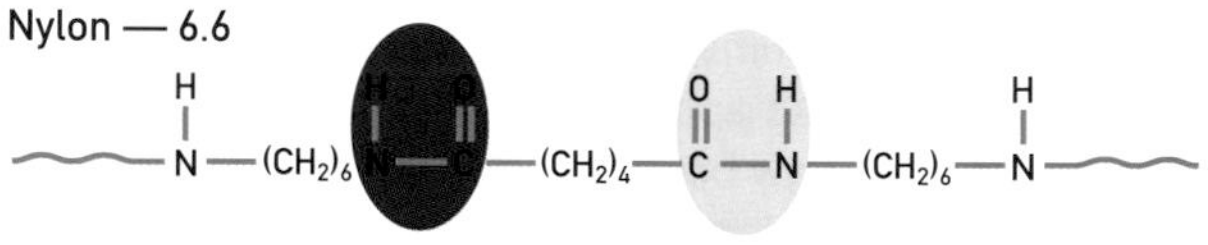

In nylon-6, all the amide groups are the same way round. In nylon-6.6, alternate amide groups point in opposite directions.

16. (a) (i) Van der Waals forces, (ii) hydrogen bonds.

    (b) Hydrogen bonds.

    So, the lower the proportion of hydrogen bonding groups to van der Waals bonding groups, the easier the molecules are to separate and the lower the melting point.

17. Hydrogen bonding groups can bond with water molecules, van der Waals bonding groups cannot. So the lower the proportion of hydrogen bonding groups to van der Waals bonding groups, the lower the water absorption.

18. The molecules of

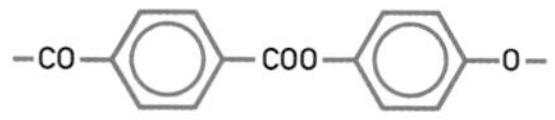

    are more linear and will pack together more closely than those of

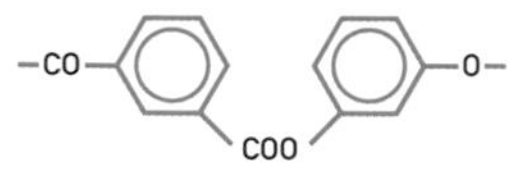

    making them harder to separate.

Electrochemistry is concerned with exploiting reactions in which electrons are transferred between chemical species, often from an electrode into a solution.

Chemical bonding is all about electrons, so it is not surprising that electron transfer reactions (also called redox reactions – see Box – *Redox reactions*) are important. Electrochemists are studying them for several reasons some of which are:

- to help them understand basic processes in chemistry and biochemistry;
- to generate electrical energy;
- to store electrical energy; and
- to provide more sensitive methods of chemical analysis.

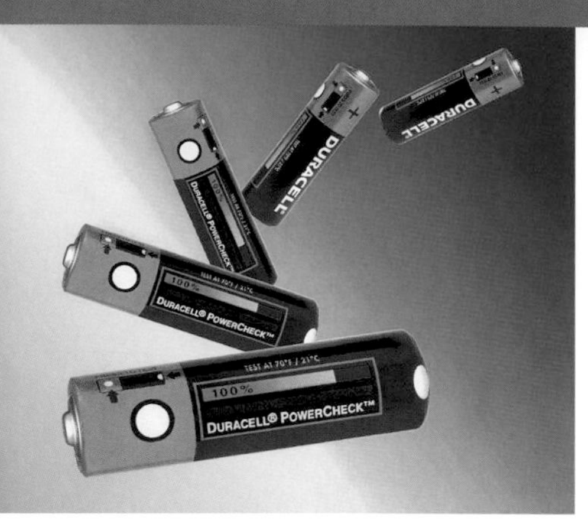

'Stick' batteries.
Courtesy of Duracell®.

Some specific examples include:

- **the chemistry of rusting and corrosion;**
- **the chemistry of extracting metals from their ores;**
- **the chemical processes that drive living cells;**
- **designing better types of batteries for everything from mobile phones to electric cars;**
- **cleaning contaminated soil; and**
- **measuring blood sugar levels in diabetics.**

In most cases, an electrochemical cell exists. This consists of a solution containing ions into which dip two **electrodes**. These carry the electric current (a flow of electrons) into and out of the cell.

Three aspects of electrochemistry are:

- **The use of chemical reactions to generate an electric current, as in batteries and fuel cells. Here a chemical reaction at one electrode produces electrons which flow through the external circuit doing useful work (and bring about a different reaction at the other electrode).**
- **The use of an electric current from an external source to bring about chemical reactions, as in electroplating or the extraction of aluminium from its ore by electrolysis. Here a difference in electrical potential from an external source (a potential difference, or more loosely, a voltage) drives ions from one electrode to the other, and also drives reactions at the electrodes.**
- **The use of electrodes as analytical probes to investigate electron transfer processes and/or to measure concentrations of substances.**

Courtesy of Electric Vehicle Association of Canada.

Courtesy of Swatch AG.

Electrode: An electrically conducting material through which electrons enter or leave an electrochemical cell and where electrochemical reactions take place.

Work: Work is done when energy is changed from one form into another, for example electrical energy to kinetic (movement) energy in a motor.

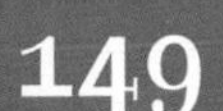

Redox reactions

Redox is short for reduction-oxidation. These types of reaction were originally defined in terms of addition and removal of oxygen but we now understand that they are essentially electron transfer reactions. **O**xidation **I**s **L**oss of electrons and **R**eduction **I**s **G**ain of electrons – easily remembered by the phrase **OIL RIG**.

1. In the reaction

$$2Mg(s) + O_2(g) \rightarrow 2MgO(s)$$

magnesium is oxidised as it gains oxygen. Write the equation ionically and explain why it is consistent with the 'OIL RIG' definition of oxidation. What has been reduced in the reaction?

Redox potentials

The ability to donate or to accept electrons can be described in terms of a so-called redox potential. A substance with a high (positive) redox potential has a strong tendency to gain electrons while a substance with a low (negative) redox potential has a strong tendency to lose them. For instance, the element fluorine has a very positive redox potential and thus readily gains electrons from other materials.

$$F_2(g) + 2e^- \rightarrow 2F^-(aq) + 2.87\ V$$

Lithium has a high negative potential and can therefore donate electrons to other materials.

$$Li(s) \rightarrow Li^+(aq) + e^- \quad -3.03\ V$$

Standard redox potentials, $E^\ominus$, are measured in volts relative to the potential of hydrogen (strictly a platinum electrode surrounded by hydrogen, at a pressure of 100 kPa and a temperature of 298 K, dipping into a 1 mol dm^{-3} solution of hydrochloric acid).

A list of selected redox potentials is given below. By convention, they are written as reversible reactions with the electron(s) on the left. The most negative are listed at the top:

Reaction	$E^\ominus$
$Li^+(aq) + e^- \rightleftharpoons Li(s)$	-3.03 V
$Na^+(aq) + e^- \rightleftharpoons Na(s)$	-2.71 V
$Zn^{2+}(aq) + 2e^- \rightleftharpoons Zn(s)$	-0.76 V
$2H^+(aq) + 2e^- \rightleftharpoons H_2(g)$	0.00 V
$Cu^{2+}(aq) + 2e^- \rightleftharpoons Cu(s)$	+0.34 V
$Ag^+(aq) + e^- \rightleftharpoons Ag(s)$	+0.80 V
$F_2(g) + 2e^- \rightleftharpoons 2F^-(aq)$	+2.87 V

This list can be a useful aid to predicting reactions as described in the main text.

2. An early battery, called a voltaic pile, consisted of cells made from strips of zinc separated from strips of silver by layers of paper soaked in salt water. Use the list of redox potentials to predict which way the electrons flow in this cell and what voltage would be produced between each pair of metals. Suggest the reactions which took place at the electrodes in this battery.

A voltaic pile given by Alessandro Volta to Michael Faraday in 1814.
The Royal Institution, London, UK
Bridgeman. Art Library London/New York.

Using electrochemistry to generate electricity – electrochemical cells

There are two types of electrochemical cell – primary and secondary. Primary cells are 'use once and throw away'. Their energy comes from the conversion of chemical energy stored within each cell; once this is spent the cell cannot be re-used. Secondary cells are rechargeable; once discharged, their chemical reactions can be reversed by applying an external source of electricity.

Did you know?

In 1780, in one of the earliest electrical experiments, Luigi Galvani discovered that placing two different metals in contact with frog muscle produced an electric current. Electricity became associated with the 'life force' which was thought to distinguish living things from dead ones. Around this time, at least two serious experiments were made to try to revive dead people using electricity. One of the subjects had been hanged and the other guillotined!

Luigi Galvani 1737 – 1798.
Reprinted by permission of McGraw-Hill, FJ Moore, A History of Chemistry.

Primary cells

An early example of a primary cell is the Daniell cell. It consists of a zinc electrode immersed in a zinc sulfate solution and a copper electrode immersed in copper sulfate solution, with the two solutions (called **electrolytes**) separated by a porous partition, Figure 1.

We can use the list of redox potentials in the Box – *Redox reactions* to see how the cell works by using the so-called anticlockwise rule. Write the half reactions for each electrode in the order they appear in the list, Figure 2. When the electrodes are connected, the upper half reaction will go from right to left, releasing electrons while the lower will accept electrons and go from left to right – hence the term anticlockwise.

If we add the two half reactions (after multiplying one or both to balance the electrons, if necessary) we get the overall reaction. In the case of the Daniell cell, this is:

$$Cu^{2+}(aq) + Zn(s) \rightarrow Cu(s) + Zn^{2+}(aq)$$

The difference in redox potentials gives the voltage of the cell (1.1 V in this case) and the electron flow will be from the more negative to the more positive (zinc to copper in this case).

A useful rule of thumb is that if the difference in potentials, $E^{\ominus}_{reaction}$, is greater than +0.4 V, then the reaction will go to completion, provided it takes place fast enough.

The principles of this rule can be applied to all cells.

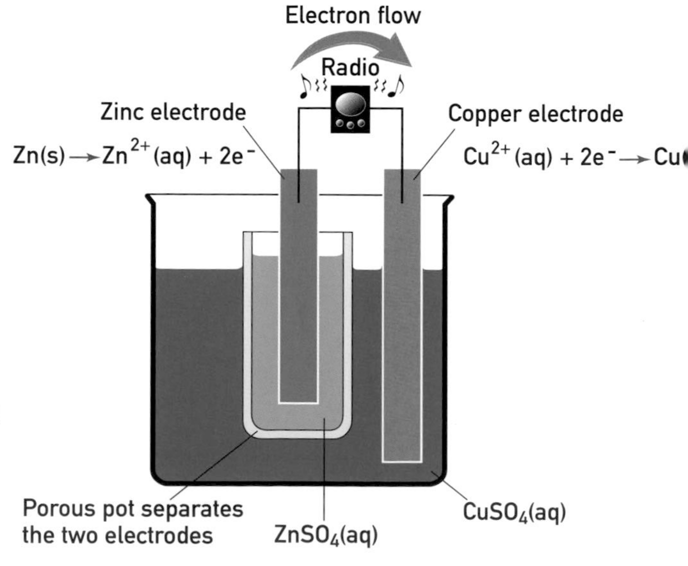

Figure 1. The Daniell cell.

Glossary

Electrolyte: A pure liquid or solution containing ions which will conduct electricity by the movement of these ions.

Courtesy of Sharp Electronics (UK) Ltd.

Courtesy of National Dairy Council.

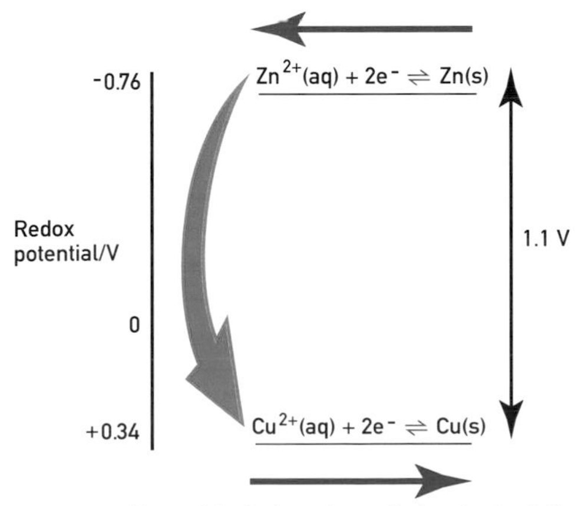

The anticlockwise rule applied to the Daniell cell.
The half reactions which occur:

$$Zn(s) \longrightarrow Zn^{2+}(aq) + 2e^-$$
$$Cu^{2+} + 2e^- \longrightarrow Cu(s)$$
$$Zn(s) + Cu^{2+}(aq) \longrightarrow Zn^{2+}(aq) + Cu(s)$$

Figure 2. The anticlockwise rule.

Secondary cells (rechargeable batteries)

The Daniell cell is a primary or single-use cell: once all the zinc has dissolved (or all the copper ions are used up), it cannot be re-used. For many applications – car starter motors, mobile phones, laptop computers – we need rechargeable or 'secondary' batteries. One of the mainstays of the rechargeable battery industry is the lead-acid battery, which is still used to power electric vehicles such as milk floats or starter motors in conventional vehicles. It was first discovered by Gaston Planté in 1860 and consists of two plates, one of lead and the other of lead coated with lead(IV) oxide, dipping into a solution of sulfuric acid ($2H^+(aq) + SO_4^{2-}(aq)$). When the battery is being used, or discharged, lead metal is oxidised at one electrode to form insoluble lead(II) sulfate, releasing two electrons per atom of lead.

$$Pb(s) + SO_4^{2-}(aq) \rightarrow PbSO_4(s) + 2e^-$$

These electrons travel around the external circuit to the other electrode where lead(IV) dioxide is reduced to lead(II) sulfate.

$$PbO_2(s) + 4H^+(aq) + SO_4^{2-}(aq) + 2e^- \rightarrow PbSO_4(s) + 2H_2O(l)$$

3. Write the overall reaction which occurs on discharge of a lead-acid battery.

When the battery is recharged, the process is reversed. The lead-acid cell can produce short bursts of high current and is still used a great deal. However, its relatively high mass compared with the amount of energy stored in the battery is a disadvantage. This has prompted research into batteries with higher 'energy densities', that is ones which store a lot of energy in a small mass. Some of the candidates include lithium batteries, nickel-metal hydride, sodium-sulfur and ZEBRA (Zero Emission Battery Research Activities) batteries. However, to compete with the lead-acid battery, their performances and costs need to improve significantly.

Sony lithium-ion battery module.
Courtesy of Electric Vehicle Association of Canada.

Courtesy of Energizer®.

Modern applications of electrochemistry

Lithium batteries

Lithium-based technology has been extensively researched, with the result that both non-rechargeable and rechargeable lithium batteries are now available. Their advantage is that, theoretically, they can store a lot of electrical charge per kilogram of battery – an energy density nearly 300 amp hours per kilogram compared with less than 90 amp hours per kilogram for a lead-acid cell (based on the materials used in the cell reaction only). (One amp hour refers to a battery supplying a current of 1 amp for one hour.)

4. If a battery has a capacity of 3000 amp hours and supplies electricity at 1.5 V, how much energy in joules does it store?

Hint: Electrical energy is given by the expression $E = VIt$ where E is the energy, V the potential difference (voltage), I the current and t the time for which it flows.

There are two reasons for this improvement: first, lithium metal has a very low density; secondly, the redox potential of lithium, see Box – *Redox reactions* is much more negative than that of other materials used in battery technology so the voltage of the cell tends to be greater. Unfortunately this second characteristic also creates a major snag. Lithium's high reactivity means that it is readily oxidised, causing oxide films to form over its surface which partially insulate the metal from the electrolyte.

Researchers have, however, overcome this problem by developing a clever idea – insertion electrodes which can accommodate the small lithium ion within their lattices. This removes the need to use free lithium metal. Two kinds of electrodes are used: one made of a metal compound such as manganese dioxide, and the other of carbon, often graphite. When the cell discharges, lithium ions from the electrolyte insert into the manganese dioxide.

$$x\mathrm{Li^+(electrolyte)} + \mathrm{MnO_2(s)} + xe^- \rightarrow \mathrm{Li}_x\mathrm{MnO_2(s)}$$

While at the other electrode a lithium ion is extracted from the carbon matrix.

$$x\mathrm{Li(electrode)} \rightarrow x\mathrm{Li^+(electrolyte)} + xe^-$$

The electrolyte is polymer-based rather than an aqueous solution, Figure 3.

Reversing the two processes recharges the cell. Such a battery can generate approximately 4 V – the voltage of nearly three normal 'stick' batteries connected in series. This type of cell is known as a lithium-ion cell. It is used in newer laptop computers and mobile phones.

Other (non-rechargeable) types of lithium cells have the same voltage as the average stick battery (1.5 V) and are available as long-life replacements. They are more environmentally-friendly because they do not contain any heavy metals and are currently used where lightness is important. However, if lithium technology is to be used in electric cars, it will need a lot more development.

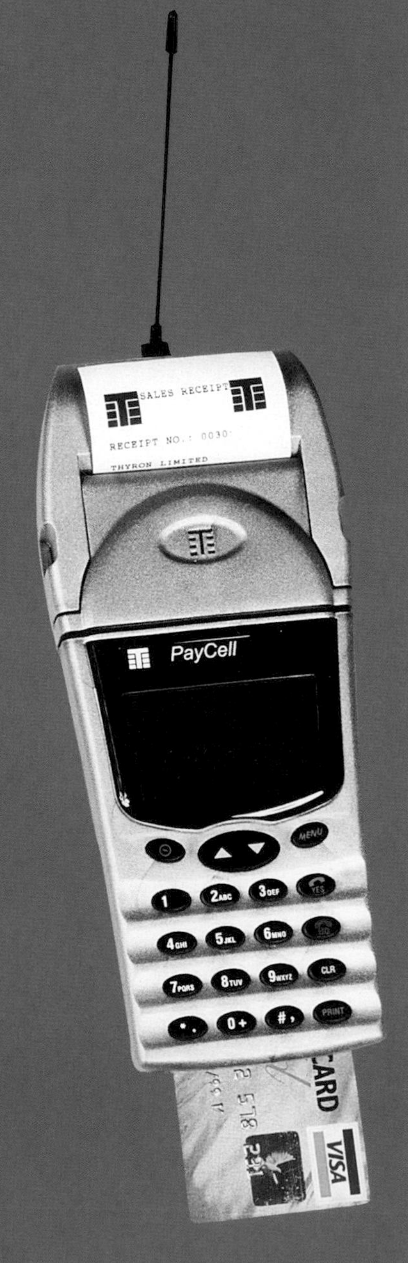

Courtesy of the Vodafone Group.

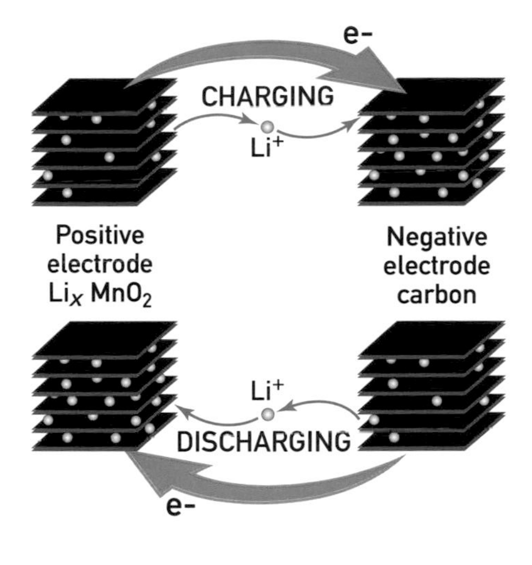

Figure 3. The rechargeable lithium battery. On charging, Li^+ ions move from the Li_xMnO_2 electrode to the carbon one. At the same time, electrons move through the external circuit. The reverse happens on discharge.

Batteries for electric cars

For vehicle applications, the main candidates are high-performance secondary batteries – nickel metal-hydride, sodium-sulfur or ZEBRA batteries.

Only nickel-metal hydride batteries are based on an aqueous electrolyte, while sodium-sulfur and ZEBRA batteries are high-temperature batteries (working at around 570 K).

Nickel-metal hydride batteries have two distinctly different electrodes. One electrode uses alloys of rare-earth metals like lanthanum that can 'soak up' hydrogen atoms. One example is $LaNi_5$. This soaks up hydrogen atoms to give the compound $LaNi_5H$ which then reacts as follows:

$$LaNi_5H(s) + OH^-(aq) \rightarrow LaNi_5(s) + H_2O(l) + e^- \text{ (anode)}$$

The other electrode is based on nickel alone.

$$NiOOH(s) + H_2O(l) + e^- \rightarrow Ni(OH)_2\ (s) + OH^-\ (aq) \text{ (cathode)}$$

To complete the cell, hydroxide ions (OH^-) migrate from the cathode, where they are generated, to the anode, where they are consumed. The cell operates with a potential of 1.35 V. Unfortunately the high cost of the nickel and the rare-earth alloy electrodes make the cell relatively expensive initially. However, these cells can be recharged many times.

Both sodium-sulfur and ZEBRA cells have a molten sodium metal electrode surrounded by a solid ceramic (aluminium oxide) through which only sodium ions can pass. The electrochemical reaction is:

$$2Na \rightarrow 2Na^+ + 2e^-$$

The difference between the sodium-sulfur battery and the ZEBRA battery lies in the materials used for the other electrode. The sodium-sulfur battery uses liquid sulfur (in a separate compartment).

$$xS + 2e^- \rightarrow S_x^{2-}\text{(electrolyte)}$$

The ZEBRA battery relies on a molten salt, sodium aluminium chloride ($NaAlCl_4$), as the electrolyte. This is heated above its melting point to become an ionically conducting liquid. Nickel chloride is dissolved in this electrolyte. The electrode reaction is:

$$NiCl_2(s) + 2e^- \rightarrow Ni(s) + 2Cl^- \text{ (electrolyte)}$$

Because both these batteries contain very hot, potentially dangerous materials, the safety implications are obvious. For this reason the ZEBRA battery has become the more popular of the two; if the ceramic aluminium oxide case is broken, then the reaction between the constituents takes place.

$$3Na(l) + NaAlCl_4(l) \rightarrow Al(s) + 4NaCl(s)$$

The solid products passivate the break and do not cause any serious problems. However, both cells have to be temperature-controlled during operation.

Panasonic nickel metal hydride HEV battery module.
Courtesy of Electric Vehicle Association of Canada.

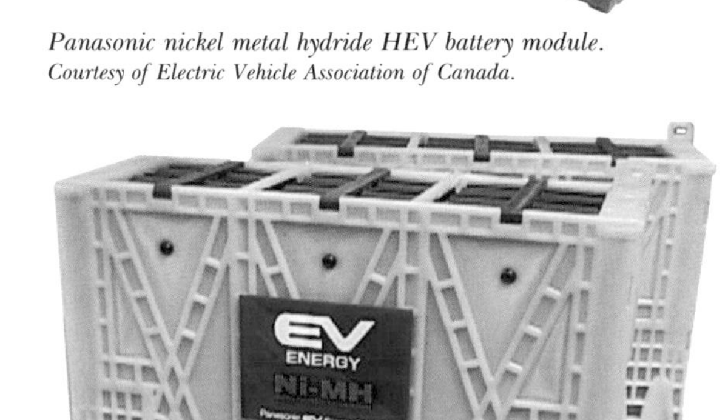

Panasonic-Toyota Nickel-metal hydride holder.
Courtesy of Electric Vehicle Association of Canada.

ZEBRA battery-powered bus.
Courtesy of Electric Vehicle Association of Canada.

The promise of fuel cells

In batteries, the chemical energy is stored inside the battery – within the materials of the electrodes and electrolyte. There is, however, another method of generating electrical power – by feeding the cell with a continuous supply of chemicals which react at the electrodes to produce electrical energy. This is called a fuel cell. Currently, car companies are very interested in them because they could be used to power environmentally friendly vehicles. Fuel and an **oxidant** are fed into the cell in the same way as petrol and air are fed into an internal combustion engine. However, in a fuel cell the oxidant and the fuel do not react directly to give heat but are electrochemically 'burnt' to generate electricity.

In the simplest case, hydrogen and oxygen gases are converted to water in two processes at the electrodes. These electrodes usually contain platinum or a related catalyst to increase the rate of the overall process, and are separated by an ion-exchange **membrane** which allows only certain ions through to complete the circuit. This particular type of fuel cell, known as a proton-exchange membrane, or PEM, fuel cell has a membrane that allows hydrogen ions (protons) to travel between the two electrodes while keeping the gases apart. Figure 4 shows how the system operates. Hydrogen and oxygen gases are fed to the two gas-permeable platinum electrodes. At the hydrogen electrode, protons and electrons are produced, while at the oxygen electrode, oxygen reacts with protons and electrons, to produce water.

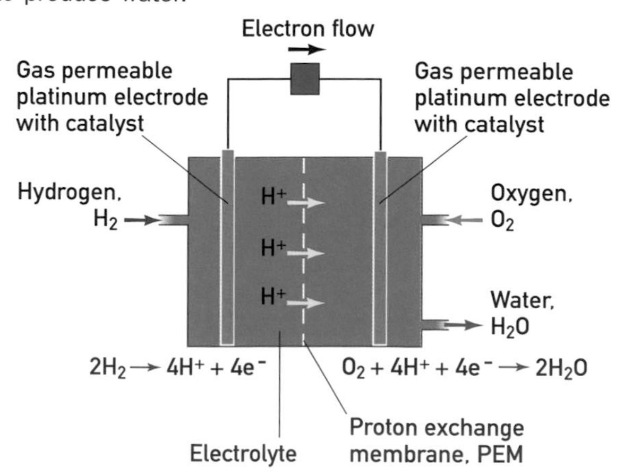

Figure 4. The PEM fuel cell.

The half reactions are:

$$2H_2(g) \rightarrow 4H^+(aq) + 4e^-$$

$$O_2(g) + 4H^+(aq) + 4e^- \rightarrow 2H_2O(l)$$

The overall reaction is:

$$2H_2(g) + O_2(g) \rightarrow 2H_2O(l)$$

The electrons travel around the circuit and do work. To complete the circuit within the cell, protons migrate from the hydrogen to the oxygen electrode. The overall reaction produces electrical energy, water and some heat.

Fuel cells of this type are ideal for vehicles aiming at zero emissions: water vapour is the only exhaust gas. However, in fairness, the whole energy cycle has to be considered. The source of hydrogen and oxygen may well have been **electrolysis** of water, using electricity that probably comes from power stations burning **fossil fuels**. Nevertheless, this clean technology is ideally suited for urban areas, and various companies have developed prototype electric buses running on fuel cell power. The only other drawback lies in the storage and transportation of liquid hydrogen. The danger of explosion must be considered, although hydrogen is perhaps less dangerous than a liquid hydrocarbon fuel. This is because it is a gas which is less dense than air so it tends to disperse rapidly.

5. Suggest a source of (a) hydrogen, (b) oxygen other than the electrolysis of water. What environmental problems might each source cause?

What emissions come from a power station burning a fossil fuel such as natural gas (methane)?

6. What advantages would a PEM fuel cell have if used on the Space Shuttle?

Glossary

Electrolysis: The breakdown of an ionically conducting liquid brought about by electricity passing through it.

Fossil fuels: Hydrocarbon fuels such as coal, oil and natural gas, which were formed from the remains of living organisms.

Membrane: A sheet of material that separates two liquids or a liquid and a gas.

Oxidant: (Also called an oxidising agent.) A sustance that oxidises other species by removing electrons from them.

To avoid the problems of transport and storage of hydrogen, chemists are working on two alternative fuel cells. One uses methanol as the fuel (called a Direct Methanol Fuel Cell, DMFC) and the other uses a conventional hydrocarbon fuel which is first converted into hydrogen and carbon dioxide before the hydrogen is 'burnt' in the cell (called a reforming fuel cell). Both cells need further development before they will be of practical use. Problems include cost, efficiency and poisoning of the catalysts, warm up times and operating temperatures and pressures. Fuel cells and batteries are clear candidates for electrical vehicles, and perhaps a combination power system will eventually be developed. Such a hybrid would have the instant power of a battery while the fuel cell warms up, and the long range of the fuel cell where large amounts of fuel can be stored without it weighing too much. It is interesting to note that the world's first production-line electric car, the General Motors EV1, still uses a lead-acid battery system. The battery has a mass of about 540 kilograms. This mass of fossil fuel, would drive the car for thousands of kilometres, while the battery has a range of only about 110 km (70 miles). Clearly, a substantial improvement is needed.

Ballard fuel cell.
Courtesy of Electric Vehicle Association of Canada.

7. What gas is likely to be produced by DMFC and reforming fuel cells? What environmental problem does it cause?

8. This question is about the range of a petrol driven car compared to the General Motors EV1 electric car whose battery weighs 540 kg.

The density of petrol is about 0.77 kg dm^{-3}.

(a) How many dm^3 of petrol would have a mass of 540 kg?

(b) The petrol tank of a typical family car holds about 100 dm^3. How many times could you fill up the tank with 540 kg of petrol?

(c) A car's petrol consumption might be 10 km dm^{-3}. How far would such a car travel on 540 kg of petrol?

(d) How does the range of the car compare with the range of the General Motors EV1?

An electric car.

Harnessing solar energy

A third way of generating electricity chemically is via a photovoltaic cell, which converts sunlight into electricity. In one form, developed at the Swiss Federal Institute of Technology by Michael Grätzel and colleagues, a coloured dye is used which is **adsorbed** onto a titanium dioxide electrode placed in a solution containing iodide ions. The cell works as follows.

- **Light is absorbed by the dye and excites electrons in the dye molecules to higher levels.**
- **Some of these electrons are transferred to the electrode and oxidise the dye (oxidation is loss of electrons).**
- **The iodide ions reduce the dye back to its original form, themselves being oxidised to iodine in the process.**
- **The electrons flow from the titanium dioxide electrode to a second electrode (doing work in the process). Here they are reduced back to iodide ions.**

This type of solar cell is quite efficient (converting 10% of the energy of the sunlight falling onto it into electricity) and relatively cheap. It looks likely to be competitive with existing types of solar cell.

Michael Grätzel in his laboratory.

9. In the photovoltaic cell described, write equations for: (a) Loss of an electron by a dye molecule (use D to represent the dye). (b) Reduction of the oxidised dye by iodide ions. (c) Reduction of the iodine molecules back to iodide ions at the electrode.

Glossary

Adsorption: The formation of a layer of atoms or molecules of a gas or liquid on a (usually) solid surface by the formation of weak bonds (hydrogen bonds, van der Waals bonds or dipole-dipole bonds) between the surface and the adsorbed particle.

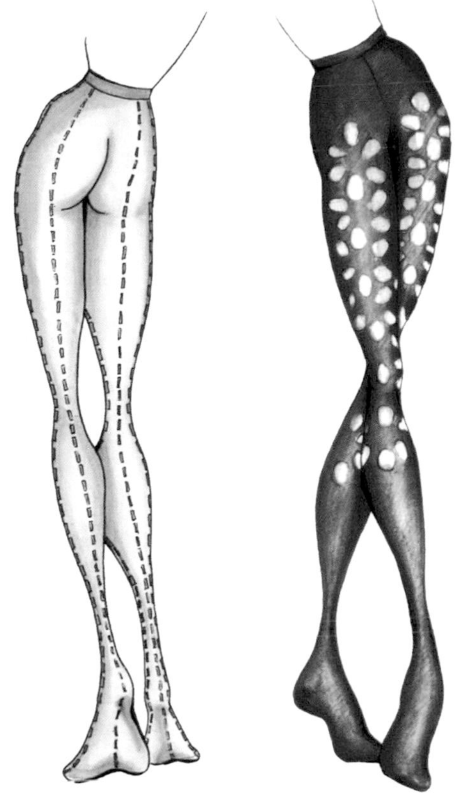

Courtesy of DuPont.

Electrochemistry in industry

Electrochemical processes account for a staggering 5% to 10% of an industrialised nation's electricity consumption, most of it being used to extract aluminium from its ore, and for the chlor-alkali industry which produces chlorine, sodium hydroxide and hydrogen from brine.

A less-familiar electrochemical process is that by which hexanedinitrile (adiponitrile) is produced from propenenitrile (acrylonitrile). Hexanedinitrile is one of the starting materials for the manufacture of nylon, and 270 000 tonnes of it is produced each year worldwide. The process is as follows:

$$2CH_2{=}CHCN + 2H_2O + 2e^- \rightarrow NCCH_2CH_2CH_2CH_2CN + 2OH^-$$

propenenitrile — hexanedinitrile

10. In the manufacture of adiponitrile, is the acrylonitrile being oxidised or reduced? Explain your reasoning.

Nylon has many uses ranging from tights to hot air balloons.

Soil remediation – treating 'brown land'

There are many potential building sites in the UK (and elsewhere) where the soil is polluted because of a previous use of the site – *eg* as a gasworks or a chemical factory. This is because in the past chemical companies did not operate under such strict legislation about waste emissions as they do today. Such sites are sometimes called 'brown land' and are unsuitable for many uses, such as building a school or for housing, until the pollution has been removed.

Electrochemistry can be used to remove some of the pollutants. The process works like a battery in reverse as in Figure 5. Electrical energy, supplied from an external power source, is passed through a cell which in this case is the contaminated soil. If the soil contains, for example, cadmium(II), a heavy metal poison often found on the sites of old gasworks, then the positive cadmium ions (Cd^{2+}) move towards the negative electrode where they can be removed, while negative ions move towards the positive electrode.

In practice, the system operates with several pairs of electrodes which are covered with appropriate ion exchange membranes and placed in the ground. An acid electrolyte is then circulated through the membrane compartment. An electrical potential is applied to the two electrodes and this breaks down water into hydrogen and oxygen at the electrode surfaces.

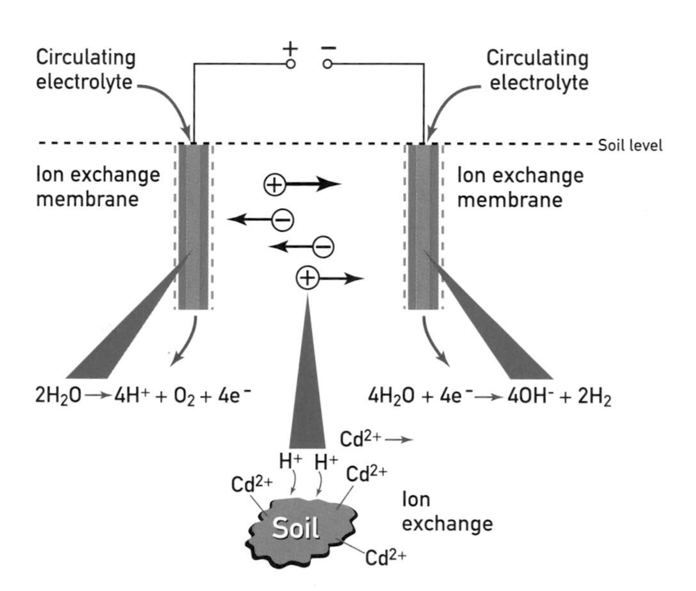

Figure 5. Removing cadmium ions from polluted soil.

At the cathode:

$$2H_2O(l) + 2e^- \rightarrow 2OH^-(aq) + H_2(g)$$

At the anode:

$$2H_2O(l) \rightarrow O_2(g) + 4H^+(aq) + 4e^-$$

As the hydrogen ions (H^+) and hydroxide ions (OH^-), which are also produced, pass through the soil towards opposite electrodes, the hydrogen ions exchange with metal cations (such as Cd^{2+}) within the soil's structure, so releasing the pollutant ions from the soil (each Cd^{2+} is replaced by two H^+ ions to maintain the balance of charges). The pollutant ions travel to the negative electrode where they pass through the membrane, dissolve in the circulating electrolyte and are removed.

This technique can be used to remove pollutants like heavy metal **cations** (copper, cadmium, iron, zinc and so on), nitrate from agricultural fertilisers, cyanide and radioactive isotopes. As well as ions, it can remove uncharged materials such as toxic organic compounds by a process called electro-osmosis whereby water molecules move from the anode to the cathode and take some of the organic toxins with them.

The economic feasibility of this process depends strongly on the local cost of electricity. Removing cadmium ions from the ground below the area of a football pitch would consume around 300 gigajoules (equivalent to the energy used by 10 one kilowatt electric fires left on for a year). This cost has to be compared with the value of the reclaimed land and the cost of cleaning the land using other techniques. Also significant is that the electrochemical process can, in most cases, be performed on site rather than having to remove the soil for processing.

Glossary

Cations: Positive ions – so-called because they move towards the cathode (the negative electrode) during electrolysis.

Displayed formula: The molecular formula of a compound written so that every atom and every bond are shown.

Interface: The boundary between two different phases of a system, for example between a solid electrode and a liquid electrolyte.

Orbital: A volume of space around an atomic nucleus where there is a high probability of finding an electron or electrons.

Electrochemistry in analysis

Scanning electrochemical microscopy

Scanning electrochemical microscopy or SECM is a relatively new technique that can investigate the **interfaces** between solids and liquids, liquids and liquids, or liquids and gases. It can be applied to study the surface of teeth and corrosion of metals, for example. It allows researchers to determine the shape of the surface as well as the chemical activity of regions on the surface.

One application of SECM is to study dentine on the surface of teeth. Figure 6 shows a hole, or pore, in the dentine. This work may enable treatments to be found for the problem of sensitive teeth.

Scanning electrochemical microscopy relies on a tiny electrode (typically a disc of a conducting material less than 50 micrometres across embedded in an insulator) known as a microelectrode or ultramicroelectrode, Figure 7. This is placed close to the surface to be examined, the two being separated by an electrolyte. The microelectrode is then moved back and forth across the surface. The microelectrode is held at a certain electrical potential that can oxidise (or reduce) electro-active molecules dissolved in the electrolyte. These molecules include ones based on a compound called ferrocene (shorthand symbol Fc) which consists of an atom of iron sandwiched between two organic rings, Figure 8.

As the microelectrode scans the interface, ferrocene molecules diffuse towards it, carrying a current between the surface and the microelectrode. This current can then be measured. If the microelectrode passes over an area that is insulating, the current decreases. On the other hand, if the surface is chemically active, the current increases. These two effects are called negative feedback and positive feedback respectively. They depend on the distance between the microelectrode and the surface. In some circumstances, the gap between the electrode and the surface can be made so small that, on average, only a single electro-active molecule is in the gap at any time.

11. Using the name to help you, write the displayed formula of a cyclopentadiene molecule. Suggest what orbitals on the iron atom might be used to form the bonds to the organic rings in ferrocene. Hint: Similar compounds are formed with the metals nickel, chromium and ruthenium but not potassium or barium, for example.

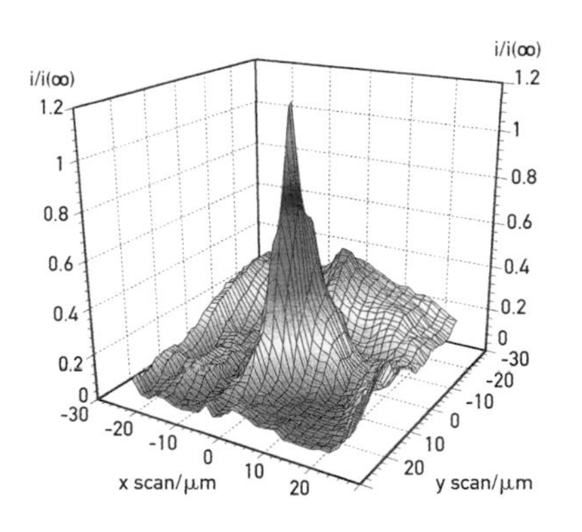

Figure 6. A SECM image of an individual hole in the dentine of a tooth.
Professor P R Unwin, University of Warwick.

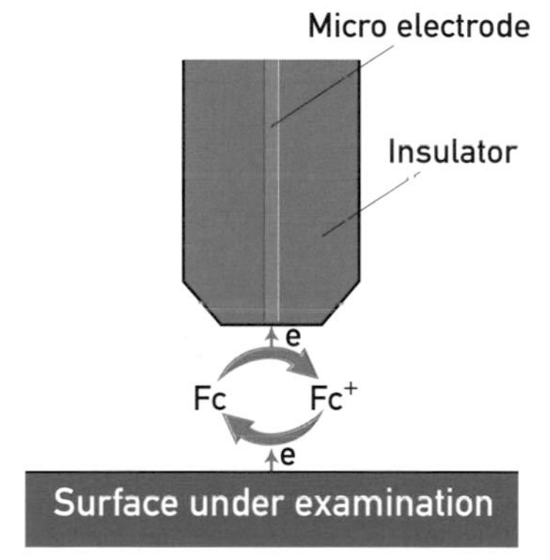

Figure 7. Scanning electrochemical microscopy.

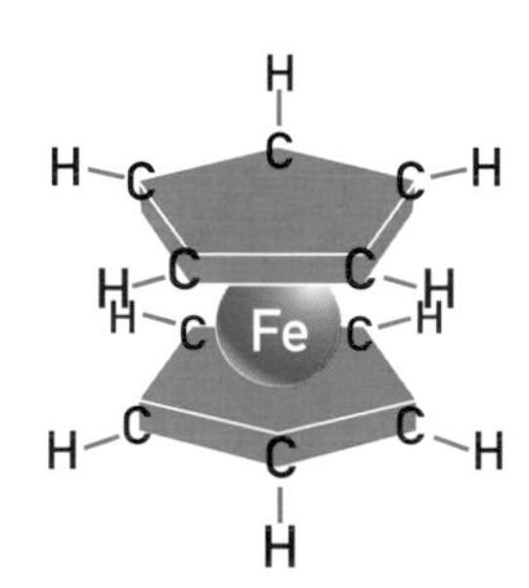

Figure 8. Ferrocene consists of an atom of iron which forms the 'meat' in a sandwich between two cyclopentadiene rings.

Courtesy of the British Diabetic Association.

Biological sensors

One way in which the redox chemistry in living cells can be exploited is in developing sensors to monitor the concentrations of important chemicals in the body, for example, the concentration of glucose in the blood. Diabetics need to measure their blood glucose levels regularly throughout the day, so they require a cheap, portable device that gives accurate measurements quickly and often.

Researchers have developed a faster and more mobile glucose biosensor which is based on an electrochemical system involving interaction between an electrode and the enzyme glucose oxidase. This enzyme oxidises glucose and is itself reduced in the process. The system works by re-oxidising the glucose oxidase in a small blood sample using an electrode – the greater the current flow, the greater the concentration of the reduced glucose oxidase and hence the greater the original glucose concentration.

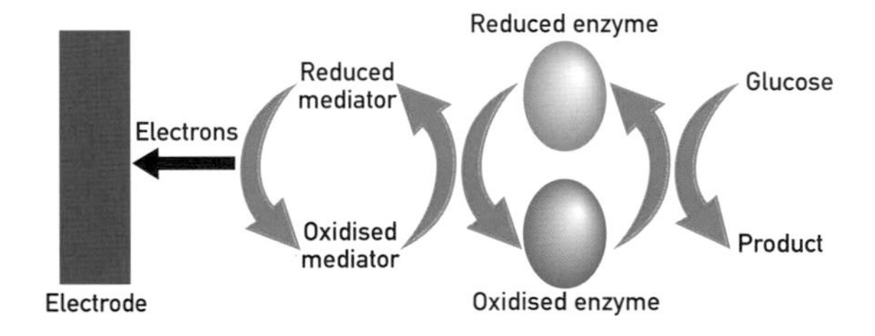

Figure 9. Small molecules carry charge from an electrode to an enzyme in the blood glucose sensor.

Unfortunately, proteins such as enzymes are very large and the electron transfer to and from the part of the bulky, convoluted protein structure which has been reduced to the electrode is very slow. However, Allen Hill and his colleagues at the Universities of Oxford and Cranfield have employed an imaginative approach to the problem.

Hill added to the enzyme solution small redox-active molecules (in the case of the glucose system, a derivative of ferrocene) that shuttle electric charge between the active centre of the enzyme and the surface of an electrode – a process he called mediation. Essentially the highly mobile and relatively small mediator molecule acts as a 'courier' in taking charge to or from the electrode surface to the enzyme redox centre where this cannot be done directly, Figure 9.

The Oxford team's work has led to a new blood glucose sensor which is now available. The market for such sensors is extremely lucrative.

Researchers have employed the mediator technique with a huge variety of enzymes and mediator combinations. Electrochemical sensors that detect more than one chemical are now available. A good example is the i-STAT sensor which can measure the level of six important chemicals in one drop of blood!

A MediSense glucose pen.
Courtesy of MediSense.

The future of electrochemistry

Investigating the structure of the electrode surface is vital if we are to understand the processes occurring on it and be able to design more effective electrode surfaces. Today, the often complex chemistry which occurs at electrodes is being investigated using sophisticated microscopic and spectroscopic techniques coupled to electrochemical systems.

The level of detail which can be seen is illustrated by the micrograph, Figure 10. This shows iodine atoms being adsorbed onto the surface of a gold, single-crystal electrode as its potential is changed. In the lower half of the picture the gold atoms can be seen. In the upper half of the image the electrode potential has been switched and iodine atoms are now seen adsorbed onto the surface. Such studies are vital for new developments in electrochemistry as chemists can actually see what is going on on the surfaces of electrodes at the level of individual atoms.

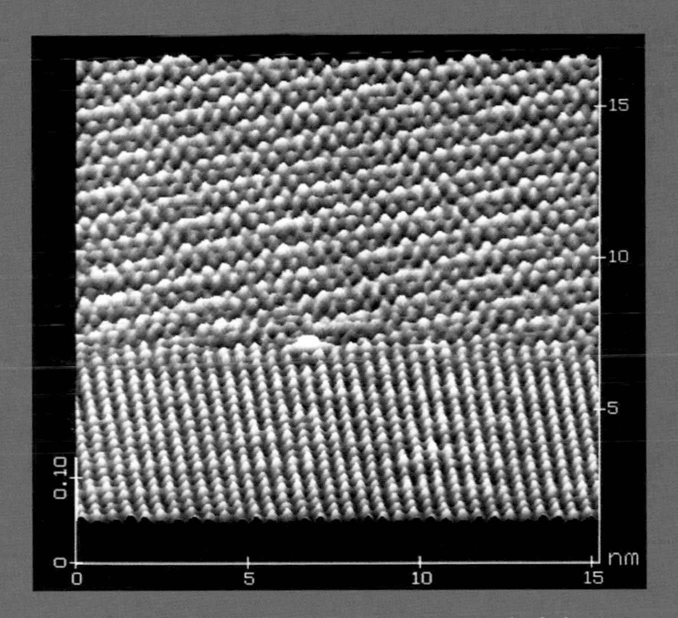

Figure 10. Micrograph showing iodine atoms (top) adsorbed onto a gold surface (bottom).
Reprinted with permission from Michael J Weaver, J Phys Chem, 1996, 100, 13079-13089. ©2000 American Chemical Society.

Electrochemical advances in battery, fuel cell and photovoltaic technology will provide new, clean, efficient and powerful energy storage and generation systems, while new electrochemical industrial processes will be cleaner and more efficient than those used now. Improved sensors and biosensors will allow important substances to be analysed much faster and more accurately. Consequently, electrochemistry has a bright future in the 21st century and beyond.

Answers

1. $2Mg(s) + O_2(g) \rightarrow 2[Mg^{2+}, O^{2-}](s)$
   Magnesium has lost two electrons in going to Mg^{2+}.
   Oxidation is loss of electrons.
   Oxygen has been reduced to oxide ions by gain of electrons.

2. From zinc (more negative) to silver (more positive).

   1.56 V.
   $Zn(s) \rightarrow Zn^{2+}(aq) + 2e^-$
   $2H_2O(l) + 2e^- \rightarrow H_2(g) + 2OH^-(aq)$
   (or $Ag^+(aq) + e^- \rightarrow Ag(s)$ could be another suggestion)

3. $Pb(s) + PbO_2(s) + 4H^+(aq) + 2SO_4^{2-}(aq) \rightarrow 2PbSO_4(aq) + 2H_2O(l)$

4. 16.2 MJ.

5. (a) Hydrogen: from cracking of petroleum fractions.
   (b) Oxygen: from air (possibly by fractional distillation).
   To produce them, both ultimately involve the burning of fossil fuels to provide energy, and result in emissions of carbon dioxide and nitrogen oxides into the atmosphere.
   Methane produces carbon dioxide (and water) when burnt. The burning process also produces nitrogen oxides (NO_x).

6. The water could be used by the astronauts for washing, drinking *etc*. The heat produced might be useful. Hydrogen and oxygen are used as launch fuels and small quantities of these could be used in the fuel cell.

7. Carbon dioxide. It is a greenhouse gas and therefore contributes to global warming.

8. (a) 700 dm^3.
   (b) 7 times.
   (c) 7000 km.
   (d) 64 times greater.

9. (a) $D(electrode) \rightarrow D^+(electrode) + e^-$

   (strictly $D(electrode) \xrightarrow{light} D^*(electrode)$, then $D^*(electrode) \rightarrow D^+(electrode) + e^-$)
   (b) $2D^+(electrode) + 2I^-(aq) \rightarrow 2D(electrode) + I_2(aq)$
   (c) $I_2(aq) + 2e^- \rightarrow 2I^-(aq)$

10. Reduced as it gains electrons.

11. d-Orbitals.

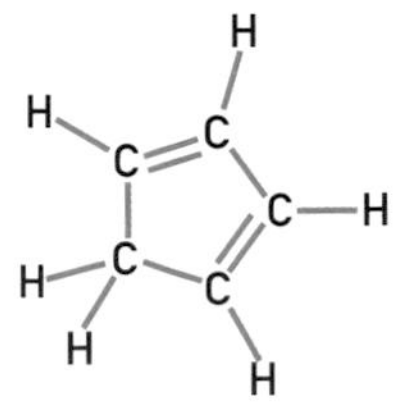

Introductory page pictures: Solectria Corporation/Photo by Sam Ogden.

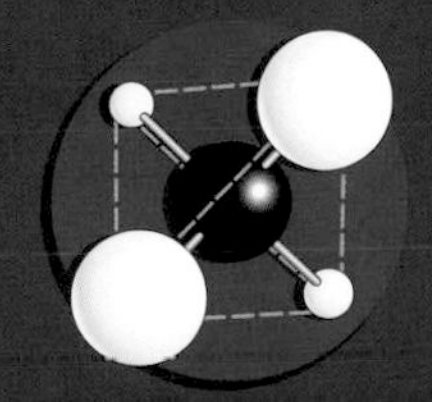

Computational chemistry and the virtual laboratory

The rapid development of powerful computers over the past 20 years has allowed chemists to investigate a wide range of chemical behaviour without using a single test-tube. Easy-to-use software packages are now available based on a wide variety of ingenious computational methods. These can be used to build computer models that probe the intimate details of simple reactions, predict the characteristics of materials such as catalysts and polymers, and visualise the interactions of biological molecules and drugs.

Computational chemistry, sometimes called 'molecular modelling' or 'theoretical chemistry' is now an indispensable tool in chemical research.

The most visible face of molecular modelling is undoubtedly computer graphics. With current software packages it is possible to produce images of all kinds of complex molecules. However, molecular modelling is much more than just pretty pictures. Today's programmes can build up a description – usually presented in the form of graphics – that reveals details of chemical behaviour, for example:

- **what happens during the course of a reaction as one molecule reacts with another;**
- **how simple organic molecules form weak bonds with metal surfaces (important in catalysis); and**
- **the way a protein embedded in a cell membrane can change its shape (important in designing drug molecules to interact with cellular receptors).**

There are three main reasons why computational chemistry is so widely used. First, it can help us to understand the behaviour of a system at a molecular level – information that may be difficult if not impossible to obtain by any experimental technique. Secondly, it enables researchers to evaluate many possible choices, in terms of reactants, reaction conditions and so on, before undertaking any experimental work. Thirdly, computational methods can be used to investigate systems under extreme conditions – such as very high pressures and/or temperatures – which cannot be reproduced in the laboratory but which exist in the Earth's core or in stars, for example.

The key to the widespread use and acceptance of molecular modelling is undoubtedly the phenomenal rate of growth in the power and availability of computers. In recent times, computational power has doubled every couple of years. Even more astonishing is how the cost of a given amount of computational power has fallen, Figure 1. The personal computers that are so widespread both at work and in the home are as powerful as the expensive mainframe computers of only a few years ago.

The other reason for the spread of molecular modelling is the availability of off-the-shelf software packages that are easy to use by non-computer experts.

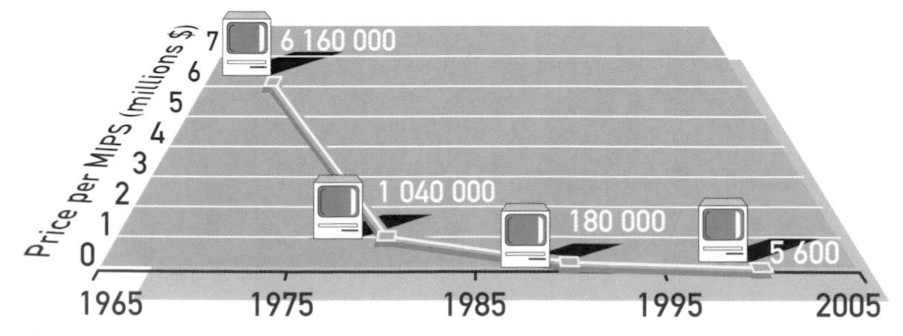

Figure 1. How the cost of computing power has fallen – the price per MIPS (Million Instructions Per Second) of an entry level IBM manufactured over the past three decades. Data courtesy of Computer Wire plc.

Glossary

Cell membrane: A biological sheet-like tissue which surrounds living cells.

Quantum theory: A theory which describes objects of sizes comparable to those of atoms and smaller. It is based on the idea that energy exists in small 'packets' called quanta and that matter also behaves like waves.

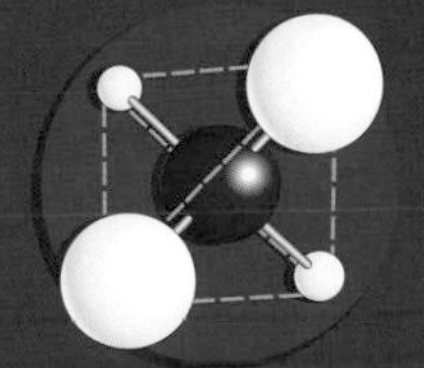

Isis facility, Rutherford Appleton Laboratory.

How does computer modelling work?

Modellers study systems ranging from single isolated molecules to proteins containing thousands of atoms immersed in a sea of solvent molecules. Computational chemists have at their disposal a variety of methods but at the heart of *all* programmes is some procedure for calculating the energy of the system. The programmes then make small changes to the system and recalculate the energy after each change. They then use the principle that a chemical system tends to adopt the lowest energy which it can attain to select the most likely configuration of the system.

Energies can be calculated using two basic methods – quantum mechanics or molecular mechanics. Deciding which is the appropriate method to use depends upon the kind of questions we want to ask and on the available computing power. Quantum mechanics offers the most fundamental approach and is mostly used on simple atoms and molecules. Molecular mechanics is particularly useful for modelling large molecules and assemblies of molecules.

Cray© T94™
Courtesy of Cray Research, Inc.

Quantum mechanical methods

A quantum mechanical calculation starts with the Schrödinger equation, which fully describes any atom or molecule in terms of its wavelike, **quantum** nature. The problem is to solve the equation and thus calculate the energy levels of the system. However, this equation can only be solved *exactly* for extremely simple systems, such as the hydrogen atom, which consists of just one proton and one electron. No exact solution is possible even for the helium atom!

When written out in full, the Schrödinger equation consists of a very large number of terms which take account of the attractions and repulsions between all possible pairings of particles of the atoms in the molecule, and also their kinetic energies. Evaluating these terms is out of the question without computers, and progress to bigger molecules has been possible only because advances in computing power have allowed the methods described below to be used. These calculations can tell us how the energy of the molecule varies with the molecule's shape. They can also predict physical properties such as dipole moments (the distribution of electronic charge in the atom or molecule represented by δ^+ and δ^-).

Since the Schrödinger equation cannot be solved exactly, theorists need to think up ingenious ways of finding approximate solutions. A key component of most of these approaches is the variation principle, which states that the 'better' the solution the lower the energy. The general idea is to solve the Schrödinger equation *approximately*, make small changes to the starting conditions and solve it again, make more changes and solve it yet again and so on. Eventually a solution of minimum energy will be found – this is the 'best' solution. William Hartree did tedious calculations like these at Cambridge in the 1930s and 40s on mechanical calculators aided by his family!

Quantum mechanical methods can be divided into two types – *ab initio* methods and semi-empirical ones. *Ab initio* means 'from the beginning'; the calculations work from first principles, using as inputs only physical constants such as the charge on the electron. Semi-empirical methods use data derived from experiment, such as ionisation energies, to help out the calculations which are therefore much quicker to do and require less computing power. Despite this and other advances, quantum mechanical calculations are restricted to single molecules containing a few tens of heavy (non-hydrogen) atoms.

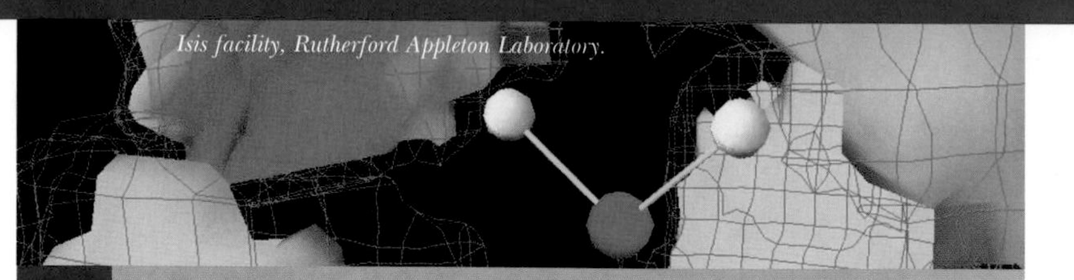
Isis facility, Rutherford Appleton Laboratory.

Simple prediction of molecular shapes – VSEPR

It is possible to make some predictions about molecular shape without doing any complex calculations. The Valence Shell Electron Pair Repulsion theory (VSEPR) is based on the idea that groups of electrons in the outer (valence) shell of an atom will repel each other so that they lie as far away as possible. Four groups of electrons will give a **tetrahedral** shape and three will form a flat shape described as **trigonal** and so on.

This theory predicts, for instance, that all the H–C–H bond angles in methane (four bonding pairs of electrons) will be 109.5° and that the H-C=O angle in methanal, HCHO, (two pairs of electrons in the C–H bonds and a group of four electrons in the C=O bond) will be about 120°.

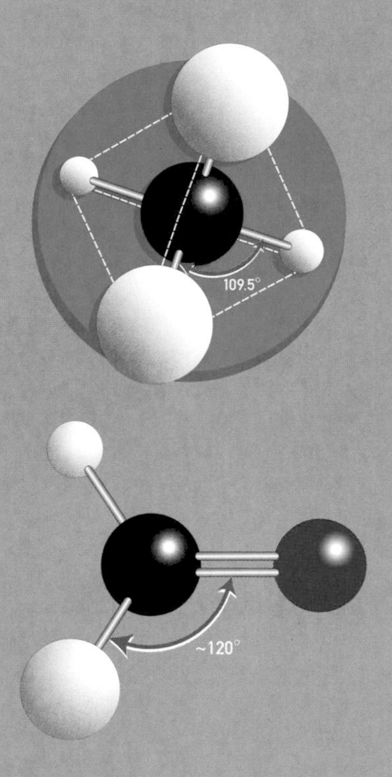

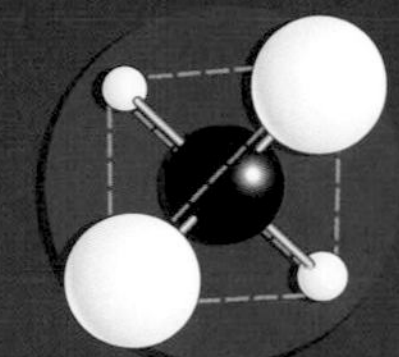

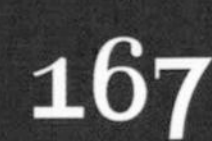

Glossary

Tetrahedral: Describes the shape of a triangular based pyramid.

Transition state: A short-lived, high energy species intermediate between reactants and products which is formed during a chemical reaction.

Trigonal: Describes the shape of an equilateral triangle.

The true test of any theoretical technique is how well its predictions agree with the values obtained from experiment. Normally, experimental data are regarded as 'definitive' but one celebrated case where theory proved to be in advance of the experiment is that of methylene (CH_2). This is an extremely simple (and unstable) species, with just two hydrogen atoms bonded to a central carbon atom, see below. The carbon atom carries a lone pair of electrons as well as the two bonding pairs. The geometry of this species had been investigated using both theoretical and experimental techniques. The early debate concentrated on whether its geometry is linear or bent.

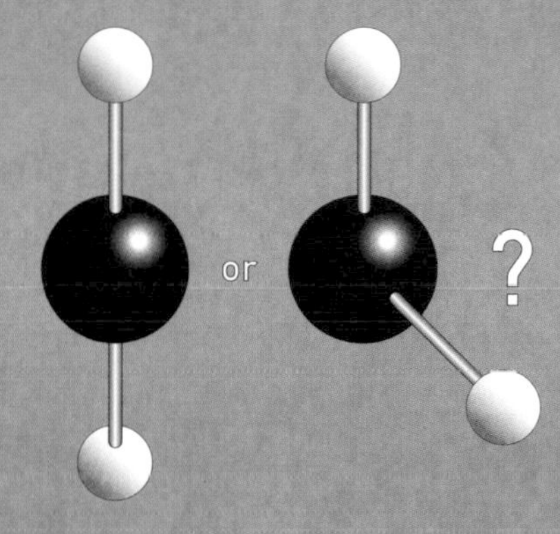

Linear and bent versions of methylene.

1. A 'dot cross' diagram for methylene is given below.

H
• x
C •
• •
H x

Use VSEPR to predict the shape of the molecule and explain your reasoning. Hint: Remember that lone pairs of electrons (which are closer to the nucleus) repel more strongly than bonding pairs.

The first theoretical studies on methylene were performed by chemists at the University of Cambridge in 1960. They concluded that CH_2 had an H-C-H angle of 129°. There was, then, no experimental data for CH_2, but in the following year one of the leading spectroscopists of the time, Gerhard Herzberg at the University of Ottawa, concluded from his experiments that the molecule was linear, *ie* the bond angle was 180°. Events came to a head when Charles Bender and Fritz Schaefer at the University of California also calculated from theory a bent geometry (with an H-C-H angle of 135.1°). Soon after Bender and Schaefer's paper, several experimental research groups reported new data on the molecule showing that the molecule was indeed bent. In the light of these findings, Herzberg re-examined his data and concluded that they were in fact consistent with a bent structure.

Quantum mechanics really scores for problems that involve the making or breaking of chemical bonds, and so it has been extensively used to investigate the detailed mechanisms of reactions – the nature of the **transition state** that forms when molecules react. Indeed it is only now, with the advent of lasers producing ultra-short pulses, that we are able to observe directly the course of a chemical reaction experimentally, and even so we can only follow simple reactions (see Chapter 4 – *Following chemical reactions*). For most of the reactions of interest to organic chemists, only indirect experimental evidence is available from which we can try to deduce features about the actual mechanism.

Using computer modelling it is possible to 'observe' the reaction all the way from the starting materials through any transition state or reaction intermediate to the products. The so-called Diels-Alder reaction for introducing cyclic structures into organic molecules is a classic example that has been studied theoretically and experimentally in detail (see Box – *The Diels-Alder reaction – one step or two?*).

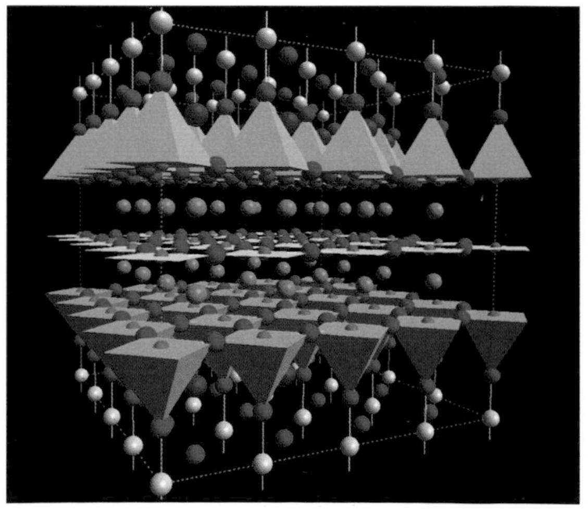

Isis facility, Rutherford Appleton Laboratory.

In the Diels-Alder reaction, a diene (a molecule with two double bonds separated by a single bond) adds on to a 'dienophile' (a molecule with an isolated carbon-carbon double bond) to form a ring-shaped structure. The simplest example is shown below.

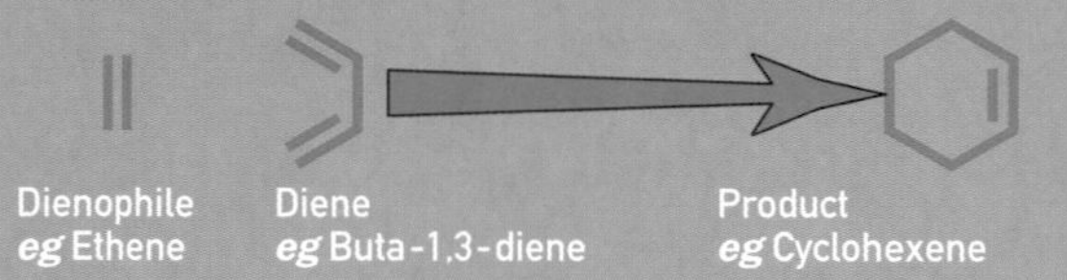

This is an extremely important reaction used to synthesise new organic molecules. Chemists have, therefore, been interested to know the details of its mechanism. A useful way to consider different mechanisms is to think about how the energy and geometry of the system might change as the reactant molecules progress through to the products. The Diels-Alder reaction involves the formation of two new bonds. If the mechanism is stepwise these bonds will be formed one at a time. The system will pass through an intermediate in which just the first of the new bonds has formed. If the mechanism is concerted, the two new bonds will be formed simultaneously and no intermediate is formed.

The situation is shown on the reaction profiles below.

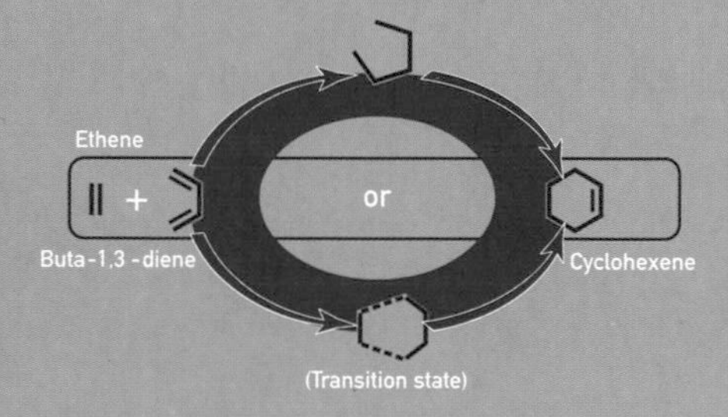

Reaction profiles for the two possible mechanisms for the Diels-Alder reaction.

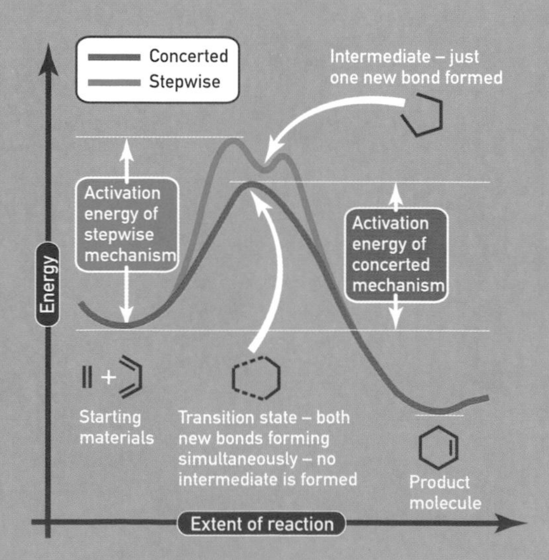

Computations indicate that the concerted mechanism has a lower activation energy than the stepwise mechanism and is therefore the one that actually occurs.

Molecular mechanics

When we wish to model larger molecules, or assemblies of molecules, quantum mechanics is usually not feasible. Such systems can be studied by the molecular mechanics, or force field method. The molecular mechanics approach considers the energy of any arrangement of atoms to be made up of several distinct parts, as shown in Figure 2. These can be calculated separately and then added up.

First, there is a contribution from the stretching or compressing of bonds. Each of the bonds in the molecule has a normal bond length. For a C-C single bond, it is around 0.154 nm, for a C=O bond it is about 0.122 nm and for a N≡N triple bond it is approximately 0.110 nm. Energy must be put in to stretch or compress the bond away from its normal length. The amount of energy depends on the amount the bond is stretched or compressed and the strength of the bond. This is just like stretching or compressing a spring. A N≡N triple bond is stronger (and therefore harder to stretch or compress) than a C-C single bond, for example.

The second contribution to the molecular mechanics energy is that due to 'angle bending'. As with bond stretching, there is an ideal value for each angle, and deviations from this ideal value require the expenditure of energy. The ideal value for atoms having a tetrahedral arrangement of bonds (such as H-C-H in a $-CH_3$ group) is around 109.5°, for example.

Most of the variation in the structure of a molecule comes not from bond-stretching or angle-bending, but from rotation about bonds. The energy of a molecule varies with its particular shape or conformation; in other words, the particular three-dimensional configuration arising from bond rotation. Perhaps the simplest example of this is ethane (C_2H_6). As the carbon-carbon bond rotates, the distance between the hydrogen atoms on the two carbon atoms changes. The hydrogen atoms repel one another slightly and the closer they are, the higher the energy of that conformation. The two extremes are the so-called eclipsed and staggered conformations which differ in energy by 12.2 kJ mol^{-1}, Figure 3.

The fourth contribution to the molecular mechanics energy arises from non-bonded forces. These are often called 'intermolecular forces' although they do work between different atoms within molecules as well. They include dipole-dipole forces and van der Waals interactions as well as hydrogen bonds. Dipole-dipole forces arise from the interaction between

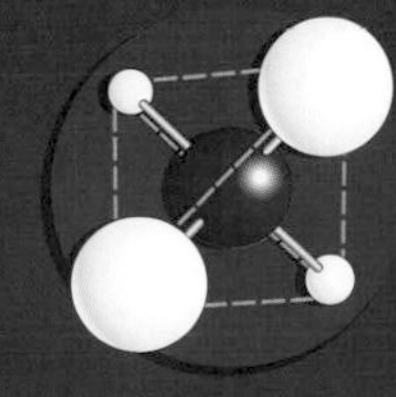

GlaxoWellcome.

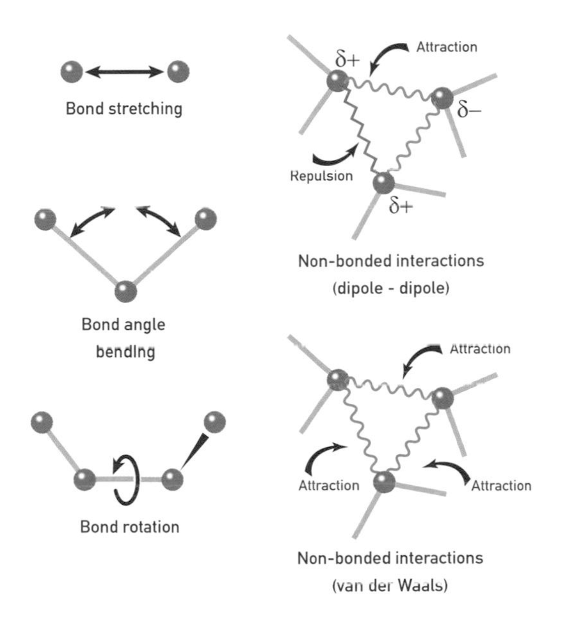

Figure 2. Contributions to the energy of an assembly of atoms.

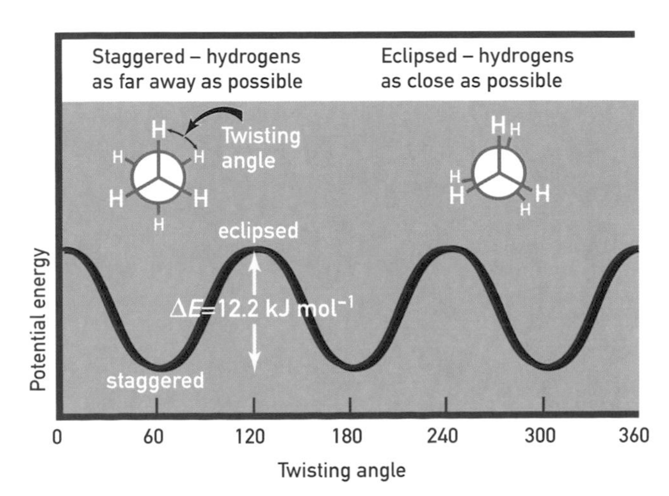

Figure 3. The energy of the different conformations of ethane. In the two graphics of ethane, we are looking along the C–C bond.

permanent partial charges (δ^+ and δ^-) within the molecule caused by uneven distributions of electron density between the atoms. The interactions between these δ^+ and δ^- charges are determined using Coulomb's law, which states that the energy between two charges is proportional to their product divided by their separation, with charges of opposite sign attracting each other and charges of like sign repelling.

Van der Waals forces are present even in systems such as the noble gases, where there cannot be any permanent dipoles. Their origin is rather more subtle than that of the straightforward electrostatic interaction. Although the electron distribution around an atom is spherically symmetrical on average, the distribution at any one instant is not uniform, leading to an instantaneous electric dipole. This instantaneous dipole can then induce a dipole in a neighbouring atom. The two dipoles now attract each other, Figure 4. As the atoms approach closer and closer they attract more and more until a position of minimum energy is reached. If they approach closer than this, they start to repel each other.

2. Explain why there are no permanent dipoles in the atoms of the noble gases.

3. Explain why the instantaneous dipole-induced dipole forces are always attractions and never repulsions.

4. Suggest why two atoms start to repel when they approach very closely.

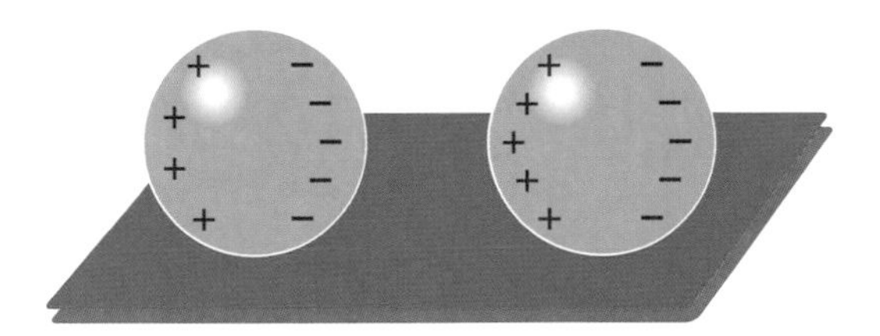

Figure 4. An instantaneous dipole of one atom (for example, in the atom on the left) can induce a dipole in a neighbouring atom, causing the two atoms to be attracted to each other.

Conformational analysis

One of the most common applications of molecular mechanics is in conformational analysis. This is the task of exploring all the ways in which the various parts of a molecule rotate in relation to each other – their so-called conformational space – in order to identify the conformations with the lowest energy. These are the conformations the molecule is most likely to have.

Even simple molecules can show some conformational variation, for example butane ($CH_3CH_2CH_2CH_3$). As the central carbon-carbon bond in butane rotates, the energy of the molecule varies as shown in Figure 5(a). There are three minimum points which correspond to the three stable arrangements. Moreover, one of these is more stable (lower in energy) than the others. The conformations corresponding to these low energy positions are shown in Figure 5(b).

5. Make a model of butane using a ball and stick molecular modelling kit. Rotate the central C-C bond to produce the conformations shown in Figure 5(b). What do you notice about the distance between the -CH_3 groups in (a) the most stable and (b) the least stable conformation?

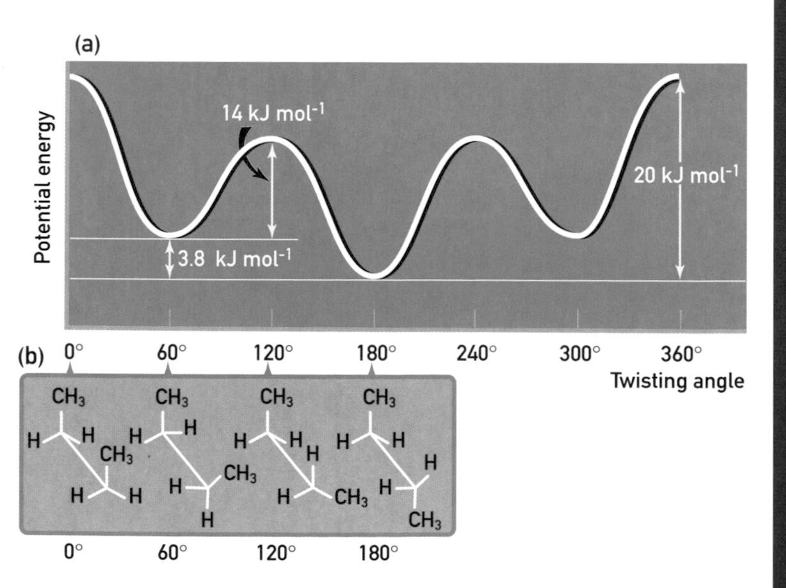

Figure 5. Energies of the different conformations of butane.

The Valence Shell Electron Pair Repulsion theory (VSEPR) – see Box – *Simple prediction of molecular shapes* – enables us to predict the bond angles of simple molecules. However, this does not completely define the shape of more complex molecules. This is because rotation is possible about single bonds. So all but the simplest molecules can adopt a range of shapes. These are called conformations. The figures above and below show just two of the many conformations of the molecule octane. In all the possible shapes, all the bond angles are 109.5°.

6. Make a model of octane ($CH_3CH_2CH_2CH_2CH_2CH_2CH_2CH_3$) with a ball and stick molecular modelling kit. Try rotating the C–C bonds. You will find that there is an enormous range of shapes possible, all obeying the rule that all the bond angles are 109.5°.

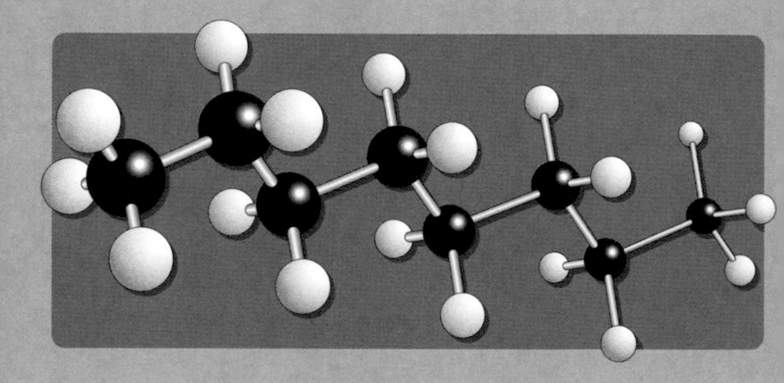

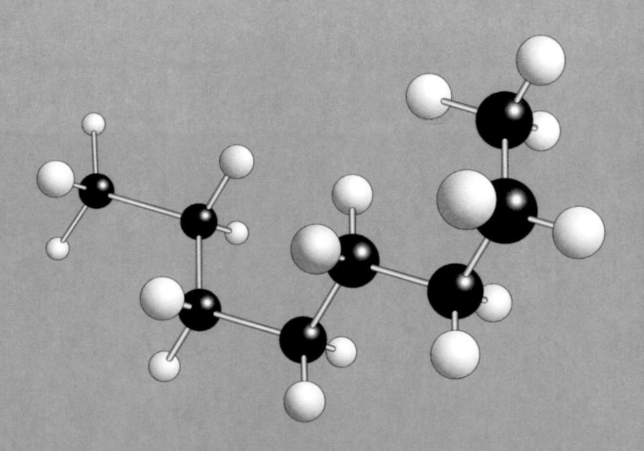

Glossary

Enthalpy: Energy measured under specified conditions – a temperature of 298 K and a pressure of 100 kPa.

Entropy: A measure of the disorder of a chemical system. It is mathematically related to the number of ways of arranging the particles and the energy of the system.

Free energy: The amount of energy in a chemical system which is available to do useful work.

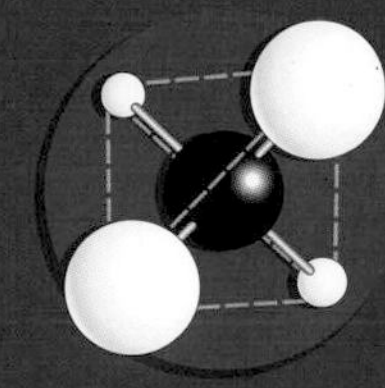

Figure 6. The structure of Enalapril.

A more typical example of the type of molecule for which conformational analysis is required is Enalapril, Figure 6, a drug used to treat high blood pressure. The conformation adopted by drug molecules is important because many of them work by fitting into gaps ('active sites') in enzymes or in protein molecules called receptors found on the surfaces of, or inside, cells. This is rather like a hand fitting a glove. Conformational analysis can be used to predict the likely shape adopted by a complex molecule and therefore to suggest whether it could fit the active site.

Some conformational analysis programmes work by selecting a conformation of the molecule and calculating its energy. A second conformation in which one of the rotatable bonds is twisted through, say, 30° is chosen and the energy calculated again. This process is repeated for all possible combinations of angles for all the rotatable bonds. Conformations with low energy values are the most stable and therefore most likely to occur.

7. (a) In the Enalapril molecule, Figure 6, which bonds can rotate and which cannot?

(b) Identify the functional groups in Enalapril.

Groups of molecules

So far, we have been concerned only with calculations performed on single molecules. In real life, the molecule of interest will be in a solution, a solid, a glass or a liquid crystal. Single isolated molecules are normally only ever observed in the gas phase. Real systems contain many molecules all interacting with each other.

An obvious practical difference between single molecules and groups of molecules is that the systems are much larger. Another key difference is that groups of molecules can be arranged in many ways. In other words, the idea of **entropy** (S) plays a much more significant role. The computational methods that we have met so far provide energies equivalent to the **enthalpy** (H) of a molecule. What really matters is the **free energy** (G). The free energy change of a reaction is related to the enthalpy change and the entropy change by the relationship $\Delta G = \Delta H - T\Delta S$. ΔG can be used to predict which reactions might take place spontaneously. It is linked to other quantities which predict this, such as the equilibrium constant, K_c, and to the emf of a cell reaction, $E^{\ominus}$.

The entropy is a measure of the number of different arrangements accessible to the system. A single molecule has contributions to its entropy from:

- **translational motion (*ie* movement from place to place);**
- **rotation of the molecule as a whole; and**
- **vibration of the bonds.**

There is also a contribution from the different conformational states available to it.

It is possible to calculate these contributions quite accurately for small systems. For a large scale system such as a solution there are vastly more states available. Even with the fast computers of today we cannot model the behaviour of systems containing numbers of molecules corresponding to everyday quantities of material. There are 6×10^{23} particles in a mole of substance, whereas the largest computers can deal with only terabytes of information ('tera' stands for one million million, or 10^{12}). Fortunately, a technique called statistical mechanics provides us with a way of calculating bulk properties by using much smaller systems – containing a manageable number of molecules – and still obtain meaningful results that can be compared with experiment. This is the realm of computer simulation.

Molecular simulation

The two major techniques used in molecular simulation are molecular dynamics and the Monte Carlo method. The molecular dynamics method solves Newton's equations of motion for the collisions between atoms in the system. The very first molecular dynamics simulations used a 'hard sphere' model which treated the atoms like a set of snooker balls moving about on a table. They enable us to predict the locations and velocities of each ball at any point in time. In this model, attractions or repulsions between the particles are ignored except during the collisions. These are treated as being perfectly elastic (*ie* no energy is lost).

Even with such a simple model it was possible to show the existence of both solid and liquid phases with a definite transition between the two. This is beautifully shown in some of the earliest examples of the use of molecular graphics in computational chemistry, Figure 7.

In reality there are both attractive and repulsive forces between the molecules due to the **non-bonded forces**, and these can be allowed for in more recent simulation methods. In current molecular dynamics simulations, the calculation is broken down into many very small steps each of about 1 femtosecond (10^{-15} s). At each time-step, the forces on the atoms are determined. From this their positions at the next small time-step ahead are deduced, and so on. The restriction to small time-steps means that the length of a typical simulation is limited to picosecond (10^{-12} s) or nanosecond (10^{-9} s) timescales. Although these appear to be very short times in our everyday world, at the molecular level they are often sufficient to enable many interesting phenomena to be studied. Figure 8 shows a molecular dynamics simulation of a peptide antibiotic in a biological membrane.

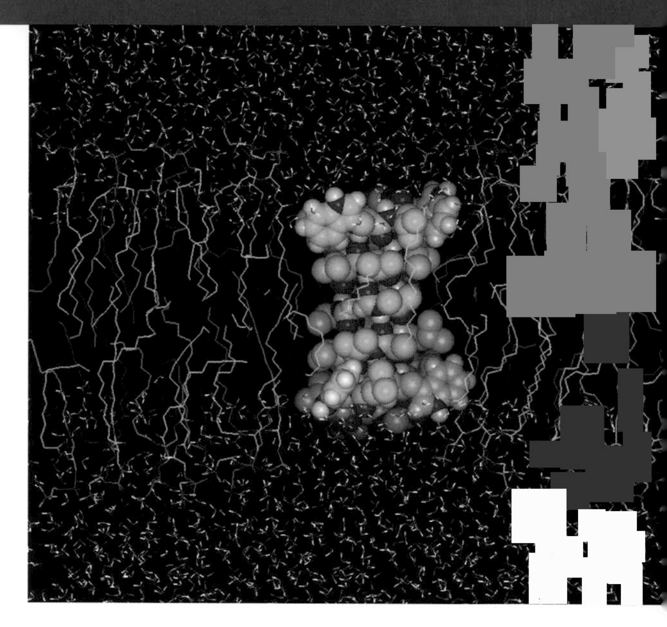

Figure 8. A molecular dynamics simulation of part of the antibiotic Gramicidin A in a biological membrane with water. The disordered nature of the long chain lipid molecules which make up the membrane is obvious.
Courtesy of Alan Robinson and Graham Richards.

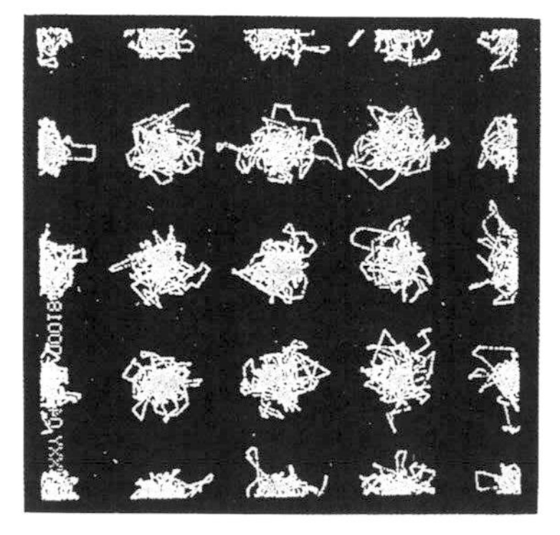

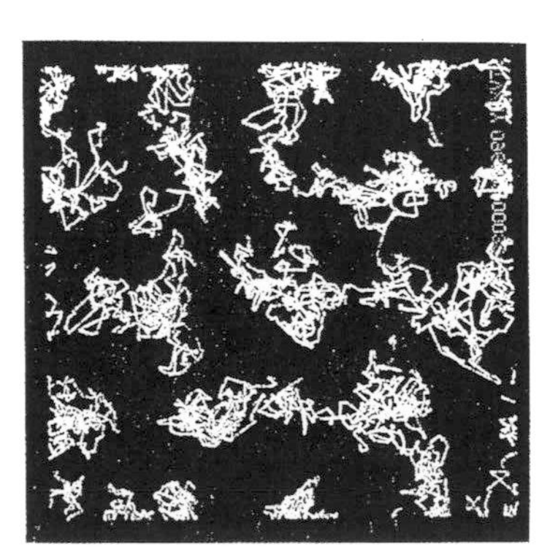

Figure 7. This early molecular graphics image shows the paths generated by 32 hard spheres (atoms) from molecular dynamics calculations. The picture on the left corresponds to the solid phase and that on the right to the liquid phase. As can be seen the particles move further away from the original position and are more randomly arranged in the liquid phase.
Reprinted with permission from B.J. Alder and T.E. Wainwright, Journal of Chemical Physics, 31, 459-466, (1959). ©2000 American Institute of Physics.

Glossary

Functional group: A reactive group in an organic compound such as alcohol, carboxylic acid, amine *etc.*

Non-bonded forces: Another name for intermolecular forces – hydrogen bonding, dipole-dipole forces and van der Waals forces.

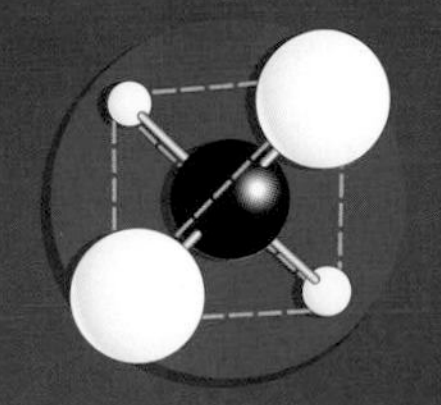

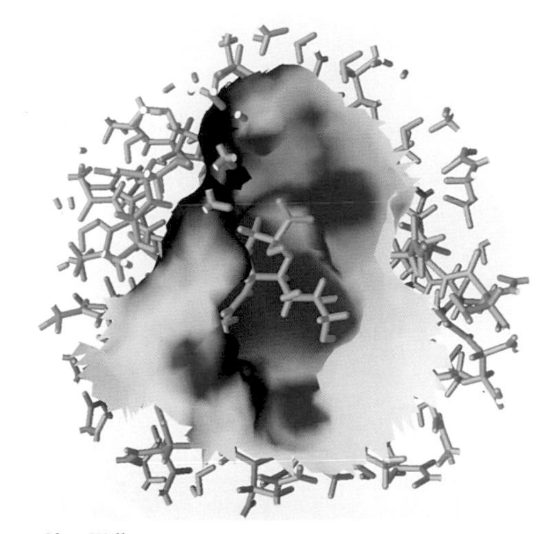

GlaxoWellcome.

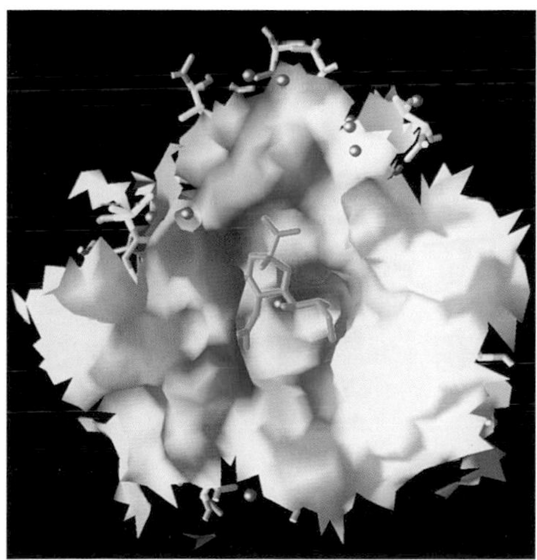

GlaxoWellcome.

Monte Carlo methods

Monte Carlo programmes (named from the random number generator at the heart of them, which acts rather like a roulette wheel) add up the interactions between all the molecules under consideration rather than dealing with individual molecules and their collisions. They can calculate bulk properties such as temperature, pressure and heat capacity and can deal with assemblies of up to tens of thousands of atoms.

Uses of molecular modelling

As well as examples of interest to theoretical chemists such as the mechanism of the Diels-Alder reaction (see Box – *The Diels-Alder reaction – one step or two?*), molecular modelling has many practical applications. Among the major users are pharmaceutical chemists who are trying to develop new molecules for use as medicines. Many drugs work by interacting with a large biological molecule, often a protein such as an enzyme or receptor (but sometimes DNA). The drug molecule locks into and blocks the active site on the protein structure so changing or inhibiting its normal biological action. The binding power and thus activity of the drug depends on its conformation and the types of **functional groups** it has.

In an increasing number of cases, the three-dimensional structure of the target protein is known from X-ray crystallography or from nuclear magnetic resonance measurements (see Chapter 2 – *Analysis and structure of molecules*). If so, using the protein structure as a guide, we can use computer modelling to design small molecular structures (called ligands) that bind with the protein's active site. These can be used to help us identify potential inhibitor molecules. One programme constructs a regular grid around the protein. At each of the points of the grid, a number of small molecular fragments (such as -OH, -COOH and $-NH_3^+$) are positioned to act as 'probes'. These are chosen to represent the functional groups commonly found in drug molecules. The interaction energy between each fragment and the protein is calculated using molecular mechanics and presented in graphic form. In drug design we are particularly interested in regions corresponding to low energy as these represent regions where that particular functional group could bind.

8. Suggest what groups present in a protein might interact with -OH, -COOH and $-NH_3^+$. For each group, say what sort of interaction is likely – *eg* 'chemical reaction leading to covalent bonding', 'chemical reaction leading to ionic bonding', 'hydrogen bonding' *etc.*

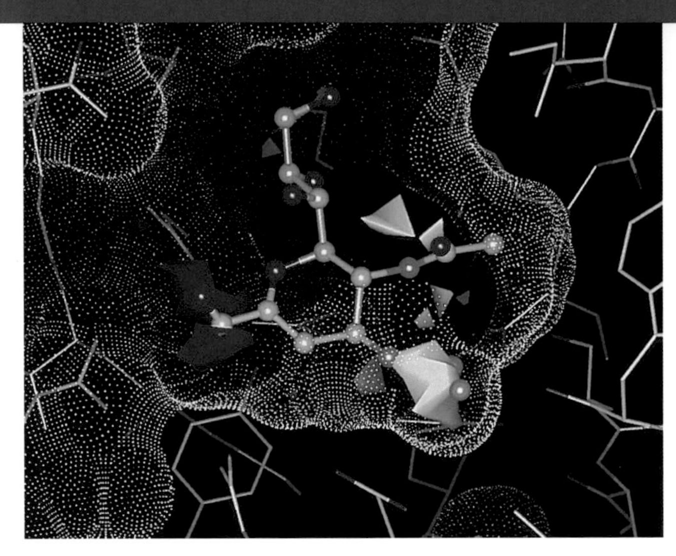

Figure 9. Analysis of the binding site of the flu virus enzyme, neuraminidase. Molecular fragments such as $-COO^-$ (red) and $-NH_3^+$ (blue) are used to probe parts of the enzyme that might bind to a potential inhibitor molecule.
Andrew Leach, Molecular Modelling: Principles and Applications. Reprinted with permission of Addison Wesley Longman Ltd.

A good example of this approach is in the design of inhibitors of the enzyme, neuraminidase. This enzyme is involved in the mechanism by which a flu virus passes from one host cell to another. Inhibitors of this enzyme are therefore potential antiflu drugs. From X-ray diffraction, the structure of the enzyme with an inhibitor was known, although the inhibitor was not very effective. When the binding site was analysed, Figure 9, it was suggested that a relatively simple modification would lead to a much more potent compound. When the suggested compound was made, it did indeed turn out to be much more effective.

In many cases we do not know the structure of the receptor protein. Nevertheless, it is still possible to deduce some useful information about the nature of its binding site. The process called pharmacophore mapping takes a series of molecules that bind to the site and attempts to derive an abstract model called a 3-D pharmacophore – the arrangement of functional groups and other atoms required for the drug to bring about a response. A simple example of a 3-D pharmacophore is shown in Figure 10.
3-D pharmacophores are often defined in terms of general features, such as a basic group, a hydrogen-bond donor or the centre of an aromatic ring and the distances between them. To generate meaningful 3-D pharmacophores is a difficult problem because of the many different possible conformations and functional groups which could be involved.

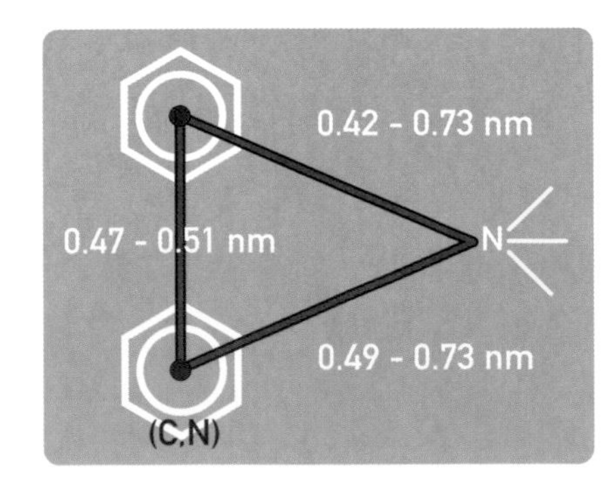

Figure 10. The 3-D pharmacophore for compounds with antihistamine properties.

There are several uses that can be made of a 3-D pharmacophore once it has been derived. A major application is in searching a database of three-dimensional structures for molecules with similar biological properties. Such molecules may represent the starting point for a new drug development programme. An approach favoured by many pharmaceutical companies is to derive a 3-D pharmacophore from known molecules (from the chemical literature or from competitors' compounds) and then search the company's database of compounds to identify new molecules 'owned' by the company. The critical feature of the 3-D databases used for such searches is that they take the three-dimensional, or conformational, properties of the molecules into account. This enables us to identify a series of molecules with similar three-dimensional properties but not necessarily similar chemical structures. A key development for 3-D database systems was the introduction of programmes that could take the computer equivalent of the two-dimensional chemical drawing and predict the most likely three-dimensional structure for the molecule. This is important because even a small molecule has a vast number of possible conformations which it can adopt. Crystal structures, which measure the actual conformation adopted, are not routinely determined for all compounds. These conversion programmes can process tens of thousands of structures per day, enabling a large database to be converted rapidly and automatically.

A 3-D database system matches the three-dimensional structures of the molecules it contains to the suspected 3-D pharmacophore. Most systems used today examine several conformations not just a single pre-stored one. Pharmaceutical companies now consider 3-D searching based on pharmacophores as a valuable tool which enables them to identify molecules apparently quite different from those already known.

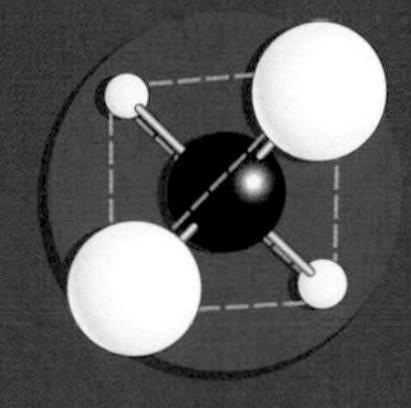

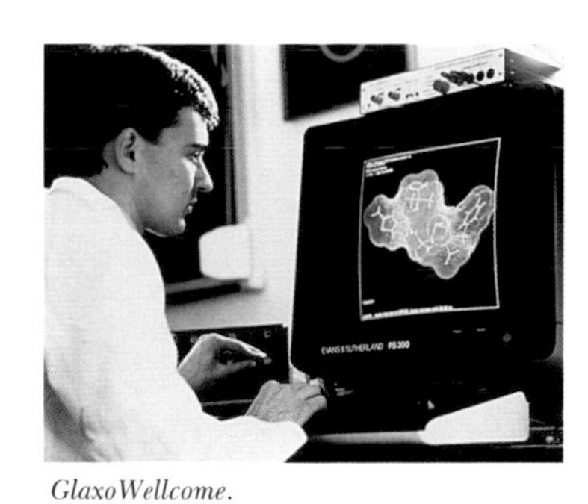

GlaxoWellcome.

Aromatic ring 0.85 - 1.2 nm Aromatic ring

0.35 - 0.65 nm 0.35 - 0.65 nm 3-D pharmacophore

H-bond donor or acceptor

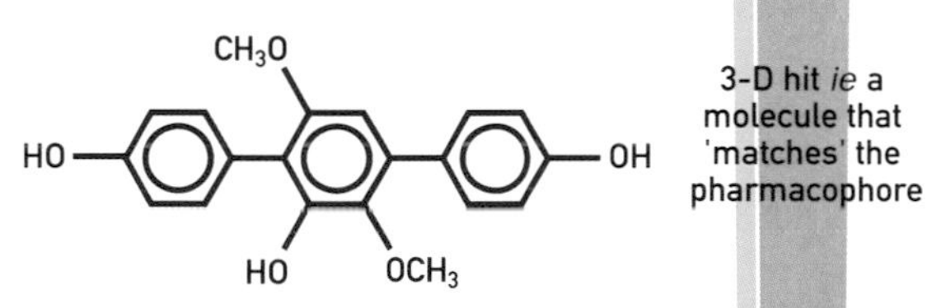

3-D hit *ie* a molecule that 'matches' the pharmacophore

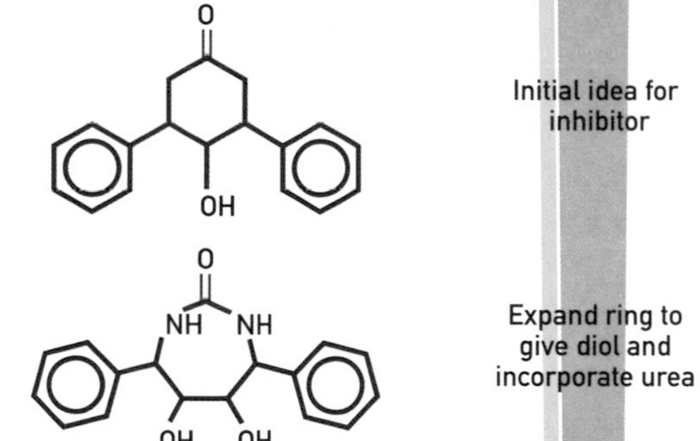

Initial idea for inhibitor

Expand ring to give diol and incorporate urea

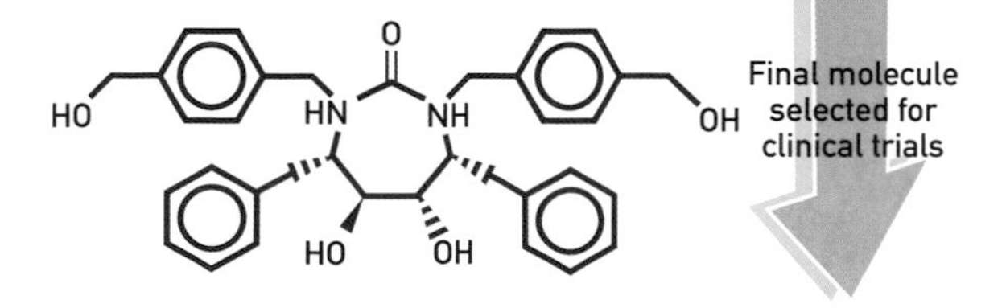

Final molecule selected for clinical trials

Figure 11. The key steps involved in the development of an anti-AIDS drug.

HIV drugs

The search for drugs useful against the HIV virus is a high priority. The DuPont Merck company in the US searched for potential inhibitors of HIV protease, the enzyme involved in creating the proteins and enzymes of the AIDS virus from polypeptides. The upper part of Figure 11 shows the 3-D pharmacophore. It consists of two hydrophobic (water-hating) functional groups (the aromatic rings) and a group that interacts with aspartate amino acids in the active site of the enzyme (the H-bond donor or acceptor).

One of the structures found not only contained these features but also other chemically useful ones. After several rounds of chemical synthesis and molecular modelling, a series of compounds with seven-membered rings was identified that were yet more potent. The sequence is shown in the flow chart in Figure 11. These compounds also had the key property of remaining active when taken by mouth. Later X-ray crystallographic analysis showed that the final molecule does indeed bind in the manner predicted. This story illustrates how a close partnership between experimental and theoretical approaches led to a very successful outcome, far in excess of what either could achieve in isolation.

9. Why are drug companies keen to make drugs that can be taken by mouth?

10. Why might some drugs be effective when injected but not when taken by mouth?

The future

Computational chemistry has come a long way since the days when programmes had to be input to the computer using punched cards, and a machine with 1 Kb of memory was considered to be state-of-the-art. Certainly, increases in computer performance have enabled major strides to be made in our understanding of chemical systems at the molecular level. However, there still remain some important problems for which a solution is sought, such as the famous 'protein folding problem', in which attempts are made to predict the three-dimensional structure of a protein solely from its amino acid sequence. It may be that ever-faster computers will enable us to solve these problems using current approaches, but history has shown us that it is often the combination of technology with a fertile mind that provides the breeding ground for the most significant advances.

Answers

1. The H-C-H bond angle is predicted by VSEPR to be somewhat less than 120°. Three pairs of electrons around the carbon atom would indicate a bond angle based on 120° but the extra repulsion of the lone pair compared with the bonding pairs would 'squeeze' the hydrogens closer together and reduce this angle slightly.

   Note for teachers. Methylene (also called carbene) actually exists in two forms – a high energy singlet state in which the two non-bonding electrons are paired and a lower energy triplet state in which these electrons are unpaired. The triplet state is almost linear and the singlet state (to which the question refers) has a bond angle of 103°. The large reduction from the predicted value is because the non-bonding electrons are in an orbital with a large proportion of p-character.

2. Permanent dipoles require the presence of an electronegative atom in a molecule. This cannot be the case in noble gases which consist of separate atoms.

3. The positive end of an instantaneous dipole will attract the electrons in the other atom towards it so that the induced dipole will form as shown, leading to attraction.

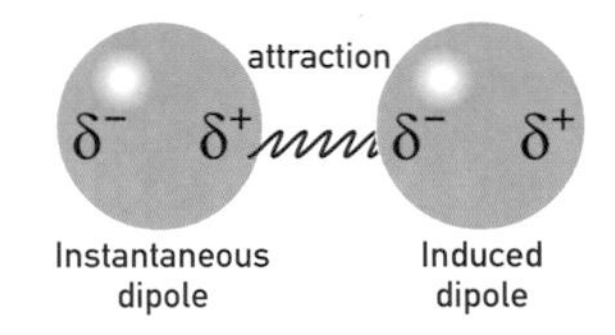

4. Their electron clouds will start to interpenetrate thus causing repulsion.

5. The $-CH_3$ groups are at their furthest distance apart in the most stable conformation and at their closest in the least stable.

6. Examination of the model shows a vast range of shapes.

7. (a) None of the bonds in the rings can rotate. The single bonds can all rotate, the double ones cannot.
   (b) From left to right: ester, benzene ring, secondary amine, amide, carboxylic acid.

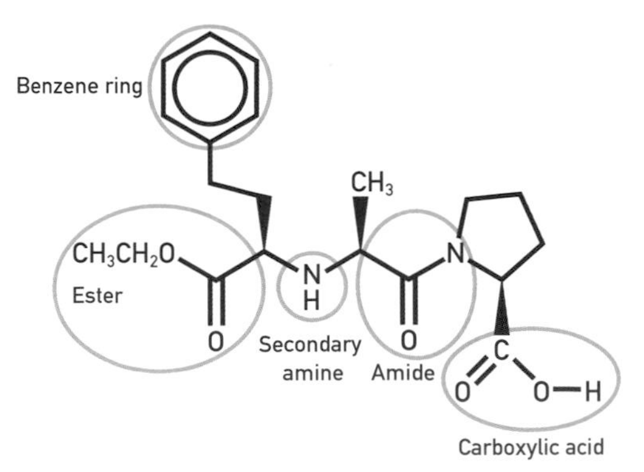

8. -OH: this could form an ester with a free -COOH group or act as both a hydrogen donor and a hydrogen acceptor in hydrogen bonding with appropriate groups. -COOH could form an ester with a free -OH group or an amide with a free $-NH_2$. It can also participate in hydrogen bonding. $-NH_3^+$ could interact ionically with a $-COO^-$ group, for example, or act as a hydrogen donor in hydrogen bonding.

9. This is the easiest method for medicines to be taken by patients at home and therefore means that the potential market is increased.

10. Medicines taken by mouth are exposed to a harsh chemical environment in the gut. This includes acids and enzymes which might decompose the medicines before they can be absorbed.

Introductory page pictures: GlaxoWellcome.

The chemistry of life

Life is chemistry at its most complex. Over the past 50 years, chemists and biologists have been uncovering the extraordinary molecular processes by which living organisms function. This new area of knowledge is helping to improve many aspects of our lives – from radically improved healthcare to cheaper food and other everyday essentials.

One of the triumphs of recent decades has been the unravelling of the complexities of the chemistry of life at the molecular level, and this has spawned a brand new area of science called molecular biology. This new discipline is not only fascinating, but can also be used to help in designing new medicines, **agrochemicals**, and even **biodegradable** plastics, that will further enhance the quality of our lives. Currently underway is perhaps the biggest undertaking of all, the Human Genome Project, which will determine the sequence of the **DNA** in human genes.

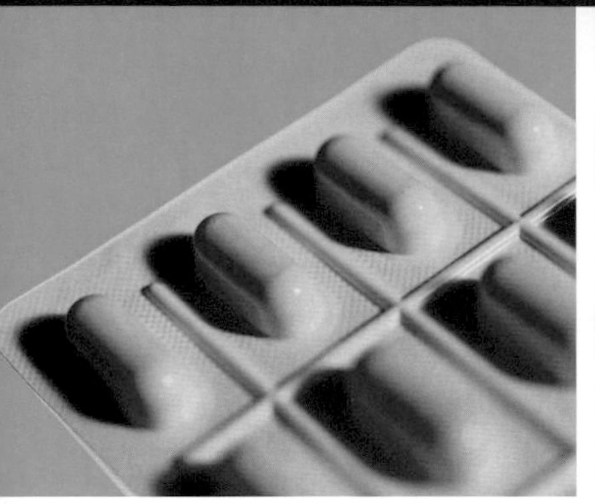

Friedrich Wöhler showed there was no 'vital force' associated with chemicals of natural origin.
Reproduced courtesy of the Library and Information Centre, Royal Society of Chemistry.

The beginning of molecular biology

We have to go back to 1828 to find how chemistry first connected with biology. In this year, the German chemist, Friedrich Wöhler, heated the 'inorganic' chemical ammonium cyanate and converted it into the 'organic' chemical urea that was identical with the material isolated from urine.

$$(NH_4)CNO(s) \rightarrow CO(NH_2)_2(s)$$

Ammonium cyanate → Urea

This proved that there was no essential difference or 'vital force' that distinguished substances, like urea, made by living things, from other chemicals, like ammonium cyanate, derived from rocks, water and air, *etc.* Wöhler wrote to a fellow chemist Jöns Berzelius: 'I must tell you that I can prepare urea without requiring a kidney of an animal, either man or dog.'

1. Ammonium cyanate and urea are in fact a pair of structural isomers. Explain what this term means.

Meanwhile, Wöhler's contemporaries were busy isolating and investigating biologically interesting molecules, such as morphine, atropine and quinine, which were produced by plants. These materials had been used for centuries in folk medicine, witchcraft and so on, but of course nothing was known about how they worked at the molecular level. Nevertheless, these early adventures in natural product chemistry (the chemistry of substances produced by living things), which spanned the period 1800 to 1930, set the scene for all the amazing current developments in the design of new medicines and therapies.

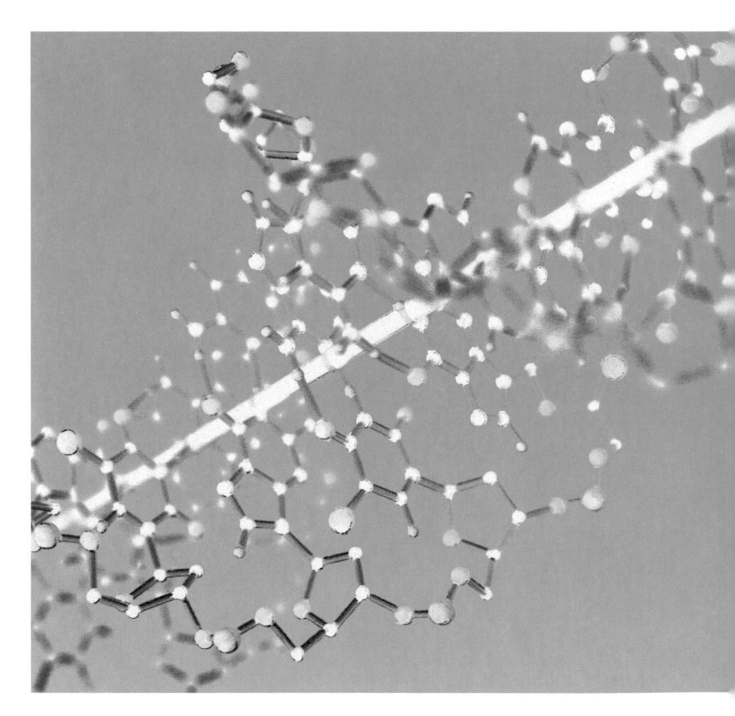

Glossary

Agrochemicals: Chemicals, such as pesticides, used to treat crops.
Biodegradable: Describes chemicals which break down to harmless products in the environment.
DNA: Deoxyribonucleic acid. The double-helix molecule which stores genetic information.
Drug: A chemical which affects how the body works.

Medicine before the 20th century

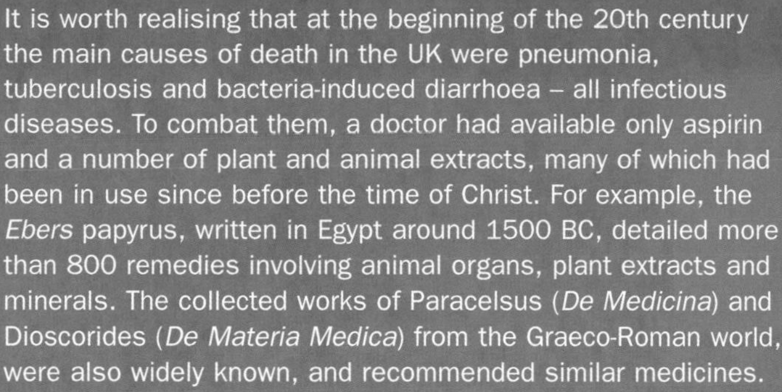

It is worth realising that at the beginning of the 20th century the main causes of death in the UK were pneumonia, tuberculosis and bacteria-induced diarrhoea – all infectious diseases. To combat them, a doctor had available only aspirin and a number of plant and animal extracts, many of which had been in use since before the time of Christ. For example, the *Ebers* papyrus, written in Egypt around 1500 BC, detailed more than 800 remedies involving animal organs, plant extracts and minerals. The collected works of Paracelsus (*De Medicina*) and Dioscorides (*De Materia Medica*) from the Graeco-Roman world, were also widely known, and recommended similar medicines.

New knowledge emerged only slowly. The *London Pharmacopoeia* of 1809 still listed plant extracts (for example aconite, belladonna, cinchona, colchicum, hemlock, henbane and opium), and inorganic salts such as *regulus antimonii* (metallic antimony) and *mercurius sublimatus corrosivus* (mercuric chloride), and formed the basis of prescribing by doctors and pharmacists at this time.

A variety of plants used in herbal medicine.
Sheila Terry/Science Photo Library.

An old pharmacy.

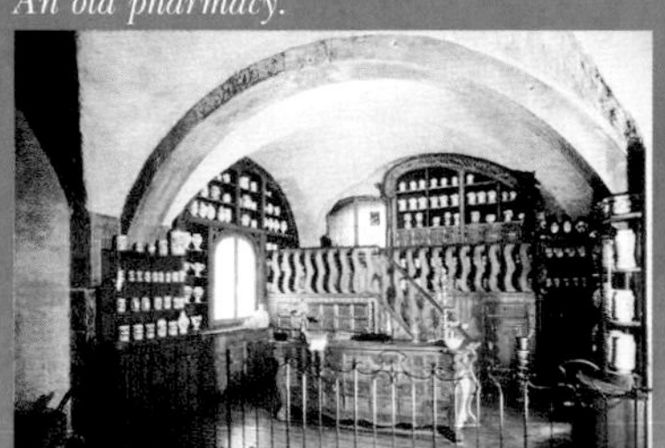

The first synthetic drugs

At the start of the 20th century, the chemical industry was primarily concerned with explosives, fertilisers and dyestuffs, not medicines. Paul Ehrlich in Berlin was the first person to employ a totally synthetic chemical compound as a **drug**. He found that some dyes used to selectively stain bacteria were also able to kill them. Since these dyes were absorbed only by the bacteria, they would kill the bacteria but not a patient infected with them. This led to the use of the dye methylene blue for the treatment of malaria in 1891. Other dyes were active against the trypanosomes that cause African sleeping sickness, and Ehrlich's studies with arsenic-substituted analogues of so-called azo-dyes provided the compound arsphenamin in 1909. This had excellent activity against the organism which caused the venereal disease syphilis.

2. The functional group in an azo-dye is -N=N-. Why would it be expected that 'arsenic-substituted analogues' of azo-dyes could be made?

Ehrlich introduced for the first time the concept of a 'magic bullet' – a chemical designed to target a particular disease: 'Here we may speak of magic bullets which aim exclusively at the dangerous intruding parasites, strangers to the organism, but do not touch the organism itself and its cells.' There then followed numerous triumphs of drug discovery – the development of the sulfonamides and penicillins surely rank as two of the supreme achievements of the 20th century. These were the first antibacterial compounds.

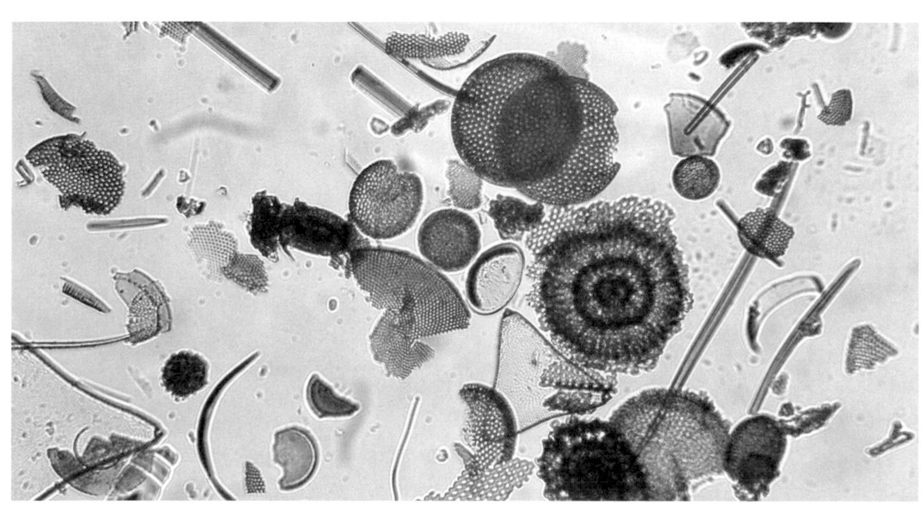

Some dyes were used to kill bacteria in the early 1900s.

Paul Ehrlich was the first person to use a synthetic compound as a drug.
Reproduced courtesy of the Library and Information Centre, Royal Society of Chemistry.

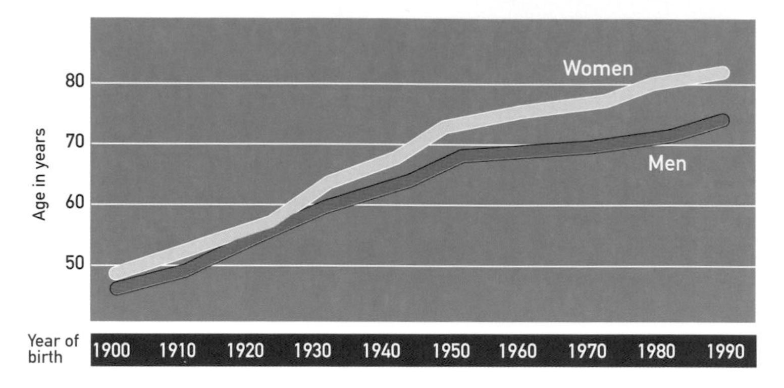

Figure 1. The life expectancy of women and men at birth in the 20th century.

It is often forgotten that the average life expectancy at the beginning of the 20th century was only about 45 years, and that around 15% of children died before their fifth birthday, Figure 1. Most of these premature deaths were a result of bacterial infections. Leading on from Ehrlich's research, the German chemist Gerhard Domagk began a systematic screening study in 1927 of the effects of gold compounds and various dyestuffs against a strain of *Streptococcus* which he had isolated from a patient who had died from septicaemia. One particular red dye, Prontosil Rubrum, Figure 2(a), displayed antibacterial activity, and in 1933, Domagk administered the dye to a baby dying from staphylococcal septicaemia. The baby survived. This was little short of miraculous, and after further tests in the UK and Germany, Prontosil Rubrum was made available for general use.

However, it was not until 1940 that a biochemist worked out the drug's mode of action. Donald Woods, working at Oxford University, found that the dye was broken down in the intestine to release a compound called 4-aminobenzenesulfonamide, Figure 2(b), and this inhibited (prevented the normal functioning of) one of the bacterium's key **enzymes**.

Bacteria make their own vitamin F (folic acid) and cannot acquire it from the environment. Folic acid is essential for the synthesis of DNA in cells and therefore for cell replication. Folic acid is biosynthesised from a compound called 4-aminobenzoic acid (often called PABA from its non-systematic name *para*-aminobenzoic acid). In the presence of either 4-aminobenzenesulfonamide or one of the many sulfonamides prepared later, the bacteria cannot produce folic acid. This is because the sulfonamide molecules are a similar shape to PABA and are mistaken for it. Humans also need folic acid but we get it from our diet, rather than from PABA. This means that the sulfonamides have selective antibacterial activity and are relatively harmless to humans – an important requirement of any drug. One of the sulfonamides, sulfapyridine, made by the UK company May & Baker and called M&B 693, Figure 3, achieved star status when it was used to cure Winston Churchill of pneumonia in 1943. Who knows what the outcome of World War II might have been without it!

H_2N NH_2 N N SO_2NH_2 small intestine H_2N SO_2NH_2

(a) Prontosil Rubrum

(b) Sulfanilamide (4-aminobenzenesulfonamide)

Figure 2. The red dye, Prontosil Rubrum (a), which is converted into 4-aminobenzenesulfonamide (b) – the first antibacterial drug.

H_2N SO_2NH N

M&B 693 (sulfapyridine)

Figure 3. The sulfonamide M&B 693 which cured Winston Churchill of pneumonia.

The sulfonamide drugs were discovered largely by chance and without understanding how they worked. It is now possible to use computer modelling to help with the design of new drugs. See Chapter 9 – *Computational chemistry and the virtual laboratory.*

3 Draw the structure of 4-aminobenzoic acid (PABA) and compare its shape with that of 4-aminobenzenesulfonamide. Look at the structure of the drug M&B 693. Suggest what structural feature the sulfonamides require to be active as antibiotics. It may help if you make models of the three structures.

Glossary

Clinical trial: A trial (of a drug, for example) carried out on actual patients.
Enzyme: A protein which acts as a biological catalyst.

The wonderful penicillins

The sulfonamides had a major impact in reducing the number of deaths of mothers after childbirth from 500 per 100 000 births in 1900 to less than 5 per 100 000 by the mid-1940s. However, sulfonamides were soon superseded by the penicillins.

Joseph Lister and others had observed that moulds like *Penicillium brevicompactum* had a little antibacterial activity. However, it was left to Alexander Fleming to herald the beginning of the antibiotic era with his accidental discovery in 1928 of the much more potent activity of the mould *Penicillium notatum*. Fleming was a bacteriologist, and his collaborators were also biologists rather than chemists, so he was not able to isolate or determine the chemical structure of the substances responsible for the antibacterial activity of the mould. This had to await the attentions of the scientists at the Sir William Dunn School of Pathology in Oxford.

During the early 1940s, the chemists Edward Abraham and Ernst Chain, under the leadership of the bacteriologist Howard Florey, carried out their ground-breaking studies that culminated in the isolation of penicillins. The first **clinical trials** were carried out in 1941, at a time when the UK was at war, and produced some miraculous results. Several patients who were close to death through advanced septicaemia made complete recoveries. The need for large quantities of this new wonder drug to treat battlefield casualties suddenly became of paramount importance. The penicillin was produced by fermentation in a process rather like brewing.

Howard Florey who isolated the first penicillin compound.
©The Nobel Foundation.

Alexander Fleming in his laboratory.
MIG, Audio Visual Services, ICSM (St Mary's).

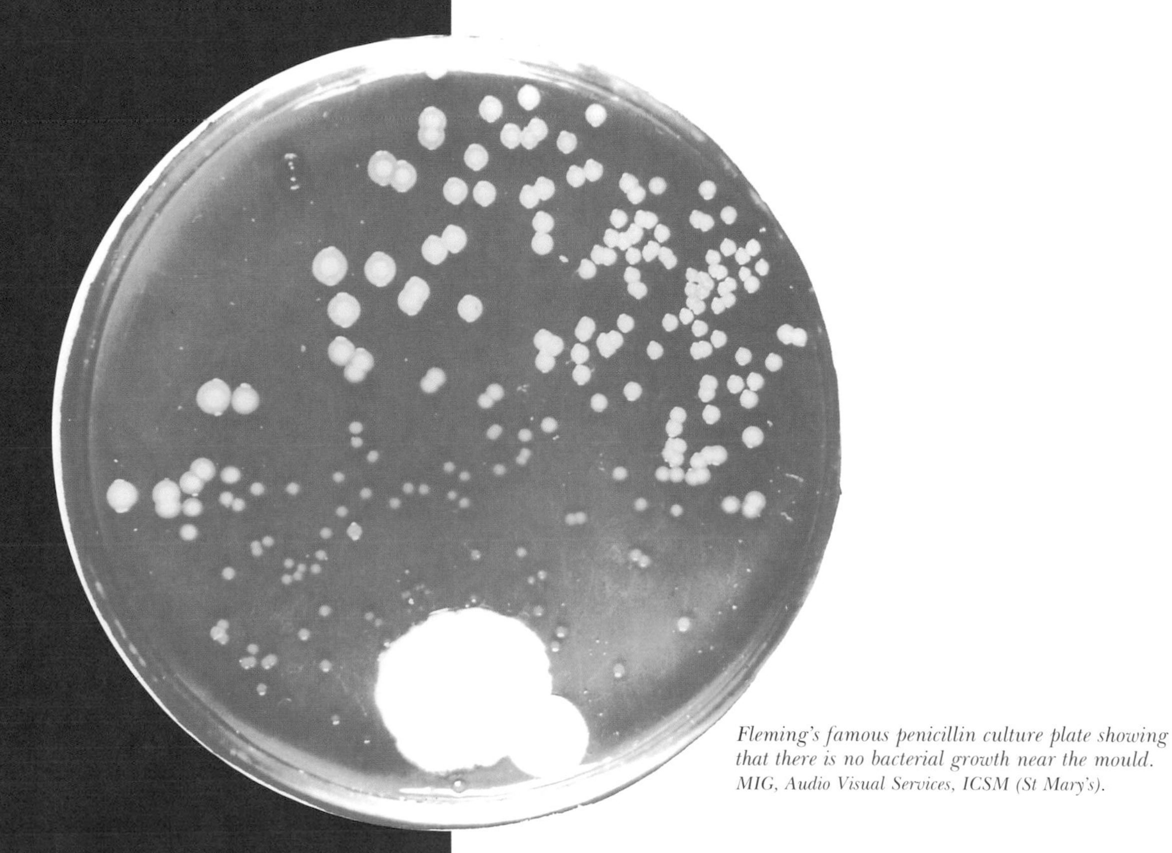

Fleming's famous penicillin culture plate showing that there is no bacterial growth near the mould.
MIG, Audio Visual Services, ICSM (St Mary's).

Dorothy Hodgkin who solved the structure of penicillin.
©The Nobel Foundation.

Figure 4. The structure of penicillin G was determined by Dorothy Hodgkin.

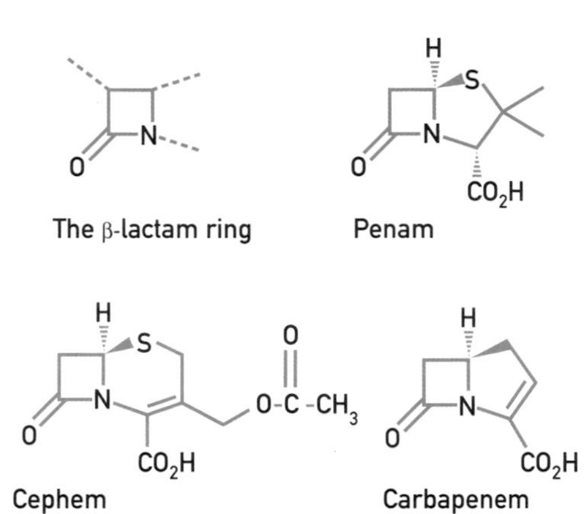

Figure 5. Some β-lactam structures.

In early trials, penicillin was in such short supply that the doctors extracted excess penicillin from patients' urine for re-use.

With hindsight, these fantastic advances are all the more extraordinary since the chemical structure of penicillin was then completely unknown. In fact, the structure of penicillin G, Figure 4, (the major constituent of this mould) was finally solved by Dorothy Hodgkin in 1945, using X-ray crystallography, a technique that can locate the positions of atoms.

The core structure of penicillin is a four-membered ring, called a β-lactam, consisting of a nitrogen atom and three carbon atoms, one of which is bound to oxygen in a carbonyl bond (C=O), Figure 5. This ring is bound to various side-chains and shares an edge with a five-membered sulfur-containing ring. Because the β-lactam ring is chemically unstable to acid or alkali, it was another 12 years before a penicillin was first synthesised chemically.

In the interim, researchers at the pharmaceutical company Eli Lilly had found that they could vary the structure of the side-chain of the penicillin, by adding the side-chain of their choice to the fermentation brew. In this way, a range of new, semi-synthetic penicillins could be produced. Penicillin V (Figure 6(a)) was probably the most successful drug to arise from this method, because it could be administered by mouth – unlike penicillin G which had to be injected.

4. What chemical conditions must a drug be able to withstand if it is to be taken by mouth?

Another major breakthrough was the discovery by Beechams that 6-aminopenicillanic acid, which was the most important intermediate for the production of semi-synthetic penicillin, could actually be isolated from the fermentation broth. This allowed them to prepare an almost unlimited array of structures. One of these, methicillin (Figure 6(b)), was possibly the most important since it was active against strains of bacteria that had become resistant to the earlier penicillins.

(a) Phenoxymethylpenicillin – penicillin V

(b) Methicillin

Figure 6. The semi-synthetic penicillins, penicillin V (a) and methicillin (b).

5. Explain in your own words what is meant by the term 'semi-synthetic'.

6. Look at the formula of penicillin V which is shown in Figure 6(a) in skeletal notation. Draw it as a displayed formula (showing every atom and every bond). List the organic functional groups you recognise.

Since the discovery of penicillin, the pharmaceutical industry has mounted a non-stop campaign to isolate new antibacterial substances to provide drugs which are more effective and to stay one step ahead of the bacteria. Bacteria have been evolving for billions of years, and have now had a further 50 years to evolve new means of becoming resistant to antibiotics.

Glossary

Amino acid: A molecule containing both a carboxylic acid (-COOH) and an amine ($-NH_2$) functional group.

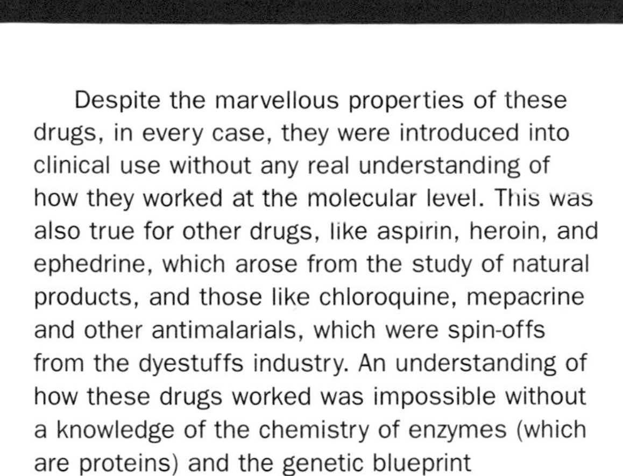

Despite the marvellous properties of these drugs, in every case, they were introduced into clinical use without any real understanding of how they worked at the molecular level. This was also true for other drugs, like aspirin, heroin, and ephedrine, which arose from the study of natural products, and those like chloroquine, mepacrine and other antimalarials, which were spin-offs from the dyestuffs industry. An understanding of how these drugs worked was impossible without a knowledge of the chemistry of enzymes (which are proteins) and the genetic blueprint deoxyribonucleic acid (DNA), the molecules that control every aspect of our lives.

Proteins – the workhorses of life

Until the 1930s, proteins were thought to be no more than random chains of **amino acids** linked together by amide groups, sometimes called peptide linkages in this context, Figure 7.

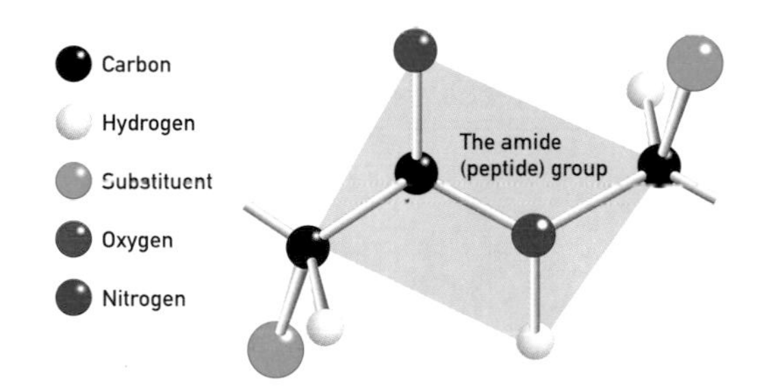

Figure 7. The amide (peptide) link between two amino acids.

The development of a method of determining the amino acid sequence of proteins, by Frederick Sanger at Cambridge University in 1945, showed that proteins were much more ordered than had been originally believed.

However, it was only with the advent of X-ray crystallography that a particular kind of geometrical structure common to many proteins, the α-helix, was first identified by Linus Pauling and Robert Corey in 1950.

The X-ray crystal structures of the proteins myoglobin from muscle, Figure 8, and haemoglobin from blood, were solved by John Kendrew and Max Perutz at Cambridge in the late 1950s. They revealed fascinating three-dimensional structures made up of regions with particular geometries, or folds – for example, the α-helix and β-pleated sheet, Figure 9.

Figure 8. The structure of myoglobin.

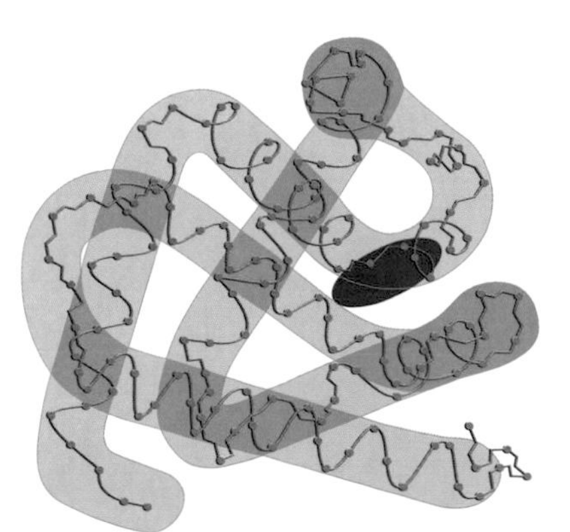

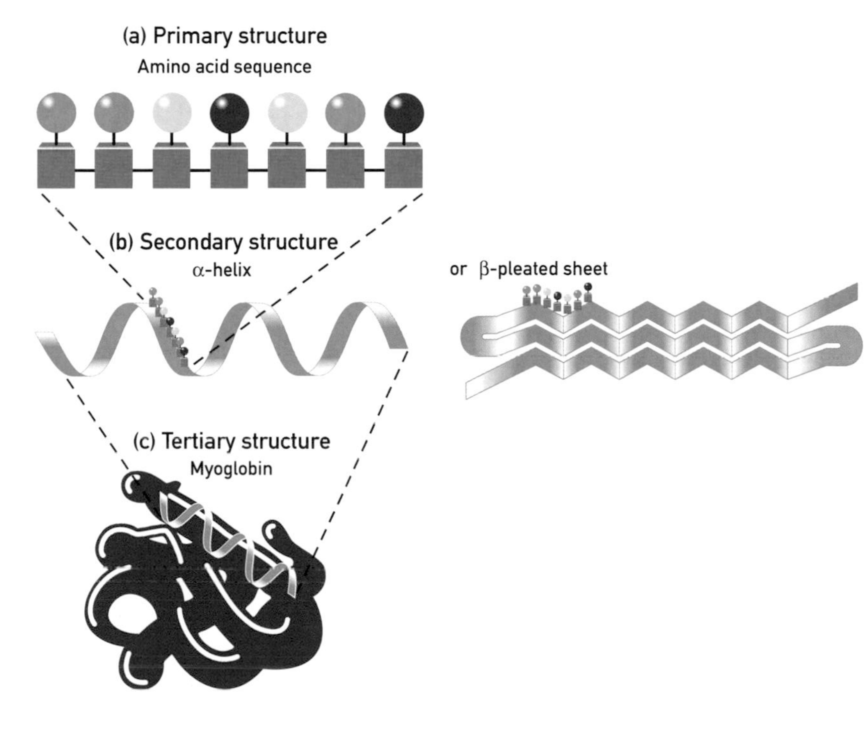

Figure 9. A protein has several layers of structure: its amino acid sequence (a); the types of folds in the structure (b); and the overall folded configuration (c).

In 1965 the X-ray structure of the first enzyme, lysozyme, the antibacterial agent found in tears, was worked out. This allowed investigators to demonstrate how lysozyme destroys the carbohydrate coats of bacteria and thus exerts its antibacterial effect.

7. State three types of bonding between protein molecules by which their three-dimensional stuctures are maintained.

Bruce Merrifield who won the Nobel Prize for chemistry in 1984. Courtesy of Ingbert Grüttner, The Rockefeller University.

Synthesising proteins

Once chemists had determined the amino acid sequence of a number of small polypeptides (molecules of fewer than 50 amino acids strung together), like insulin and human growth **hormone**, it was quite natural that they would want to synthesise them. To achieve this, Bruce Merrifield working at the Rockefeller University, New York developed the technique of solid-phase synthesis which bears his name.

The technique involves attaching an amino acid to an insoluble polystyrene support, and then chemically linking further amino acids in turn to form a growing chain. It is not necessary to purify the growing polypeptide chain; the unreacted reagents and solvents just have to be washed off. Another advantage is that more of the **reagents** can be used than is actually needed. This forces the equilibria of any reversible reactions to the right and ensures maximum yield of product at each step. At the end of the synthesis, the complete polypeptide is chemically removed from the support. It then usually folds spontaneously to attain the required natural three-dimensional form of the protein. In 1969, using this technology, Merrifield was able to prepare the enzyme ribonuclease, consisting of 124 amino acids, in an overall **yield** of 17% – a staggering achievement. The method is now routinely used to prepare both natural and synthetic polypeptides and proteins.

8. A yield of 17% may not seem impressive, but the following calculation should put it in context. Imagine each stage of the reaction has a 90% yield. The yield after step 1 is 0.90 (or 90%), after 2 steps, 0.90 x 0.90 (81%) and after 3 steps 0.90 x 0.90 x 0.90 (72.9%) and so on. The yield after *n* stages is 0.90^n. How many stages would give a yield of 17%. Hint: You will need to take logarithms. What does this tell you about the yield of each step in the Merrifield synthesis?

As the complexities of protein structures were revealed, chemists started to appreciate that their three-dimensional shapes governed their biological functions. Almost every one of life's processes depends upon the interaction of proteins with small organic molecules such as hormones, neurotransmitters or vitamins. Proteins may be enzymes catalysing biochemical reactions in the aqueous environment inside a cell, or receptors sitting in the membranes of cells. In most instances, the protein folds around these small molecules to bind to them and bring about a chemical reaction. But how does this translate into a biological response, and how does an understanding of these complex interactions help with drug design? To illustrate some of these interactions and responses, we shall look at two examples – adrenaline, and the oestrogen hormones, as used in oral contraceptives.

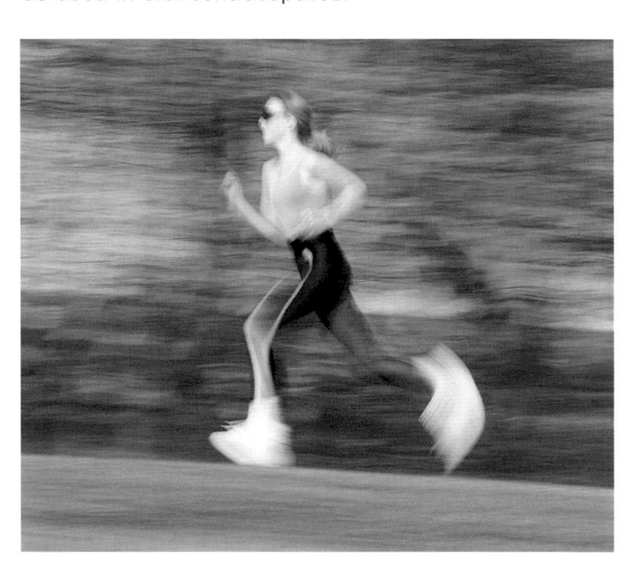

The adrenaline 'cascade'

Adrenaline is the 'fear, flight and fight' hormone. It is released from the adrenal glands in response to a wide variety of stimulating situations. It has a major effect on the heart and leads to an increase in blood pressure, heart rate and output, and thus prepares the body to react to the stimuli. These effects on the body are the result of a complex chain of chemical events that follow the binding of adrenaline to its protein receptor on the surface of heart muscle cells.
This is the classic lock and key type of binding, first described by Emil Fischer in 1894, where the shape of a so-called **active site** on the protein fits the shape of the molecule being bound to it, Figure 10, in this case adrenaline, Figure 11.

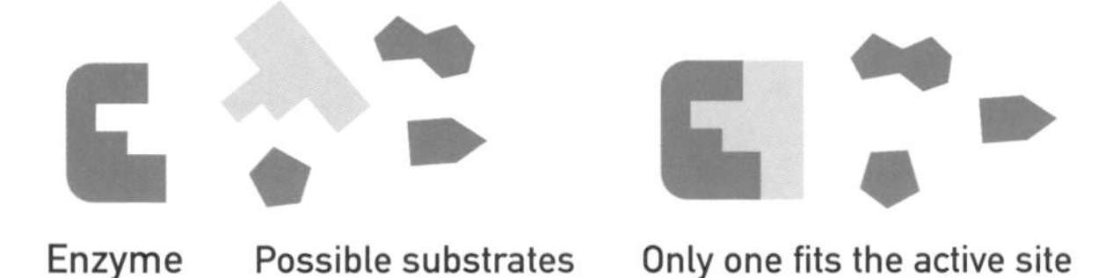

Figure 10. Lock-and-key binding: receptor proteins bind only to molecules (substrates) with a specific shape (just as a key is specific for one lock).

Glossary

Active site: The site on an enzyme where the substrate fits.

Hormone: A 'chemical messenger' molecule secreted into the bloodstream by a gland in the body and which controls a process elsewhere in the body.

Reagent: A chemical substance that reacts with another chemical.

Yield: The amount of product obtained from a chemical reaction expressed as a percentage of the theoretical maximum amount predicted by the chemical equation.

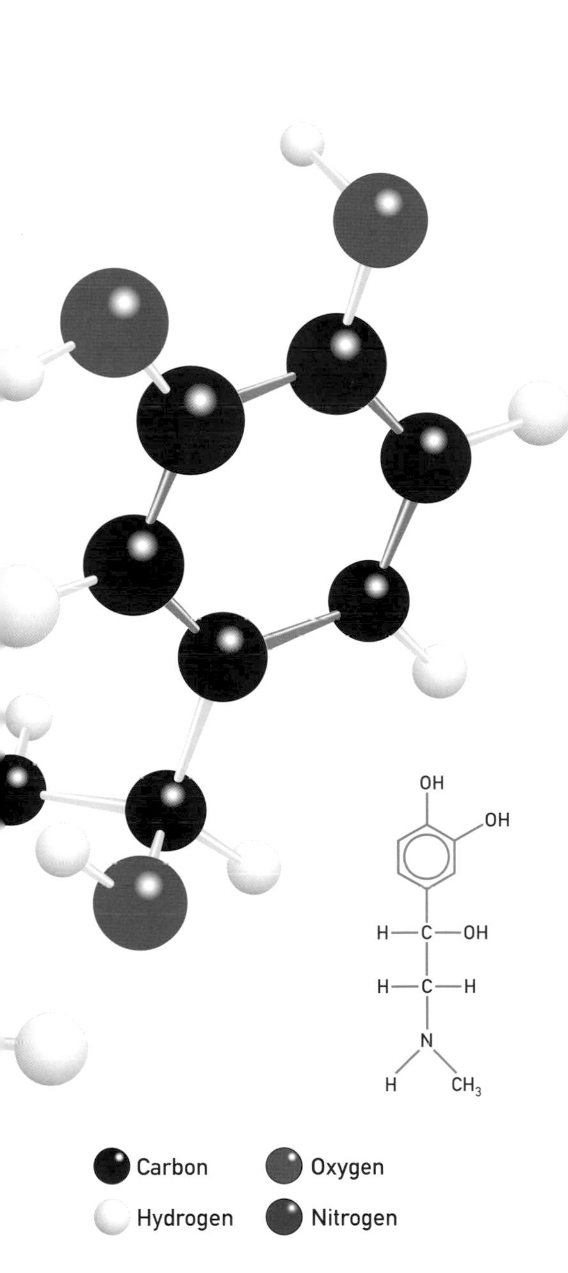

Figure 11. Adrenaline.

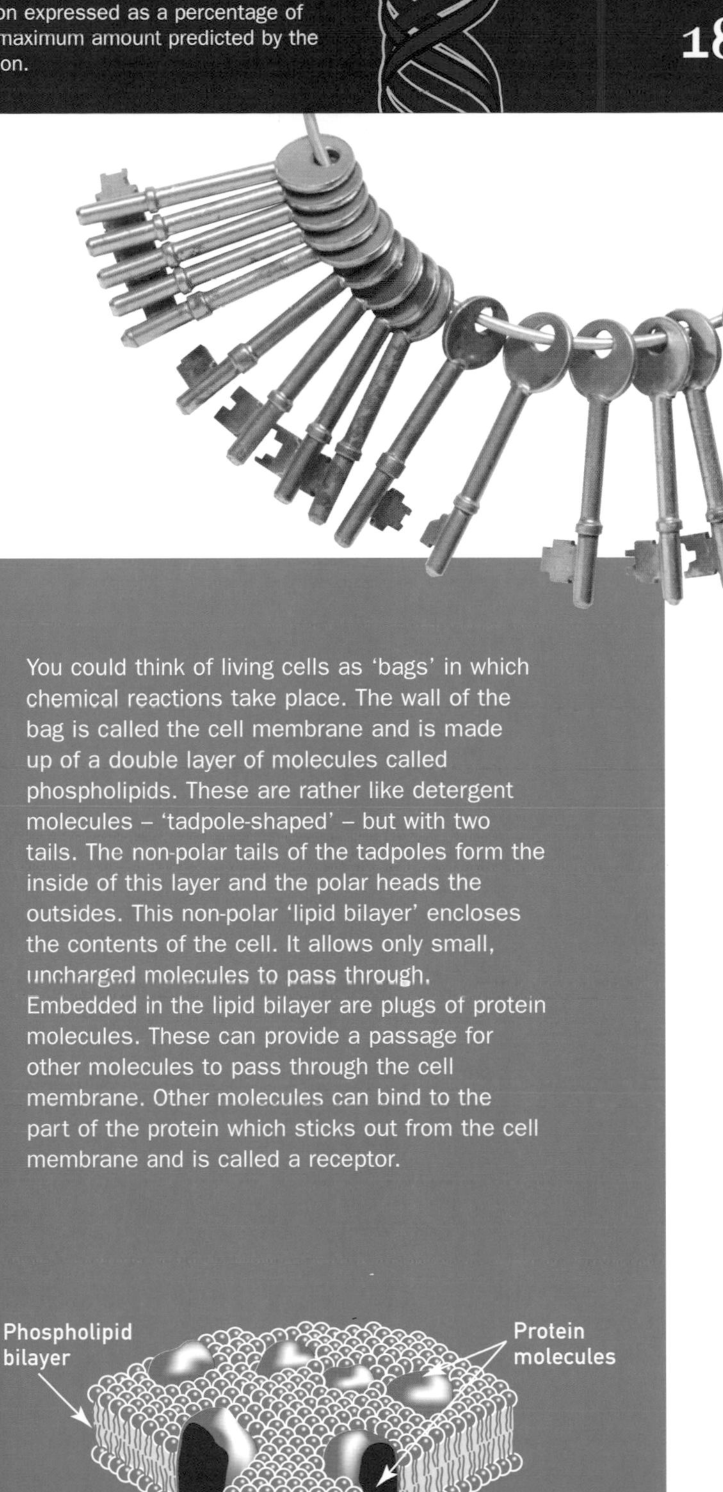

Cell membranes

You could think of living cells as 'bags' in which chemical reactions take place. The wall of the bag is called the cell membrane and is made up of a double layer of molecules called phospholipids. These are rather like detergent molecules – 'tadpole-shaped' – but with two tails. The non-polar tails of the tadpoles form the inside of this layer and the polar heads the outsides. This non-polar 'lipid bilayer' encloses the contents of the cell. It allows only small, uncharged molecules to pass through. Embedded in the lipid bilayer are plugs of protein molecules. These can provide a passage for other molecules to pass through the cell membrane. Other molecules can bind to the part of the protein which sticks out from the cell membrane and is called a receptor.

Part of a cell membrane.

The adrenaline receptor protein straddles the cell membrane. The binding of an adrenaline molecule to the outer part of the receptor causes part of the the receptor protein on the inside of the cell to break off. This fragment then sets off a cascade of further enzyme-catalysed chemical reactions inside the cell. At each stage, one molecule of starting material gives rise to several molecules of product. The overall effect is an enormous amplifying effect.

One major effect of the cascade of events caused by binding of adrenaline is the release of calcium ions from storage sites within the cell, and this leads to activation of muscles, and thus to increased heart rate.

All this also helps us to understand how heart drugs called β-blockers (for example, propanolol, Figure 12) function.

Figure 12. The heart drug, propanolol.

These were first prepared by a research team at ICI. The team was looking for molecules that would prevent the excitation of the heart caused by adrenaline and a similar molecule noradrenaline. This was an important medical goal, because these compounds can precipitate an attack of chest pain in patients suffering from heart disease. The drug molecules compete with adrenaline and noradrenaline in binding to so-called β-adrenergic receptors on the surface of heart muscle cells. When bound, however, they do not set off the cascade of reactions inside the cell and therefore inhibit or block the resulting process. The drugs are called β-blockers from the name of the receptors that they inhibit.

The above account is a classic example of so-called rational drug design – *ie* drug design based on a knowledge of how the drug works. This is now an important activity in the pharmaceutical industry. It involves:

- **identifying a key receptor implicated in a disease;**
- **identifying the molecule that normally bonds to it to set off a biochemical response – this molecule is called the agonist; and**
- **searching for another agent (usually a molecule not found in nature) that binds to the receptor better than the agonist but does not set off the response – this molecule is called an inhibitor, or antagonist.**

Glossary

Nucleus: The nucleus (of a cell) is the body in the cell which contains the genetic material.
Steroid: One of a group of organic molecules based on a four-ring system.

The contraceptive pill

Not all drugs, however, act through this type of chemical cascade. The **steroid** hormones like the oestrogens (female sex hormones), progesterone (the hormone of pregnancy), and the androgens (male sex hormones) all pass into cells without interacting with a surface receptor.

9. The steroid hormones can pass through the lipid bilayer of cells. What does this fact suggest about the stuctures of these molecules?

Once inside the cell, they attach to special protein receptors, and then the whole steroid-receptor complex passes into the **nucleus** of the cell, where it interacts with DNA.

The contraceptive pill.

The steroid hormones

In the mid-17th century, the herbalist Nicholas Culpeper had reported that some plant extracts could be used to improve fertility and others as a method of contraception. However, it was not until the 1930s when chemists isolated and found the structures of the sex hormones that people appreciated that these substances could really control fertility. Later these compounds were synthesised in large quantities, and the first oral contraceptive, ethisterone, was in restricted use from around 1937.

However, the major breakthrough, in terms of supply, was made by Russell Marker in the late 1940s who showed that the plant steroid diosgenin, isolated from the Mexican yam *Dioscorea mexicana*, could be efficiently converted into progesterone. Eventually the contraceptive norethindrone (Norlutin), Figure 13, was synthesised. This could be taken by mouth, and was marketed in the early 1960s. It was soon followed by a similar compound, norethynodrel (Enovid). For the first time in human existence, women had the opportunity to control their own fertility and in many countries the birthrate dropped dramatically.

An additional benefit of these synthetic steroids is their use in hormone replacement therapy (HRT) in menopausal women. In this context, the drugs not only help to reduce the symptoms associated with the menopause but also dramatically reduce the incidence of osteoporosis (brittle bone disease).

Researchers also noted that the growth of tumours in around one-third of breast cancer patients depended on a supply of oestrogen, so an obvious form of therapy was to deny these steroids to the tumour cells. The drug tamoxifen (Novaldex) developed by ICI does just this. It binds to the oestrogen receptors in breast cancer cells and thus prevents natural oestrogens from attaching. The whole receptor-tamoxifen complex now moves into the cell nucleus but does not bring about a burst of DNA activity as does the oestrogen complex. Cell division and growth are no longer possible and the tumour regresses.

OH
C≡CH
O

Figure 13. Norethindrone (Norlutin).

DNA

The double helix structure of DNA (deoxyribonucleic acid) which carries genetic information is well known, Figure 14. It consists of two chains of nucleotides. Each nucleotide itself consists of a molecule of a sugar (deoxyribose), a phosphate group and a base (adenine, guanine, cytosine or thymine). The two chains are held together by hydrogen bonding between the bases (cytosine, C, to guanine, G and thymine, T, to adenine, A) called base pairing. The **hydrogen bonds** are weak enough to allow the two chains to unravel and each can then act as a template for the building of a copy of the other. James Watson and Francis Crick first suggested this structure in 1953 – the culmination of a world-wide effort involving many researchers in chemistry and biology in the 75 years leading up to their discovery. Some of the key events are listed in the Box – *Key events leading to the discovery of the DNA structure*.

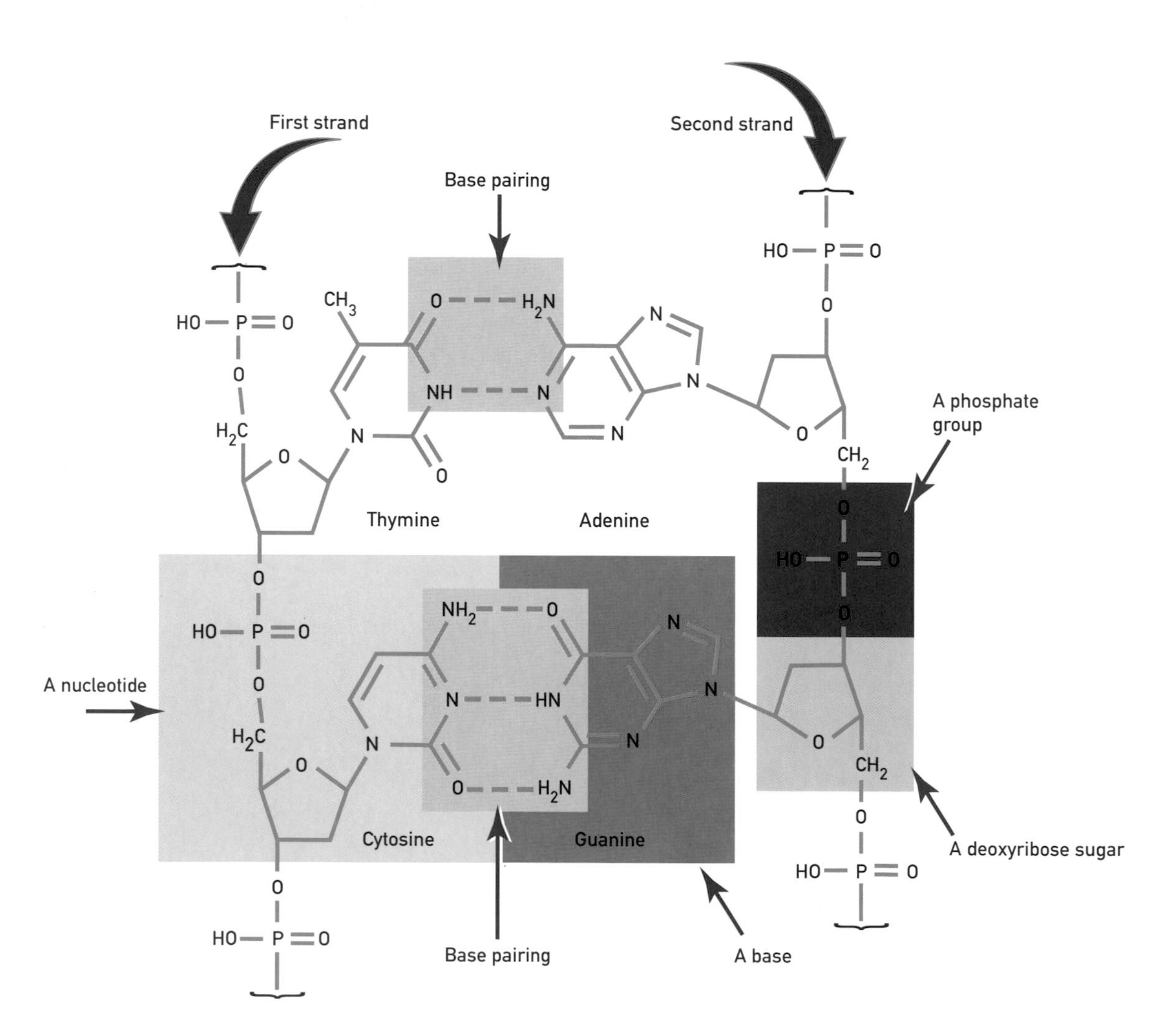

Figure 14(a). Base pairing in DNA.

Glossary

Hydrogen bonds: The strongest type of intermolecular force. They form between slightly positively charged hydrogen atoms (which are covalently bonded to an oxygen, nitrogen or fluorine atom) and another oxygen, nitrogen or fluorine atom which has a slight negative charge.

Pathogenic: Disease-causing.

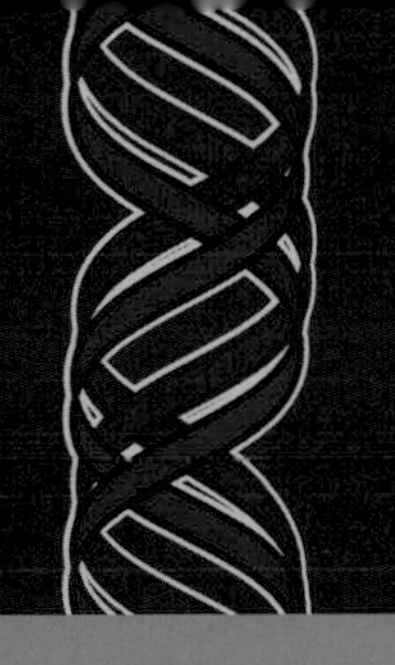

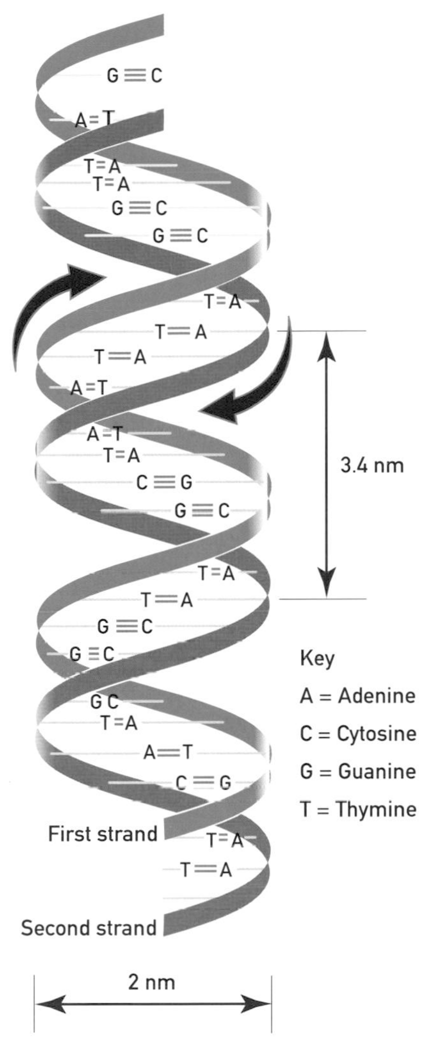

Figure 14(b). The DNA double helix.

James Watson and Francis Crick with their model of DNA.
A Barrington Brown/Science Photo Library.

Key events leading to the discovery of the DNA structure

- **The bases were identified by Albrecht Kossel and his colleagues at Heidelberg University, Germany in the early years of this century.**
- **Phoebus Levene (a former chemistry student of the Russian composer Borodin) showed that 2'-deoxy-D-ribose provided the sugar component of DNA. He also synthesised the bases.**
- **Synthesis from simple molecules of the nucleotide monomers which make up DNA was completed in the 1940s by Alexander (later Lord) Todd and his research group at Manchester and later Cambridge.**
- **In 1944, evidence for the biological significance of DNA was first produced by Oswald Avery at Rockefeller University, New York when he identified DNA as the 'transforming principle' that could turn non-pathogenic bacteria into pathogenic ones.**
- **William Astbury and Florence Bell at Leeds University conducted some X-ray studies on fibres of DNA, and concluded that the bases were stacked face-to-face like a pile of coins in groups of 8 or 16. The three-dimensional structure thus resembled parallel ladders standing out from a central axis.**
- **John Gulland at the University of Nottingham predicted that the hydroxyl (OH) groups of the bases formed hydrogen bonds with one another – the first time that this key form of bonding had been suggested as a stabiliser of DNA structure.**
- **Edwin Chargaff of Columbia University, New York, discovered Chargaff's rules (1950). They stated that: (i) the total of the bases adenine and guanine always equalled the total of bases thymine and cytosine; (ii) the number of moles of adenine equalled that of thymine, and the number of moles of guanine equalled that of cytosine. Thus: A + G = T + C and A = T and G = C.**
- **Linus Pauling in the US found that many protein molecules have helical shapes and proposed (incorrectly) a triple helix shape for DNA.**
- **X-ray diffraction patterns produced by Maurice Wilkins and Rosalind Franklin of King's College London, suggested that the dominant form of DNA possessed a spiral (helical) structure with a diameter of 2 nanometres and one turn every 3.4 nanometres.**
- **Watson and Crick, armed with the above information, and using molecular models made in the workshop at Cambridge deduced the double helix structure.**

The genetic code

The information held on a strand of DNA is used to make proteins. The DNA first acts as a template for the formation of a molecule of a closely related substance – RNA (ribonucleic acid) in which the deoxyribose sugar is replaced by a slightly different sugar, ribose, and the base thymine is replaced by a similar base, uracil, Figure 15.

Uracil (U)   Ribose

Figure 15. RNA contains uracil rather than thymine and a ribose sugar instead of a deoxyribose.

10. Uracil can form hydrogen bonds with adenine in a similar way to the way which thymine can. Draw a diagram to show the hydrogen bonding between uracil and adenine.

The sequence of bases in RNA governs what proteins are made in the cell. Each sequence of three bases (U, C, A and G) represents an amino acid which is incorporated into a new protein. UUC, for example produces the amino acid phenylalanine (Phe). This idea is known as the genetic code, Figure 16.

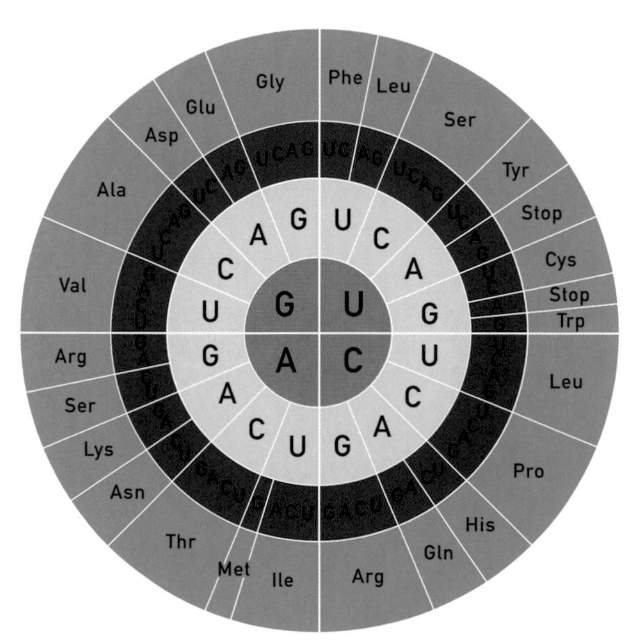

Figure 16. A representation of the genetic code. Reading from inside to out gives the amino acid represented by any three letter code. For example, AGG represents arginine. There are also codes such as UGA to stop a protein chain. Most amino acids have more than one code.

11. What amino acid is produced by the code AAA? What other code will also produce this amino acid?

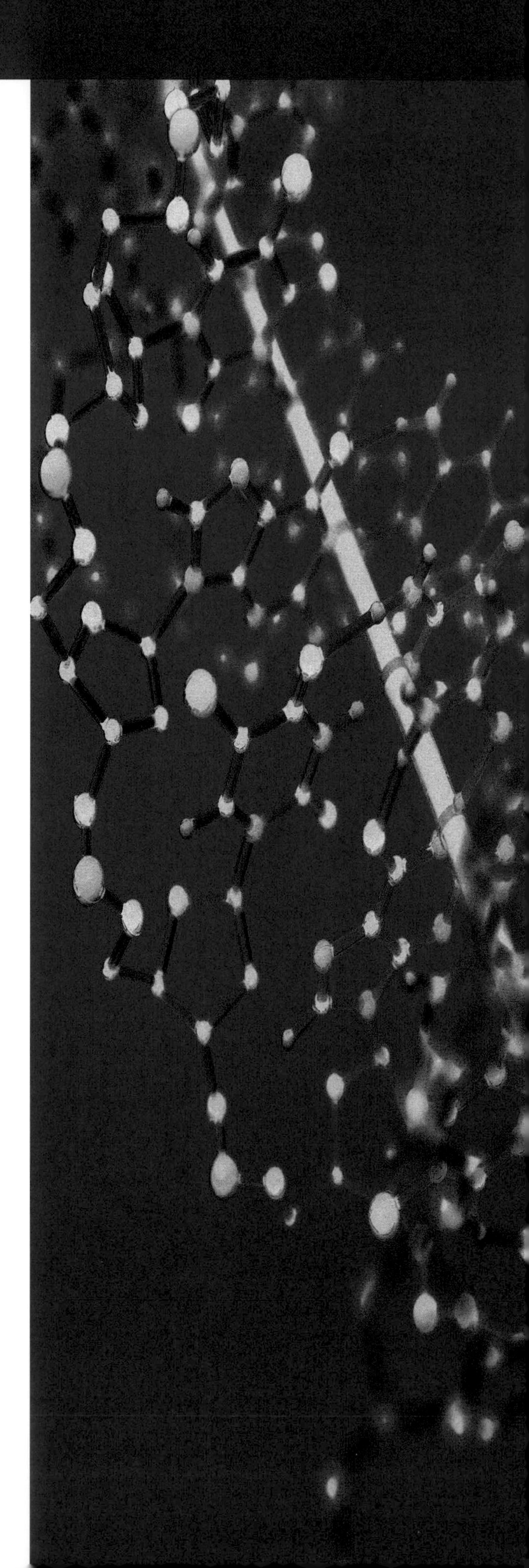

Genetic engineering

Since DNA (via RNA) produces specific proteins, it should be possible, by altering their DNA, to tailor organisms such as bacteria to produce different proteins to order. In the 1960s enzymes were discovered which could join two sections of DNA and ones which could cut DNA at specified points. This made it possible to 'cut and splice' DNA from different organisms to produce so-called recombinant DNA and do genetic engineering. This means that, for example, a bacterium could be engineered which would make a particular chemical required for use as a drug, food or fuel. To do this, of course, we need to know the sequence of bases in natural DNA. Methods of doing this were developed in the 1970s and work is in progress to discover the entire human genome – the complete sequence of human DNA.

More recently, it has become possible to produce enormous numbers of copies of a particular DNA, using an enzyme called DNA polymerase. All that is needed is a minute quantity of a master sequence, for instance the DNA from one cell or from a small amount of DNA produced synthetically. This polymerase chain reaction (PCR) technology has significantly accelerated developments in molecular biology, and has revolutionised the field of genetics as well.

The technique of DNA fingerprinting, developed by Alec Jeffreys of Leicester University in 1985, also depends upon this technique. It allows a genealogist to trace a family's heritage, or a palaeontologist to map the evolutionary tree of a long-extinct species, or a forensic scientist to identify a criminal by using small samples of DNA.

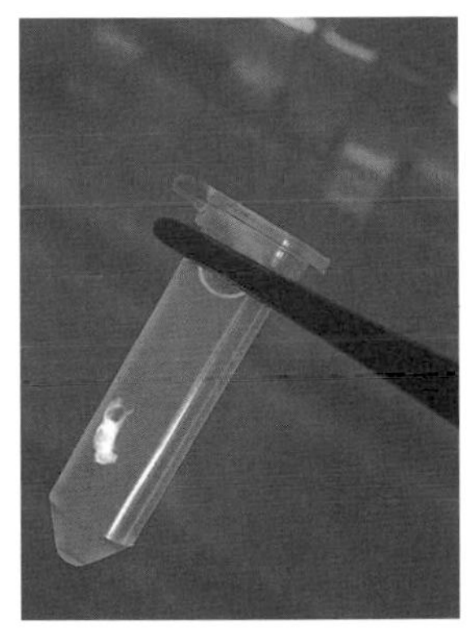

Sample tube containing pellet of human DNA (white).
Philippe Plailly/Science Photo Library.

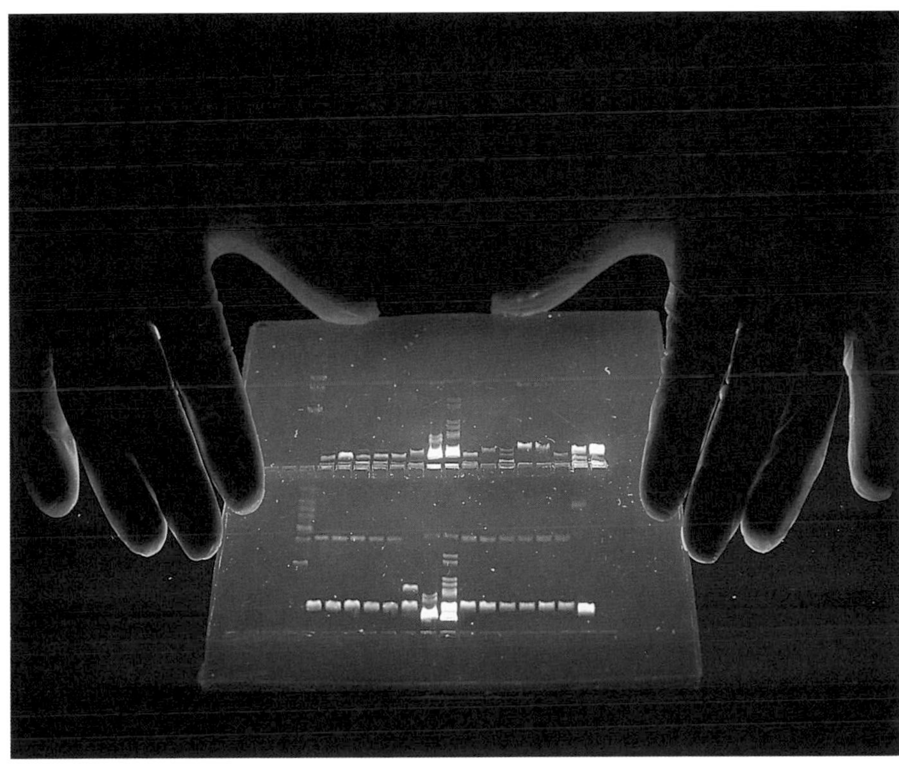

A typical DNA fingerprint.
J C Revy/Science Photo Library.

The role of DNA in drug design

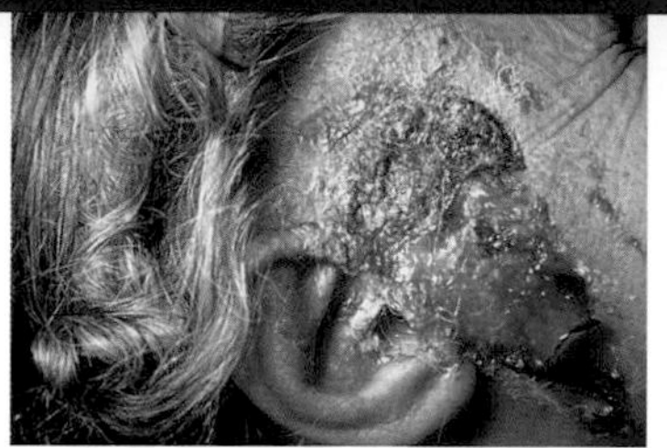

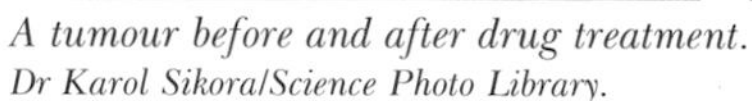

A tumour before and after drug treatment.
Dr Karol Sikora/Science Photo Library.

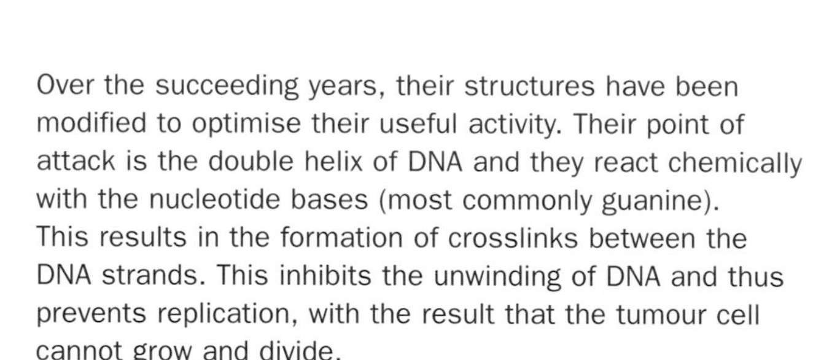

Anticancer drugs

Tumours are made up of rogue cells that have lost control over their own growth and replication. Because they are so similar to normal cells, it is very difficult to find 'magic bullet'-type agents that target cancer cells while leaving healthy tissue untouched. Many drugs used to destroy cancer cells in cancer chemotherapy show little discrimination between cancer cells and normal cells. They rely on the fact that cancer cells often grow much more rapidly than normal cells and take up nutrients (and drugs) faster. However, those healthy tissues that undergo rapid replacement, such as the lining of the digestive tract, white blood cells and hair follicles, are also seriously affected by such drugs and this leads to nausea, anaemia and hair loss during chemotherapy (drug treatment). These side effects seriously limit the dose of drug that can be administered, and this has led to increasing efforts to develop drugs and drug-delivery systems that *can* specifically target tumour cells.

The first anticancer agents were the nitrogen mustards. These were developed from mustard gas used in World War I, Figure 17, one of the few positive things to arise from what was otherwise an abuse of chemistry. Nitrogen mustards were first used clinically in 1942, with some spectacular, if short-lived, therapeutic effects.

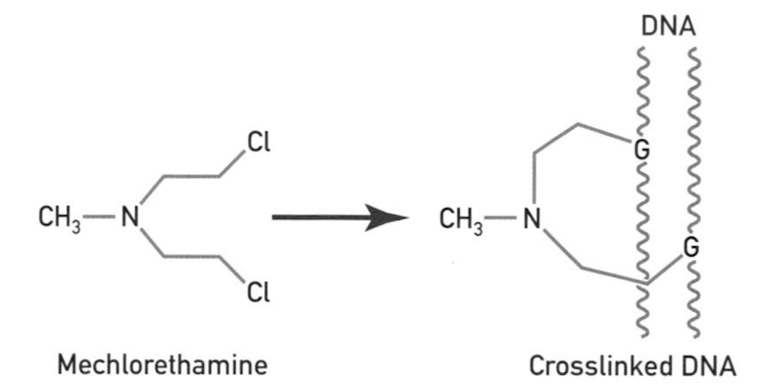

Figure 17. How anticancer agent mechlorethamine, a nitrogen mustard, crosslinks with DNA and prevents it from unwinding.

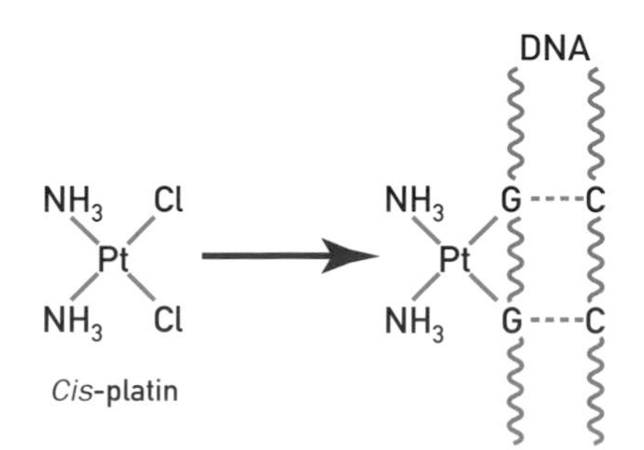

Figure 18. Cis-platin forms bonds with guanine bases in DNA and prevents replication.

Over the succeeding years, their structures have been modified to optimise their useful activity. Their point of attack is the double helix of DNA and they react chemically with the nucleotide bases (most commonly guanine). This results in the formation of crosslinks between the DNA strands. This inhibits the unwinding of DNA and thus prevents replication, with the result that the tumour cell cannot grow and divide.

Platinum anticancer drugs also work by bonding to DNA and preventing replication. The most famous of these drugs is *cis*-platin which bonds to two guanine bases on the same strand of DNA, Figure 18. *Cis*-platin was discovered in 1965. Ten years later it was revolutionising the treatment of various cancers, most spectacularly testicular teratoma, where a death sentence was translated into a survival rate of more than 90%.

? 12. Draw the structure of *trans*-platin.

However, like most anticancer drugs, the mustards and *cis*-platin are not given alone but in combination with other drugs – many of them natural products. One family of drugs stops cells dividing by interfering with the functions of an enzyme called tubulin polymerase, which controls the production and destruction of the protein tubulin. This forms the thin strands that hold dividing cells together until they are ready to lead a separate existence of their own. Members of this family include the vinca alkaloids (vinblastine and vincristine), the podophyllins, and taxol – used against breast cancer. Other natural products like doxorubicin and dactinomycin possess flat, multi-ring structures which slip in between the coils of DNA and act as a kind of molecular 'glue' preventing the separation of the two strands. A final class of drugs is metabolised to produce radicals which interact with oxygen to form highly reactive oxygen radicals which destroy the DNA.

? 13. What is a radical? Why are radicals so reactive?

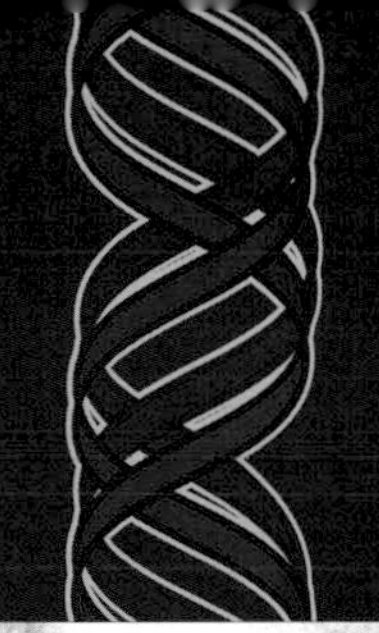

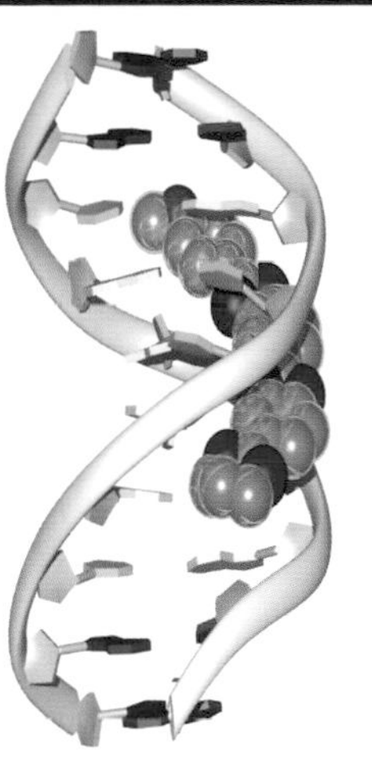

An anticancer drug in the minor groove of DNA.

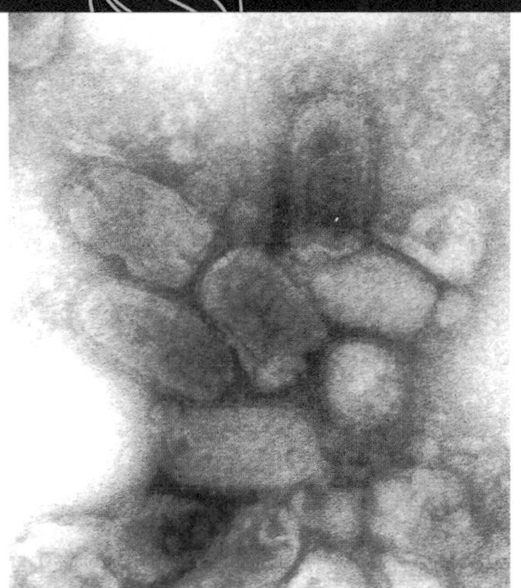

The rabies virus.
Courtesy of New York State Department of Health's Wadsworth Center.

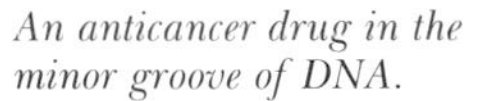

Antiviral drugs

The other area where a knowledge of the molecular biology of the disease is helping with the design of drugs is in antiviral chemotherapy. Viruses can be considered as cellular parasites. They hijack the DNA replication and protein biosynthesis systems of the host cell for the production of their own genetic material (DNA or RNA) and for their enzymes and coat proteins.

Viruses are responsible for a number of serious diseases including rabies, yellow fever, poliomyelitis, measles, rubella, influenza, AIDS, and certain cancers. For many of these diseases it has been possible to immunise people by injecting them with a modified form of the virus (a vaccine). This stimulates the immune system to produce antibodies. These then recruit other members of the white cell family that eventually destroy the invading virus. While this strategy has been immensely successful in some cases – smallpox has been completely eradicated, for example – other viruses have proved to be much more difficult to destroy.

The human immunodeficiency virus (HIV) is estimated to be present in around 30 million people worldwide, and has already caused at least 15 million deaths from acquired immune deficiency syndrome (AIDS).

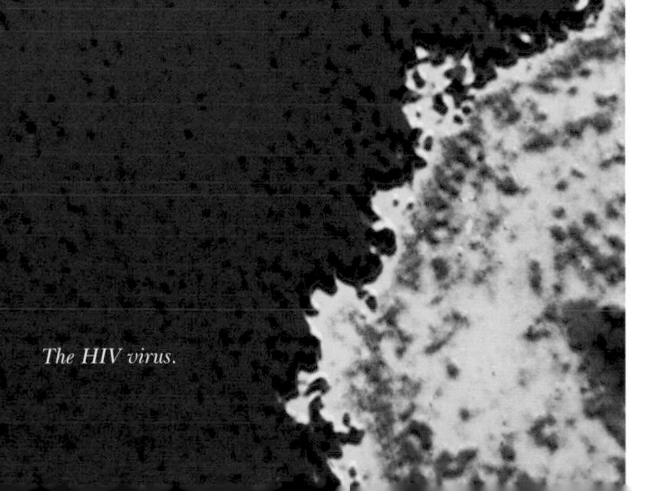

The HIV virus.

Because the virus undergoes rapid mutation, it avoids recognition by the immune system, and it is also virtually impossible to produce a vaccine. Chemotherapy is the only answer, and this requires an understanding of the biology of the virus. After considerable efforts, scientists identified several enzymes unique to the virus that could be targeted by new drugs. HIV is a retrovirus which means that it uses RNA rather than DNA as its genetic store, and requires enzymes to help it produce DNA.

A full understanding of the life-cycle of this virus, and a knowledge of the three-dimensional structures of key viral enzymes (such as reverse transcriptase and HIV protease) has allowed the pharmaceutical industry to design drugs to target these enzymes. Several major families of compounds have been identified, and these include the dideoxynucleosides like AZT (Zidovudine developed by Wellcome, 1987) and ddC (Zalcitabine from Hoffmann La Roche, 1991), Figure 19.

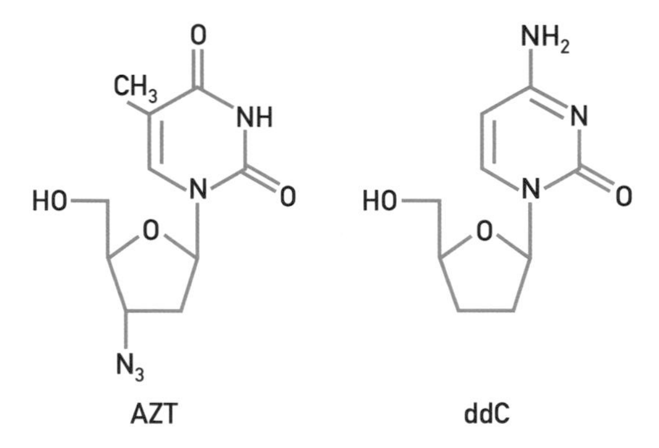

Figure 19. Two antiviral drugs for treating HIV infection.

Chemistry and gene therapy

The extraordinary efficiency with which viruses transfer their genetic material for incorporation into host cell DNA suggests that they could be used to transfer more useful genes. Indeed, clinical trials are underway to test whether viruses like *herpes simplex* (the virus that causes cold sores) and *adenoviruses* (commonly associated with colds and sore throats) can be used to transfer genes to correct the defects present in Duchenne muscular dystrophy and cystic fibrosis.

One form of cancer chemotherapy, so-called GDEPT (gene directed enzyme activated prodrug therapy) involves injecting a gene that codes for a particular enzyme into a tumour. This is followed by administration of a prodrug – a molecule which is not an active drug but which can be turned into an active drug by the enzyme. So the active drug is produced only in the tumour itself which has been injected with the enzyme. For example, injecting a herpes simplex virus-kinase gene and then administering the anti-herpes drug acyclovir (ACV), results in conversion of ACV to produce its active form ACV-triphosphate, and this then inhibits DNA replication in the dividing tumour cells.

Of more interest to the food and chemical industries are the possibilities of engineering genes in micro-organisms and plants so that they will produce large amounts of valuable chemicals, plastics or edible proteins. A growing awareness of the environmental problems caused by plastics now requires that they have both the required physical properties and also improved biodegradability. Synthetic protein-based polymers modelled on the structures of spider silk and mammalian collagen (the fibrous protein in skin, bone, tendon and cartliage) have already been produced by transferring engineered genes to the bacterium *E.coli.* But the real advances will come when similar genes can be transferred to plants to allow cheap mass production.

As molecular biologists unravel more and more of the secrets of the gene sequences of organisms from bacteria to humans, they will provide chemists with almost unlimited opportunities to design and synthesise new molecules. If the 20th century has been one in which chemists have designed drugs to treat disease, then the 21st century will surely see the advent of drugs to prevent disease and even eradicate it.

Future challenges for chemistry in biology and medicine

We have outlined some of the important chemical advances which have allowed us to probe biological systems to develop drugs to cure diseases and to help us to understand fundamental life processes. For both of these tasks chemists will be employed in the 21st century.

We are in a continuous race to develop new antibacterial agents and keep one step ahead of the 'superbugs' which threaten to plunge us back to the 'dark age' before we had antibiotics. The advent of new diseases such as AIDS or BSE/CJD which appear to have jumped the species barrier from other animals, or the spread of rare diseases such as the Ebola virus of Central Africa to areas of greater population will continually throw up new challenges in chemistry, medicine and biology. Recent advances in cancer treatment such as the use of taxol and other drugs for the treatment of breast or cervical cancer have proved to be effective but still many other cancers cannot be treated by chemotherapy.

Combinatorial chemistry (see Chapter 1 – *Make me a molecule*) allows us to produce large libraries of new synthetic compounds very rapidly, and it may be from these new reservoirs of molecules rather than natural sources that we find promising compounds in the coming years.

Nature is still capable of surprising scientists. Fifteen years ago no one would have believed that nitric oxide (nitrogen(II) oxide, NO), a gaseous compound best known as a pollutant in car exhausts, was used in many body tissues as a biological messenger causing smooth muscle to relax.

Once the biological role of nitric oxide had been established, it explained why compounds such as glyceryl trinitrate and amyl nitrite – prescribed for more than a century to treat angina (narrowing of the arteries to the heart) – worked. These compounds decompose in the body to give nitric oxide. The importance of nitric oxide has led to the development of new drugs which interfere with smooth-muscle relaxation including sildenafil (Viagra), Figure 20, which became the top selling drug of 1998 and is used temporarily to overcome penile erection problems.

Evidence for life on Mars has led to other avenues of research.

It is a sobering thought that we still do not know the structures and biological roles of more than 70% of the compounds found in human blood plasma. Our understanding of many biological processes and their regulation is still rudimentary. The completion of the Human Genome Project will provide us with the 'blueprint' for the human body. However, the role played by the majority of genes within this plan and the proteins which they encode still need to be discovered.

The recent speculation about evidence for life on Mars being present in certain meteorite fragments has led to another avenue of active research. This is prebiotic chemistry and explores how life has evolved from simple molecules such as ammonia, water, carbon dioxide and hydrogen sulfide. Why has nature chosen proteins, sugars and nucleic acids to act as catalysts, sources of energy and repositories of genetic information? Does the discovery of RNA molecules with catalytic function (ribozymes) prove that prior to cellular life as we know it there was an RNA world capable of molecular organisation and replication that only involved nucleic acids? Would life on other planets have to be carbon-based? If not how would we search for it. These are all challenges that await the chemists of the future.

O CH$_3$ CH$_3$CH$_2$O HN N N N O$_2$S N N CH$_3$ HO$_2$C CO$_2$H OH CO$_2$H

Figure 20. Viagra.

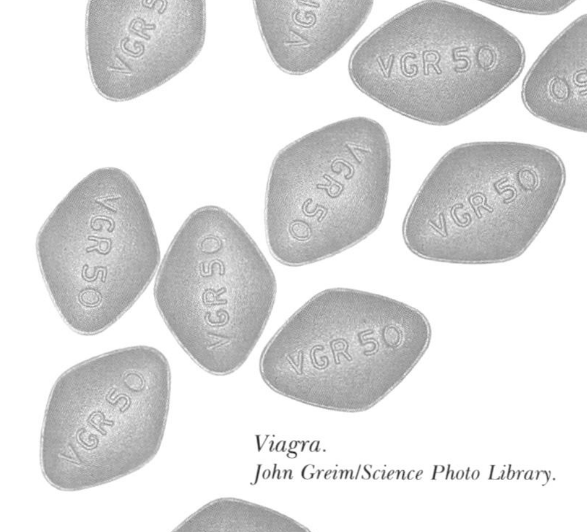

Viagra.
John Greim/Science Photo Library.

Answers

1. They have the same molecular formula but different arrangements of atoms in space.

2. Arsenic is in the same group as nitrogen in the Periodic Table and would therefore be expected to bond in the same sort of way.

3.

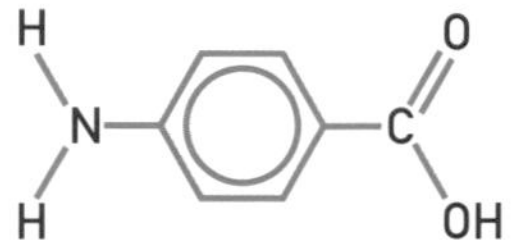

   It is similar in that it has a primary aromatic amine functional group in the 4-position.
   The primary aromatic amine group in the 4-position seems to be required. (This is in fact the case as structures of other sulfonamide drugs confirm.)

4. Acid and enzymes in the stomach.

5. Molecules made by modifying a naturally-occuring compound.

6.

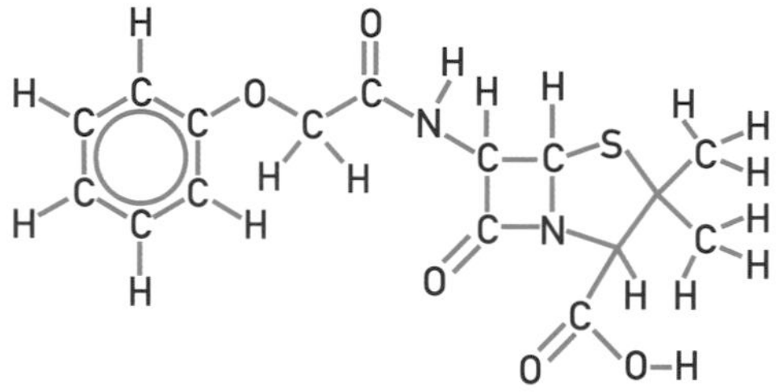

   Amide, ether, carboxylic acid, aromatic ring

7. Three from: hydrogen bonding, dipole-dipole interactions, ionic interactions such as acid-base bonds, S-S bonds.

8. 17 stages (to nearest whole no.) The yield of each step must be much greater than 90%. (In fact the average yield of each step is 98.6%.)

9. They must be relatively non-polar.

10.

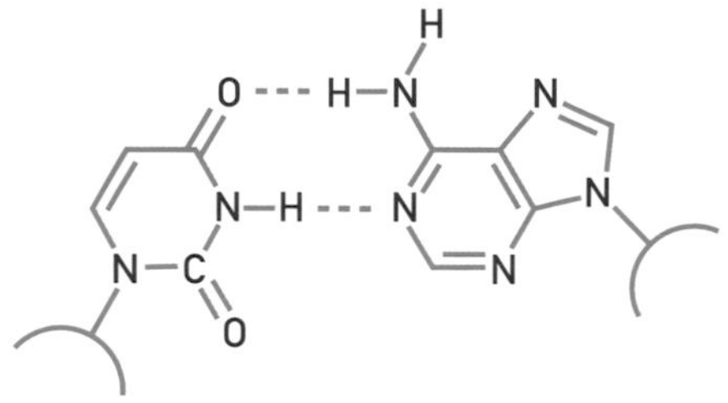

11. Lysine. AAG

12.

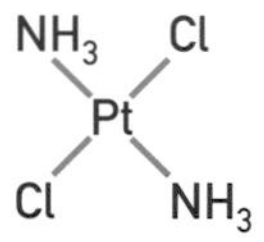

13. A species with an unpaired electron.
    Radicals react indiscriminately to pair up their unpaired electrons.